Springer Series in Statistics

Advisors:
P. Bickel, P. Diggle, S. Fienberg, K. Krickeberg,
I. Olkin, N. Wermuth, S. Zeger

Springer
New York
Berlin
Heidelberg
Barcelona
Hong Kong
London
Milan
Paris
Singapore
Tokyo

Springer Series in Statistics

Andersen/Borgan/Gill/Keiding: Statistical Models Based on Counting Processes.
Andrews/Herzberg: Data: A Collection of Problems from Many Fields for the Student and Research Worker.
Anscombe: Computing in Statistical Science through APL.
Berger: Statistical Decision Theory and Bayesian Analysis, 2nd edition.
Bolfarine/Zacks: Prediction Theory for Finite Populations.
Borg/Groenen: Modern Multidimensional Scaling: Theory and Applications
Brémaud: Point Processes and Queues: Martingale Dynamics.
Brockwell/Davis: Time Series: Theory and Methods, 2nd edition.
Daley/Vere-Jones: An Introduction to the Theory of Point Processes.
Dzhaparidze: Parameter Estimation and Hypothesis Testing in Spectral Analysis of Stationary Time Series.
Efromovich: Nonparametric Curve Estimation: Methods, Theory, and Applications.
Fahrmeir/Tutz: Multivariate Statistical Modelling Based on Generalized Linear Models.
Farebrother: Fitting Linear Relationships: A History of the Calculus of Observations 1750 - 1900.
Farrell: Multivariate Calculation.
Federer: Statistical Design and Analysis for Intercropping Experiments, Volume I: Two Crops.
Federer: Statistical Design and Analysis for Intercropping Experiments, Volume II: Three or More Crops.
Fienberg/Hoaglin/Kruskal/Tanur (Eds.): A Statistical Model: Frederick Mosteller's Contributions to Statistics, Science and Public Policy.
Fisher/Sen: The Collected Works of Wassily Hoeffding.
Good: Permutation Tests: A Practical Guide to Resampling Methods for Testing Hypotheses.
Goodman/Kruskal: Measures of Association for Cross Classifications.
Gouriéroux: ARCH Models and Financial Applications.
Grandell: Aspects of Risk Theory.
Haberman: Advanced Statistics, Volume I: Description of Populations.
Hall: The Bootstrap and Edgeworth Expansion.
Härdle: Smoothing Techniques: With Implementation in S.
Hart: Nonparametric Smoothing and Lack-of-Fit Tests.
Hartigan: Bayes Theory.
Hedayat/Sloane/Stufken: Orthogonal Arrays: Theory and Applications.
Heyde: Quasi-Likelihood and its Application: A General Approach to Optimal Parameter Estimation.
Heyer: Theory of Statistical Experiments.
Huet/Bouvier/Gruet/Jolivet: Statistical Tools for Nonlinear Regression: A Practical Guide with S-PLUS Examples.
Jolliffe: Principal Component Analysis.
Kolen/Brennan: Test Equating: Methods and Practices.
Kotz/Johnson (Eds.): Breakthroughs in Statistics Volume I.

(continued after index)

Dimitris N. Politis
Joseph P. Romano
Michael Wolf

Subsampling

Springer

Dimitris N. Politis
Department of Mathematics
University of California, San Diego
9500 Gilman Drive
La Jolla, CA 92093-0112
USA
politis@math.ucsd.edu

Joseph P. Romano
Department of Statistics
Stanford University
Stanford, CA 94305-4065
USA
romano@stat.stanford.edu

Michael Wolf
Departamento de Estadistica y Econometria
Universidad Carlos III de Madrid
Calle Madrid 126
28903 Getafe
Spain
mwolf@est-econ.uc3m.es

Library of Congress Cataloging-in-Publication Data
Politis, Dimitris N.
   Subsampling / Dimitris N. Politis, Joseph P. Romano, Michael Wolf.
     p. cm. — (Springer series in statistics)
   Includes bibliographical references and index.
   ISBN 0-387-98854-8 (alk. paper)
    1. Bootstrap (Statistics).   I. Romano, Joseph P., 1960– .
II. Wolf, Michael, 1960– .   III. Title.   IV. Series.
QA276.7.P646   1999
519.5′44—dc21   99-15017

Printed on acid-free paper.

© 1999 Springer-Verlag New York, Inc.
All rights reserved. This work may not be translated or copied in whole or in part without the written permission of the publisher (Springer-Verlag New York, Inc., 175 Fifth Avenue, New York, NY 10010, USA), except for brief excerpts in connection with reviews or scholarly analysis. Use in connection with any form of information storage and retrieval, electronic adaptation, computer software, or by similar or dissimilar methodology now known or hereafter developed is forbidden. The use of general descriptive names, trade names, trademarks, etc., in this publication, even if the former are not especially identified, is not to be taken as a sign that such names, as understood by the Trade Marks and Merchandise Marks Act, may accordingly be used freely by anyone.

Production managed by Robert Bruni; manufacturing supervised by Thomas King.
Photocomposed pages prepared from the authors' LaTeX files.
Printed and bound by Edwards Brothers, Inc., Ann Arbor, MI.
Printed in the United States of America.

9 8 7 6 5 4 3 2 1

ISBN 0-387-98854-8 Springer-Verlag New York Berlin Heidelberg   SPIN 10726682

To our families and friends

Preface

Since Efron's (1979) profound paper on the bootstrap, an enormous amount of effort has been spent on the development of bootstrap, jackknife, and other resampling methods. The primary goal of these computer-intensive methods has been been to provide statistical tools that work in complex situations without imposing unrealistic or unverifiable assumptions about the data-generating mechanism. Bootstrap methods often can achieve this goal, but like *any* statistical method, they cannot be applied without thought. While the bootstrap will undoubtedly serve as a widely used tool well into the 21st century, it sometimes can fail. Moreover, the asymptotic justification of the bootstrap is often peculiar to the problem at hand.

It was realized in Politis and Romano (1992c, 1994b) that a very general approach to constructing asymptotically valid inference procedures exists by appropriate use of subsampling. That is, the statistic of interest, such as an estimator or test statistic, is evaluated at subsamples of the data, and these subsampled values are used to build up an estimated sampling distribution. Historically, the roots of subsampling can be traced to Quenouille's (1949) and Tukey's (1958) jackknife. In fact, Mahalanobis (1946) suggested the use of subsamples to estimate standard errors in studying crop yields, though he used the term interpenetrating samples. Hartigan (1969, 1975) exploited the use of subsamples to construct confidence intervals and to approximate standard errors; he obtained some finite sample results in the symmetric location problem, as well as some asymptotic results for asymptotically normal statistics. Hartigan's constructions involve recalculating a statistic over all subsamples and all subsample sizes, but he also considers using only randomly chosen subsamples, or what he calls balanced subsamples. Of course, the use of subsamples is quite related to the delete-d jackknife, which was developed mainly in the context of variance estimation; see Wu (1986) and Shao and Wu (1989). In the time series context, Carlstein (1986) considered the use of subsamples, or what he calls subseries, to approximate the variance of a statistic. Later, Wu (1990) realized how jackknife pseudo-values can be used for distribution estimation as well, but only in the context of independent, identically distributed (i.i.d.) observations when the statistic is approximately linear so that asymptotic normality holds; his arguments were specific to this context.

The main contribution presented here can be summarized as follows: the consistency properties of estimated sampling distributions based on subsampling hold under extremely weak assumptions, and even in situations where the bootstrap fails. In contrast to the aforementioned results, we can obtain quite general results by considering subsamples of fixed size (but depending on the sample size and possibly data-dependent). In fact, it is hard to conceive of a theory for the construction of confidence intervals that works (in a first order sense) under weaker conditions. Moreover, the mathematical arguments to justify such a claim are based on fairly simple ideas, thus allowing for ease of generalization. Therefore, we will consider subsampling not just in the usual context of independent and identically distributed observations, but also in the context of dependent data situations, such as time series, random fields, or marked point processes.

The idea of using subsamples as a diagnostic tool to describe the sampling distribution of an estimator was also considered in Sherman and Carlstein (1996). In the i.i.d. context, subsampling is closely related to the usual bootstrap, the main difference being resampling the data without replacement versus with replacement. It follows that the bootstrap also enjoys general first order consistency properties, but only if one is willing to resample at sizes much smaller than the original sample size, as noted in Politis and Romano (1992c, 1993c); the benefits of using a smaller resample size when bootstrapping are further exploited in Bickel, Götze, and van Zwet (1997). Of course, the usual bootstrap can work perfectly well, and it enjoys good higher order properties in some nice problems. Subsampling, on the other hand, is more generally applicable. Ultimately, it may be used as a robust starting point toward even more refined procedures (some of which we present).

The primary goal of this book is to lay some of the foundation for subsampling methodology and related methods. The book is laid out in two parts. Chapters 1–7 provide the basic theory of the bootstrap and subsampling. Chapter 1 is devoted to developing some of the basic consistency properties of the bootstrap. Obviously, one can write volumes about the bootstrap, and there are now several books whose main subject is the theoretical development of the bootstrap. Our purpose is to provide some mathematical tools needed in studying consistency properties of the bootstrap, thereby setting the stage for studying subsampling. Thus, this chapter serves as a tutorial on the bootstrap that can be included in a graduate course on asymptotic methods. The chapter is largely based on Beran (1984) and one of the author's lecture notes in a class on the bootstrap taught by Beran at U.C. Berkeley in the fall of 1983. Chapter 2 considers the use of subsampling in the case of independent and identically distributed observations. Some comparisons are made with the bootstrap. Chapter 3 considers subsampling stationary time series, and generalizations to the nonstationary or heteroskedastic case are presented in Chapter 4; the moving blocks bootstrap is also considered. Generalizations to random fields are developed in Chapter 5, and marked point processes are discussed in Chapter 6. In Chapter 7, a general theorem on subsampling is developed that allows the parameter space of interest to be quite abstract. This concludes the first of two parts of the book. The second part of the book is concerned with extensions of the basic theory, as well as practical issues in implementation and applications. Chapter 8 addresses complex situations where the convergence rate of the estimator is unknown since it depends on parameters that are not assumed known; subsampling is shown to be generally useful here as well. The issue of choice of block size is discussed in Chapter 9. Chapter 10 is devoted to higher-order accuracy and extrapolation. Two important cases where the convergence rate is unknown are detailed in Chapters 11 and 12. Chapter 11 considers the case of the mean with heavy tails (infinite variance), while Chapter 12 considers inference for an autoregressive parameter in the possibly integrated case. In Chapter 13, we consider a financial application by using subsampling to discuss whether stock returns can be predicted from dividend yields. The appendices contain some results on mixing sequences that are used throughout the book.

This book is intended for graduate students and researchers; it can be used for an advanced graduate course in statistics that assumes a basic knowledge of theoretical statistics usually taught at the first-year Ph.D. level. Even if one does not want to devote an entire course to subsampling, Chapters 1–3 are designed so that they can be included in any course on asymptotic methods in statistics.

We have benefitted from the support of many friends and colleagues. Special thanks are due to Patrice Bertail for his instrumental help toward the original development of the ideas in Chapters 8 and 10. Some of the book was presented in graduate courses taught at Stanford University and U.C. San Diego in 1997 and 1998, and we thank our students for providing helpful comments. We also would like to thank the National Science Foundation for supporting much of this research and writing over the past few years.

Contents

Preface vii

I Basic Theory   1

1 Bootstrap Sampling Distributions   3
- 1.1 Introduction   3
  - 1.1.1 Pivotal Method   4
  - 1.1.2 Asymptotic Pivotal Method   6
  - 1.1.3 Asymptotic Approximation   6
  - 1.1.4 Bootstrap Approximation   7
- 1.2 Consistency   9
- 1.3 Case of the Nonparametric Mean 11
- 1.4 Generalizations to Mean-like Statistics 16
- 1.5 Bootstrapping the Empirical Process 19
- 1.6 Differentiability and the Bootstrap 23
- 1.7 Further Examples 30
- 1.8 Hypothesis Testing 33
- 1.9 Conclusions 38

2 Subsampling in the I.I.D. Case   39
- 2.1 Introduction 39
- 2.2 The Basic Theorem 41
- 2.3 Comparison with the Bootstrap 47
- 2.4 Stochastic Approximation 51

	2.5	General Parameters and Other Choices of Root	52
		2.5.1 Studentized Roots	52
		2.5.2 General Parameter Space	53
	2.6	Hypothesis Testing	54
	2.7	Data-Dependent Choice of Block Size	59
	2.8	Variance Estimation: The Delete-d Jackknife	62
	2.9	Conclusions	64

3 Subsampling for Stationary Time Series — 65

3.1	Introduction	65
3.2	Univariate Parameter Case	67
	3.2.1 Some Motivation: The Simplest Example	67
	3.2.2 Theory and Methods for the General Univariate Parameter Case	68
	3.2.3 Studentized Roots	74
3.3	Multivariate Parameter Case	74
3.4	Examples	77
3.5	Hypothesis Testing	90
3.6	Data-Dependent Choice of Block Size	92
3.7	Bias Reduction	93
3.8	Variance Estimation	95
	3.8.1 General Statistic Case	95
	3.8.2 Case of the Sample Mean	97
3.9	Comparison with the Moving Blocks Bootstrap	98
3.10	Conclusions	100

4 Subsampling for Nonstationary Time Series — 101

4.1	Introduction	101
4.2	Univariate Parameter Case	102
4.3	Multivariate Parameter Case	106
4.4	Examples	108
4.5	Hypothesis Testing and Data-Dependent Choice of Block Size	117
4.6	Variance Estimation	117
4.7	Conclusions	119

5 Subsampling for Random Fields — 120

5.1	Introduction and Definitions	120
5.2	Some Useful Notions of Strong Mixing for Random Fields	121
5.3	Consistency of Subsampling for Random Fields	123
	5.3.1 Univariate Parameter Case	123
	5.3.2 Multivariate Parameter Case	127
5.4	Variance Estimation and Bias Reduction	129
5.5	Maximum Overlap Subsampling in Continuous Time	132
5.6	Some Illustrative Examples	134

Contents xiii

	5.7	Conclusions	137
6	**Subsampling Marked Point Processes**		**138**
	6.1	Introduction	138
	6.2	Definitions and Some Different Notions on Mixing	140
	6.3	Subsampling Stationary Marked Point Processes	144
		6.3.1 Sampling Setup and Assumptions	144
		6.3.2 Main Consistency Result	146
		6.3.3 Nonstandard Asymptotics	149
	6.4	Stochastic Approximation	151
	6.5	Variance Estimation via Subsampling	154
	6.6	Examples	155
	6.7	Conclusions	158
7	**Confidence Sets for General Parameters**		**159**
	7.1	Introduction	159
	7.2	A Basic Theorem for the Empirical Measure	160
	7.3	A General Theorem on Subsampling	164
	7.4	Subsampling the Empirical Process	166
	7.5	Subsampling the Spectral Measure	168
	7.6	Conclusions	170

II Extensions, Practical Issues, and Applications 171

8	**Subsampling with Unknown Convergence Rate**		**173**
	8.1	Introduction	173
	8.2	Estimation of the Convergence Rate	177
		8.2.1 Convergence Rate Estimation: Univariate Parameter Case	177
		8.2.2 Convergence Rate Estimation: Multivariate Parameter Case	182
	8.3	Subsampling with Estimated Convergence Rate	184
	8.4	Conclusions	187
9	**Choice of the Block Size**		**188**
	9.1	Introduction	188
	9.2	Variance Estimation	189
		9.2.1 Case of the Sample Mean	189
		9.2.2 General Case	192
	9.3	Estimation of a Distribution Function	193
		9.3.1 Calibration Method	194
		9.3.2 Minimum Volatility Method	197
	9.4	Hypothesis Testing	200
		9.4.1 Calibration Method	200

	9.4.2 Minimum Volatility Method	201
9.5	Two Simulation Studies	202
	9.5.1 Univariate Mean	203
	9.5.2 Linear Regression	204
9.6	Conclusions	206
9.7	Tables	208

10 Extrapolation, Interpolation, and Higher-Order Accuracy — 213

10.1	Introduction	213
10.2	Background	216
10.3	I.I.D. Data: The Sample Mean	218
	10.3.1 Finite Population Correction	218
	10.3.2 The Studentized Sample Mean	219
	10.3.3 Estimation of a Two-Sided Distribution	220
	10.3.4 Extrapolation	221
	10.3.5 Robust Interpolation	224
10.4	I.I.D. Data: General Statistics	225
	10.4.1 Extrapolation	225
	10.4.2 Case of Unknown Convergence Rate to the Asymptotic Approximation	226
10.5	Strong Mixing Data	228
	10.5.1 The Studentized Sample Mean	228
	10.5.2 Estimation of a Two-Sided Distribution	231
	10.5.3 The Unstudentized Sample Mean and the General Extrapolation Result	232
	10.5.4 Finite Population Correction in the Mixing Case	235
	10.5.5 Bias-Corrected Variance Estimation for Strong Mixing Data	236
10.6	Moderate Deviations in Subsampling Distribution Estimation	244
10.7	Conclusions	251

11 Subsampling the Mean with Heavy Tails — 253

11.1	Introduction	253
11.2	Stable Distributions	254
11.3	Extension of Previous Theory	257
11.4	Subsampling Inference for the Mean with Heavy Tails	260
	11.4.1 Appealing to a Limiting Stable Law	260
	11.4.2 Using Self-Normalizing Sums	263
11.5	Choice of the Block Size	265
11.6	Simulation Study	266
11.7	Conclusions	267
11.8	Tables	269

12 Subsampling the Autoregressive Parameter — 270
  12.1 Introduction — 270
  12.2 Extension of Previous Theory — 272
    12.2.1 The Basic Method — 272
    12.2.2 Subsampling Studentized Statistics — 274
  12.3 Subsampling Inference for the Autoregressive Root — 278
  12.4 Choice of the Block Size — 284
  12.5 Simulation Study — 286
  12.6 Conclusions — 287
  12.7 Tables — 289

13 Subsampling Stock Returns — 291
  13.1 Introduction — 291
  13.2 Background and Definitions — 292
    13.2.1 The GMM Approach — 293
    13.2.2 The VAR Approach — 294
    13.2.3 A Bootstrap Approach — 296
  13.3 The Subsampling Approach — 297
  13.4 Two Simulation Studies — 298
    13.4.1 Simulating VAR Data — 299
    13.4.2 Simulating Bootstrap Data — 301
  13.5 A New Look at Return Regressions — 302
  13.6 Additional Looks at Return Regressions — 303
    13.6.1 A Reorganization of Long-Horizon Regressions — 303
    13.6.2 A Joint Test for Multiple Return Horizons — 305
  13.7 Conclusions — 307
  13.8 Tables — 308

Appendices — 315

A Some Results on Mixing — 315

B A General Central Limit Theorem — 321

References — 327

Bibliography — 327

Index of Names — 341

Index of Subjects — 345

Part I
Basic Theory

1
Bootstrap Sampling Distributions

1.1 Introduction

The bootstrap was discovered by Efron (1979), who coined the name. In this chapter, the bootstrap is developed as a general method to approximate the sampling distribution of a statistic, a pivot, or a root (defined below), in order to construct confidence regions for a parameter of interest. The use of the bootstrap to approximate a null distribution in the construction of hypothesis tests is also considered. Much of the theoretical foundations of the bootstrap are laid out in Bickel and Freedman (1981), Singh (1981), and Beran (1984). The development begins by focusing on the independent, identically distributed (i.i.d.) case.

Let $x_n = (X_1, \cdots, X_n)$ be a sample of n independent and identically distributed (i.i.d.) random variables taking values in a sample space S and having unknown probability distribution P, where P is assumed to belong to a certain collection $\mathbf{P}$ of distributions. The collection $\mathbf{P}$ may be "parametric," "semiparametric," or "nonparametric," in the usual sense utilized throughout much of the statistical literature. Of course, any family $\mathbf{P}$ is parametrized by the collection of probabilities P in $\mathbf{P}$, and there is no need to place any restriction on $\mathbf{P}$ at this point.

The interest lies in constructing a confidence region for some parameter $\theta(P)$, whose range $\{\theta(P) : P \in \mathbf{P}\}$ will be denoted Θ. Typically, Θ is a subset of the real line, but we also consider more general parameters. For example, the problem of estimating the entire cumulative distribution function (c.d.f.) of real-valued observations will be treated, so that Θ is an appropriate function space.

This leads to considering a *root* $R_n(x_n, \theta(P))$, a term first coined by Beran (1984), which is just some functional depending on both x_n and $\theta(P)$. The idea is that confidence intervals or hypothesis tests could be constructed if the distribution of the root were known. For example, an estimator $\hat{\theta}_n$ of a real-valued parameter $\theta(P)$ might be given so that a natural choice is $R_n(x_n, \theta(P)) = [\hat{\theta}_n - \theta(P)]$, or alternatively $R_n(x_n, \theta(P)) = [\hat{\theta}_n - \theta(P)]/s_n$, where s_n is some estimate of the standard deviation of $\hat{\theta}_n$. Unless otherwise stated, the root $R_n(x_n, \theta(P))$ will be assumed to be a real-valued function of the data x_n and the parameter $\theta(P)$.

When **P** is suitably large so that the problem is nonparametric in nature, a natural construction for an estimator $\hat{\theta}_n$ of $\theta(P)$ is $\hat{\theta}_n = \theta(\hat{P}_n)$, where $\hat{P}_n$ is the empirical distribution of the data. Of course, this construction implicitly assumes that $\theta(\cdot)$ is defined on the empirical measures so that $\theta(\hat{P}_n)$ is at least well defined. Alternatively, in regular parametric problems for which **P** is indexed by a parameter ψ belonging to a subset Ψ of $\mathbb{R}^p$ so that $\mathbf{P} = \{P_\psi : \psi \in \Psi\}$, then $\theta(P)$ can be described as a functional $t(\psi)$. Hence, $\hat{\theta}_n$ is often taken to be $t(\hat{\psi}_n)$, where $\hat{\psi}_n$ is some desirable estimate of ψ, such as a maximum likelihood estimator, a one-step maximum likelihood estimator, or a minimum distance estimator.

Let $J_n(P)$ be the distribution of $R_n(x_n, \theta(P))$ when x_n is a sample of size n from P, and let $J_n(\cdot, P)$ be the corresponding cumulative distribution function defined by

$$J_n(t, P) = Prob_P\{R_n(x_n, \theta(P)) \leq t\}.$$

In order to construct a confidence region for $\theta(P)$ based on the root $R_n(x_n, \theta(P))$, the sampling distribution $J_n(P)$ or its appropriate quantiles must be known or estimated. Some standard methods, based on pivots and asymptotic approximations, are now briefly reviewed. Note that in many of the examples when the observations are real valued, it is more convenient and customary to index the unknown family of distributions by the cumulative distribution function F rather than P. We will freely use both, depending on the situation.

1.1.1 Pivotal Method

In certain exceptional cases, $J_n(P)$ does not depend on P, and hence an exact confidence region for $\theta(P)$ can be constructed. In this case, the root $R_n(x_n, \theta(P))$ is called a *pivot*. In general, if $R_n(x_n, \theta(P))$ is a pivot, its distribution may be determined to a desired level of accuracy by simulating the distribution of $R_n(x_n, \theta(P))$ under P for any choice of P in **P**. The following is a more concrete construction.

For $i = 1, \ldots, B$, let $x_{n,i}$ be a sample of size n from any P in $\mathbf{P}$. Here, we are tacitly assuming one can easily accomplish this sampling. Of course, when $\mathbf{P}$ consists of a class of cumulative distribution functions F on the real line, one can usually just obtain observations from F by $F^{-1}(U)$, where U is a random variable having the uniform distribution on $(0, 1)$. This construction assumes an ability to calculate an inverse function $F^{-1}(\cdot)$. A sample $x_{n,i}$ of n i.i.d. variables can then be obtained from n i.i.d. uniform $(0,1)$ observations. This is repeated B times to get samples $x_{n,1}, \ldots, x_{n,B}$. Then, the true sampling distribution $J_n(P)$ can be approximated by the empirical distribution of the B values $R_n(x_{n,i}, \theta(P))$; specifically,

$$\hat{J}_{n,B}(t) = B^{-1} \sum_{i=1}^{B} 1\{R_n(x_{n,i}, \theta(P)) \leq t\};$$

here and throughout, $1\{E\}$ denotes the indicator of the event E. For large B, $\hat{J}_{n,B}(t)$ will be a good approximation to the true sampling distribution $J_n(t, P)$. One way (though perhaps crude) of quantifying the closeness of this approximation is the following. By the Dvoretsky, Kiefer, Wolfowitz inequality (Serfling, 1980, p. 51), there exists a universal constant C so that

$$Prob_P\{\sup_t |\hat{J}_{n,B}(t) - J_n(t, P)| > d\} \leq C \exp(-2Bd^2).$$

Note that Hu (1985) shows that we can take $C = 4 \cdot 2^{1/2}$ (which is a marked improvement over the original value). Hence, if we desire the probability of the supremum distance between $\hat{J}_{n,B}(\cdot)$ and $J_n(\cdot, P)$ to be greater than d with probability less than ϵ, all we need to do is ensure that B is large enough so that $C \exp(-2Bd^2) \leq \epsilon$. Since B, the number of simulations, is determined by the statistician (assuming enough computing power), the problem is solved. Note, however, we have ignored any error from the use of a pseudo-random number generator, which presumably would be needed to generate the Uniform $(0,1)$ variables.

Classical examples where confidence regions may be formed from a pivot are the following.

Example 1.1.1 (Location and scale families). Consider the situation where $x_n = (X_1, \cdots, X_n)$ is a sample of n real random variables having a distribution function of the form $F[(x - \theta)/\sigma]$, where F is known, θ is a location parameter, and σ is a scale parameter. In general, if $\hat{\theta}_n$ is a location and scale equivariant estimator and if s_n is a location invariant and scale equivariant estimator, then the root $R_n(x_n, \theta(P)) = n^{1/2}[\hat{\theta}_n - \theta(P)]/s_n$ is a pivot. In the case where F is the standard normal distribution function, $\hat{\theta}_n$ is the sample mean, and s_n^2 is the usual unbiased estimate of variance, then R_n has a t-distribution with $n-1$ degrees of freedom. When F is not normal, exact distribution theory may be difficult, but one may resort to simulating the distribution of $R_n(x_n, \theta(P))$ under any P; the accuracy of the approximating distribution will depend on the simulation technique and can be analyzed without having to worry about $\theta(P)$ being unknown. This example can be generalized to a class of parametric problems where group invariance considerations apply.

Example 1.1.2 (Kolmogorov–Smirnov uniform confidence bands). Suppose $x_n = (X_1, \cdots, X_n)$ is a sample of n real random variables having a distribution function F, and the problem is to construct a confidence band for $\theta(F) = F$ based on the root

$$R_n(x_n, F) = n^{1/2} \sup_x |F(x) - \hat{F}(x)|,$$

where $\hat{F}(x)$ is the empirical distribution of the data x_n. In this case, if F is continuous, then the distribution of $R_n(x_n, F)$ under F does not depend on F and its quantiles have been tabled. Without the assumption that F is continuous, the distribution of $R_n(x_n, F)$ under F does depend on F.

1.1.2 Asymptotic Pivotal Method

In general, the above construction breaks down because $R_n(x_n, \theta(P))$ has a distribution $J_n(P)$ which depends on the unknown probability distribution P generating the data. An alternative approach is based on asymptotic considerations. Indeed, it is sometimes the case that $J_n(P)$ converges weakly to a limiting distribution J that is independent of P, in which case the quantiles of J may be used to construct a confidence region for $\theta(P)$.

Example 1.1.3 (Univariate mean). Suppose $x_n = (X_1, \ldots, X_n)$ is a sample of n real-valued random variables having distribution function F. Interest focuses on $\theta(F) = E_F(X_i)$, the mean of the observations. Assume X_i has a finite nonzero variance $\sigma^2(F)$. Let the root R_n be the usual t-statistic defined by $R_n(x_n, \theta(F)) = n^{1/2}[\bar{X}_n - \theta(F)]/s_n$, where $\bar{X}_n$ is the sample mean and s_n^2 is the (unbiased version of the) sample variance. Then, $J_n(F)$ converges weakly to J, the standard normal distribution.

1.1.3 Asymptotic Approximation

More typically, $J_n(P)$ converges to a limiting distribution $J(P)$ that depends on P. An approximation of the asymptotic distribution is $J(\hat{P}_n)$, where $\hat{P}_n$ is some estimate of P. For example, $J(P)$ is often a normal distribution with mean zero and variance $\sigma^2(P)$. The approximation then consists of a normal approximation based on an estimated variance $\sigma^2(\hat{P}_n)$. In any case, the quantiles of $J_n(P)$ may be approximated by those of $J(\hat{P}_n)$. Of course, this approach depends very heavily on knowing the form of the asymptotic distribution as well as being able to construct consistent estimates of unknown parameters upon which $J(P)$ depends. Moreover, the method essentially consists of a double approximation; first, the finite sampling distribution $J_n(P)$ is approximated by an asymptotic approximation $J(P)$, and then $J(P)$ is in turn approximated by $J(\hat{P}_n)$.

Example 1.1.3 (continued) In the previous example, consider instead using the non-studentized root $R_n(x_n, \theta(F)) = n^{1/2}[\bar{X}_n - \theta(F)]$. In this case, $J_n(F)$ converges weakly to $J(F)$, the normal distribution with mean zero and variance $\sigma^2(F)$. The resulting approximation to $J_n(F)$ is the normal distribution with mean zero and variance s_n^2. Alternatively, one can estimate the variance by any consistent estimator, such as $\sigma^2(\hat{F}_n)$, where $\hat{F}_n$ is the empirical distribution function.

Example 1.1.4 (Trimmed mean). Suppose $x_n = (X_1, \ldots, X_n)$ is a sample of n real-valued random variables with distribution function F. Assume that F is symmetric about some unknown value $\theta(F)$. Let $\hat{\theta}_{n,\alpha}(X_1, \ldots, X_n)$ be the α-trimmed mean; specifically,

$$\hat{\theta}_{n,\alpha} = \frac{1}{n - 2\lfloor \alpha n \rfloor} \sum_{i=\lfloor \alpha n \rfloor + 1}^{n - \lfloor \alpha n \rfloor} X_{(i)},$$

where $X_{(1)} \leq X_{(2)} \leq \cdots \leq X_{(n)}$ denote the order statistics and $\lfloor \alpha n \rfloor$ is the greatest integer less than or equal to αn. Consider the root $R_n(x_n, \theta(F)) = n^{1/2}[\hat{\theta}_{n,\alpha} - \theta(F)]$. Then, under reasonable smoothness conditions on F and assuming $0 \leq \alpha < 1/2$, it is well known that $J_n(F)$ converges weakly to $J(F)$, where $J(F)$ is the normal distribution with mean zero and variance $\sigma^2(\alpha, F)$, where

$$\sigma^2(\alpha, F) = \frac{1}{(1-2\alpha)^2} \left[\int_{F^{-1}(\alpha)}^{F^{-1}(1-\alpha)} (t - \theta(F))^2 dF(t) + 2\alpha (F^{-1}(\alpha) - \theta(F))^2 \right]. \tag{1.1}$$

Then, a very simple first-order approximation is to use $J(\hat{F}_n)$, where $\hat{F}_n$ is the empirical distribution; indeed, $J(\hat{F}_n)$ is merely the normal distribution with mean zero and variance $\sigma^2(\alpha, \hat{F}_n)$. Note that $\sigma^2(\alpha, \hat{F}_n)$ is asymptotically equivalent (to a reasonably high order) to

$$s_{n,\alpha}^2 = \frac{1}{n(1-2\alpha)^2} \left[\sum_{i=k+1}^{n-k} (X_{(i)} - \hat{\theta}_{n,\alpha})^2 + k(X_{(k+1)} - \hat{\theta}_{n,\alpha})^2 + k(X_{(n-k)} - \hat{\theta}_{n,\alpha})^2 \right].$$

1.1.4 Bootstrap Approximation

Clearly, none of the above approaches is general enough to cover all applications. Moreover, except for the special case when the root is a pivot, an approach based on asymptotic considerations may lead to procedures with poor properties for finite sample sizes. The bootstrap procedure discussed in this chapter is an alternative, more general, direct approach to approximate the sampling distribution $J_n(P)$. The method consists of estimating the exact finite sampling distribution $J_n(P)$ by $J_n(\hat{P}_n)$, where $\hat{P}_n$ is an estimate of P in $\mathbf{P}$. In this light, the bootstrap estimate $J_n(\hat{P}_n)$ is a simple functional estimate of $J_n(P)$. In nonparametric problems, $\hat{P}_n$ is typically taken to be the empirical distribution of the data. In parametric problems where $\mathbf{P} = \{P_\psi : \psi \in \Psi\}$, $\hat{P}_n$ may be taken to be $P_{\hat{\psi}_n}$, where $\hat{\psi}_n$ is an estimate of ψ.

In general, $J_n(x, \hat{P}_n)$ need not be continuous and strictly increasing in x, so that unique and well-defined quantiles may not exist. To get around this, simply define for any P in $\mathbf{P}$,

$$J_n^{-1}(1-\alpha, P) = \inf\{x : J_n(x, P) \geq 1 - \alpha\}. \tag{1.2}$$

Alternatively, one could define

$$J_n^{-1}(1-\alpha, P) = \sup\{x : J_n(x, P) \leq 1 - \alpha\}$$

or any number between the two. For the sake of definitiveness, we will use (1.2) as the definition. A resulting bootstrap confidence region for $\theta(P)$ of nominal level $1 - \alpha$ takes the form

$$B_n(1 - \alpha, x_n) = \{\theta \in \Theta : R_n(x_n, \theta) \leq J_n^{-1}(1 - \alpha, \hat{P}_n)\}. \tag{1.3}$$

An alternative region, making use of both tails of the bootstrap sampling distribution is

$$\{\theta \in \Theta : J_n^{-1}(\frac{\alpha}{2}, \hat{P}_n) \leq R_n(x_n, \theta) \leq J_n^{-1}(1 - \frac{\alpha}{2}, \hat{P}_n)\}.$$

Outside certain exceptional cases, the bootstrap approximation $J_n(x, \hat{P}_n)$ cannot be calculated exactly. Typically, one resorts to a Monte Carlo approximation to $J_n(P)$. Specifically, conditional on the data x_n, for $i = 1, \ldots, B$, let $x_{n,i}^* = (X_{1,i}^*, \ldots, X_{n,i}^*)$ be a sample of n i.i.d. observations from $\hat{P}_n$. Of course, when $\hat{P}_n$ is the empirical distribution, this amounts to resampling the original observations with replacement. The bootstrap approximation $J_n(\hat{P}_n)$ is then estimated by the empirical distribution of the B values $R_n(x_{n,i}^*, \hat{\theta}_n)$.

The goal of this chapter is to understand some basic consistency properties of the bootstrap approximation just introduced. In Section 1.2, a basic theorem for proving consistency is presented. In Section 1.3, this theorem is applied to the case of the nonparametric mean. This case is very important because one can easily generalize from it to more complicated situations; some generalizations are presented in Section 1.4. Section 1.5 addresses the bootstrap of the empirical process, which plays a fundamental role for bootstrapping nonparametric statistics. Nonlinear differentiable functions of the empirical process are then considered in Section 1.6. Further examples are presented in Section 1.7. Section 1.8 addresses the use of the bootstrap in the context of hypothesis testing.

1.2 Consistency

In this section, the consistency of the bootstrap approximation $J_n(\hat{P}_n)$ to the true sampling distribution $J_n(P)$ of $R_n(x_n, \theta(P))$ is discussed. Typically, one can show that $J_n(P)$ converges weakly to a limit law $J(P)$. Since the bootstrap replaces P by $\hat{P}_n$ in $J_n(\cdot)$, it is useful to study $J_n(\hat{P}_n)$ under more general sequences $\{P_n\}$. In order for the bootstrap to be valid, $J_n(P)$ must be smooth in P. Thus, we are led to studying the asymptotic behavior of $J_n(P_n)$ under fixed sequences of probabilities $\{P_n\}$ which are "converging" to P in a certain sense. Once it is understood how $J_n(P_n)$ behaves for fixed sequences $\{P_n\}$, it is easy to pass to random sequences $\{\hat{P}_n\}$.

In the theorem below, ρ_L is the Lévy metric between distribution functions. It could be replaced by any metric metrizing weak convergence of distributions on the line. The reader not familiar with metrics between distribution functions may skip the first conclusion of the theorem dealing with ρ_L without much lost.

Theorem 1.2.1. *Let $\mathbf{C_P}$ be a set of sequences $\{P_n \in \mathbf{P}\}$ containing the sequence $\{P, P, \cdots\}$. Suppose that, for every sequence $\{P_n\}$ in $\mathbf{C_P}$, $J_n(P_n)$ converges weakly to a common limit law $J(P)$ having distribution function $J(x, P)$. Let x_n be a sample of size n from P. Assume that $\hat{P}_n$ is an estimate of P based on x_n such that $\{\hat{P}_n\}$ falls in $\mathbf{C_P}$ with probability one.*

i. *Then,*

$$\rho_L(J_n(\cdot, P), J_n(\cdot, \hat{P}_n)) \to 0 \text{ with probability one.} \quad (1.4)$$

If $J(\cdot, P)$ is continuous and strictly increasing at $J^{-1}(1 - \alpha, P)$, then

$$J_n^{-1}(1 - \alpha, \hat{P}_n) \to J^{-1}(1 - \alpha, P) \text{ with probability one.} \quad (1.5)$$

ii. *The bootstrap confidence set $B_n(1 - \alpha, x_n)$ given by Equation (1.3) is asymptotically valid; that is,*

$$\mathrm{Prob}_P\{\theta(P) \in B_n(1 - \alpha, x_n)\} \to 1 - \alpha \text{ with probability one.} \quad (1.6)$$

Also, if $J(x, P)$ is continuous in x, then

$$\sup_x |J_n(x, P) - J_n(x, \hat{P}_n)| \to 0 \text{ with probability one.} \quad (1.7)$$

The conclusions of the theorem are quite obvious, given the conditions the theorem imposes. This does not detract from the usefulness of the theorem, as the assumptions can be verified in many interesting examples. To prove the theorem, the following lemma is needed.

Lemma 1.2.1. *Let $\{G_n\}$ be a sequence of distribution functions on the real line converging weakly to a distribution function G; that is, $G_n(x) \to G(x)$ at all continuity points x of G. Assume G is continuous and strictly increasing at $y = G^{-1}(1-\alpha)$. Then,*

$$G_n^{-1}(1-\alpha) = \inf\{x : G_n(x) \geq 1-\alpha\} \to G^{-1}(1-\alpha).$$

Proof. Fix $\delta > 0$. Let $y - \epsilon$ and $y + \epsilon$ be continuity points of G for some $0 < \epsilon \leq \delta$. Then,

$$G_n(y-\epsilon) \to G(y-\epsilon) < 1-\alpha$$

and

$$G_n(y+\epsilon) \to G(y+\epsilon) > 1-\alpha.$$

Hence, for all sufficiently large n,

$$y - \epsilon \leq G_n^{-1}(1-\alpha) \leq y + \epsilon,$$

and so, $|G_n^{-1}(1-\alpha) - y| \leq \delta$ for all sufficiently large n. Since δ was arbitrary, the result is proved. ∎

Proof of Theorem 1.2.1. For the proof of part (i), note that the assumptions imply

$$\rho_L(J_n(\cdot, P), J_n(\cdot, P_n)) \to 0$$

whenever $\{P_n\}$ is any sequence in $\mathbf{C_P}$. Thus, since $\{\hat{P}_n\} \in \mathbf{C_P}$ with probability one, (1.4) follows. Lemma 1.2.1 implies $J_n^{-1}(1-\alpha, P_n) \to J^{-1}(1-\alpha, P)$ whenever $\{P_n\} \in \mathbf{C_P}$; so (1.5) follows.

As far as the proof of part (ii) is concerned, to deduce (1.6), the probability on the left side of (1.6) is equal to

$$Prob_P\{R_n(x_n, \theta(P)) \leq J_n^{-1}(1-\alpha, \hat{P}_n)\}. \tag{1.8}$$

Under P, $R_n(x_n, \theta(P))$ has a limiting distribution $J(\cdot, P)$ and, by (1.5), $J_n^{-1}(1-\alpha, \hat{P}_n) \to J^{-1}(1-\alpha, P)$. Thus, by Slutsky's theorem, (1.8) tends to $J(J^{-1}(1-\alpha, P), P) = 1-\alpha$. Finally, when $J(x, P)$ is continuous in x, it follows by Polya's theorem that

$$\sup_x |J_n(x, P_n) - J(x, P)| \to 0 \tag{1.9}$$

whenever $\{P_n\} \in \mathbf{C_P}$. Then, (1.9) holds when $P_n = P$, and the convergence in (1.9) holds with probability one when P_n is replaced by $\hat{P}_n$; (1.7) now follows. ∎

Remark 1.2.1. In the theorem, one can remove the assumption that $J(\cdot, P)$ is continuous and strictly increasing. In particular, let

$$c_L(1-\alpha, P) = \inf\{x : J(x, P) \geq 1-\alpha\}$$

and
$$c_U(1-\alpha, P) = \sup\{x : J(x, P) \leq 1-\alpha\}.$$

Then, one can show that
$$J[c_L(1-\alpha, P)-, P] \leq \liminf_{n\to\infty} Prob_P[\theta(P) \in B_n(1-\alpha, x_n)]$$
$$\leq \limsup_{n\to\infty} Prob_P[\theta(P) \in B_n(1-\alpha, x_n)] \leq J[c_U(1-\alpha, P), P].$$

In particular, if $J(\cdot, P)$ is continuous at $c_L(1-\alpha, P)$, then the convergence (1.6) still holds; see Beran (1984).

Remark 1.2.2. Often, the set of sequences $\mathbf{C_P}$ can be described as the set of sequences $\{P_n\}$ such that $d(P_n, P) \to 0$, where d is an appropriate metric on the space of probabilities. Indeed, one should think of $\mathbf{C_P}$ as a set of sequences $\{P_n\}$ that are converging to P in an appropriate sense. Thus, the convergence of $J_n(P_n)$ to $J(P)$ is locally uniform in a specified sense. Unfortunately, the appropriate metric d will depend on the precise nature of the choice of root.

1.3 Case of the Nonparametric Mean

In this section, we consider the case of Example 1.1.3, the nonparametric mean. This example deserves special importance because many statistics can be approximated by mean-like statistics, so that understanding this example is fundamental. Given a sample $x_n = (X_1, \cdots, X_n)$ from a distribution F on the real line, consider the problem of constructing a confidence interval for $\theta(F)$ equal to the mean of F. Let $\sigma^2(F)$ denote the variance of F. For now, we assume that $\sigma^2(F)$ is finite. Consider the root $R_n(x_n, \theta(F)) = n^{1/2}[\bar{X}_n - \theta(F)]$, where $\bar{X}_n = \sum_i X_i/n$. The conditions for Theorem 1.2.1 are verified in the following.

Proposition 1.3.1. *Let F be a distribution on the line with finite, nonzero variance $\sigma^2(F)$. Let $\mathbf{C_F}$ be the set of sequences $\{F_n\}$ such that F_n converges weakly to F, $\theta(F_n) \to \theta(F)$, and $\sigma^2(F_n) \to \sigma^2(F)$.*

i. *Then, $J_n(F_n)$ converges weakly to $J(F)$, where $J(F)$ is the normal distribution with mean zero and variance $\sigma^2(F)$.*
ii. *Hence,*
$$\rho(J_n(F), J_n(\hat{F}_n)) \to 0$$
with probability one, and bootstrap confidence intervals are asymptotically valid in the sense of (1.6).

Remark 1.3.1. In Proposition 1.3.1, the assumption that $\theta(F_n) \to \theta(F)$ actually follows from the other assumptions.

12   1. Bootstrap Sampling Distributions

In the proof of Proposition 1.3.1, we will need the following lemma.

Lemma 1.3.1. *Suppose $\{Z_n\}$ is a sequence of real, nonnegative random variables such that the distribution of Z_n converges weakly to the distribution of Z and $E(Z_n) \to E(Z)$, with $E(Z) < \infty$.*

Then, $\{Z_n\}$ satisfies

$$\lim_{\beta \to \infty} \limsup_{n \to \infty} E[Z_n 1(Z_n > \beta)] = 0. \qquad (1.10)$$

Proof. Since the distribution of Z_n converges weakly to the distribution of Z, we can assume, by Skorohod's Almost Sure Representation theorem (see Billingsley [1986], Theorem 25.6), that the Z_n are all defined on the probability space $(\Omega, \mathbf{F}, \mu)$, where Ω is the unit interval, $\mathbf{F}$ is the Borel sets, and μ is Lebesgue measure; furthermore, Z_n converges almost surely to Z. Assume the lemma were false; by monotonicity in β of $E[Z_n 1(Z_n > \beta)]$, we must then have

$$\lim_{\beta \to \infty} \limsup_{n \to \infty} E[Z_n 1(Z_n > \beta)] = \epsilon,$$

where $\epsilon > 0$. Fix $\delta = \epsilon/2$; if $\epsilon = \infty$, take $\delta = 1$. Choose c such that $E[Z 1(Z < c)] > E(Z) - \delta$. Then,

$$E(Z_n) = E[Z_n 1(Z_n < c)] + E[Z_n 1(Z_n \geq c)]. \qquad (1.11)$$

By dominated convergence, $E[Z_n 1(Z_n < c)] \to E[Z 1(Z < c)]$. Thus, the lim sup over n of the right side of equation (1.11) is greater than or equal to $E(Z) - \delta + \epsilon > E(Z)$; on the other hand, the left side of (1.11) tends to $E(Z)$, yielding a contradiction. ∎

Remark 1.3.2. In fact, we can actually conclude from the lemma that $\{Z_n\}$ is uniformly integrable; that is, $\lim_{\beta \to \infty} \sup_n E[Z_n 1(Z_n > \beta)] = 0$.

Proof of Proposition 1.3.1. For purposes of the proof of (i), construct $X_{n,1}, \cdots, X_{n,n}$, which are independent with identical distribution F_n, and set $\bar{X}_n = \sum_i X_{n,i}/n$. We must show that the law of $n^{1/2}(\bar{X}_n - \mu(F_n))$ converges weakly to $J(F)$. It suffices to verify the Lindeberg condition for $Y_{n,i}$, where $Y_{n,i} = X_{n,i} - \mu(F_n)$. This entails showing that, for each $\epsilon > 0$,

$$\lim_{n \to \infty} E[Y_{n,1}^2 1(Y_{n,1}^2 > n\epsilon^2)] = 0. \qquad (1.12)$$

Fix any $\delta > 0$. We must show the expected value on the left side of equation (1.12) is less than or equal to δ for all sufficiently large n. By Lemma 1.3.1, the $Y_{n,1}^2$ satisfy: there exists a β_0 such that for all $\beta > \beta_0$,

$$\limsup_{n \to \infty} E[Y_{n,1}^2 1(Y_{n,1}^2 > \beta_0)] < \delta.$$

But, as soon as $n\epsilon^2 \geq \beta_0$,

$$E[Y_{n,1}^2 1(Y_{n,1}^2 > n\epsilon^2)] \leq E[Y_{n,1}^2 1(Y_{n,1}^2 > \beta_0)],$$

which is less than δ for all sufficiently large n, implying Lindeberg's condition holds.

In order to prove (ii) by applying Theorem 1.2.1 in this example, an appropriate resampling distribution $\hat{F}_n$ must be specified. Let $\hat{F}_n$ be the empirical distribution of the data. Then, $\{\hat{F}_n\} \in \mathbf{C_F}$ with probability one: By the Glivenko–Cantelli theorem, $\sup_x |\hat{F}_n(x) - F(x)| \to 0$ with probability one; also, by the strong law of large numbers, $\theta(\hat{F}_n) \to \theta(F)$ with probability one and $\sigma^2(\hat{F}_n) \to \sigma^2(F)$ with probability one. Thus, bootstrap confidence intervals for the mean based on the root $R_n(x_n, \theta(F)) = n^{1/2}(\bar{X}_n - \theta(F))$ are asymptotically valid in the sense of the theorem. ∎

Remark 1.3.3. The following extension of Proposition 1.3.1 holds. Let $\mathbf{D_F}$ be the set of sequences $\{F_n\}$ such that F_n converges weakly to a distribution G and $\sigma^2(F_n) \to \sigma^2(G) = \sigma^2(F)$. Then, Proposition 1.3.1 holds with $\mathbf{C_F}$ replaced by $\mathbf{D_F}$. (Actually, one really only needs to define $\mathbf{D_F}$ so that and sequence $\{F_n\}$ is tight and any weakly convergent subsequence of $\{F_n\}$ has the above property.) Thus, the possible choices for the resampling distribution are quite large in the sense that the bootstrap approximation $J_n(\hat{G}_n)$ can be valid even if $\hat{G}_n$ is not at all close to F. For example, the choice where $\hat{G}_n$ is normal with mean $\bar{X}_n$ and variance equal to a consistent estimate of the sample variance results in consistency. Therefore, the normal approximation can in fact be viewed as a bootstrap procedure with a perverse choice of resampling distribution.

Remark 1.3.4. Let F and G be two distribution functions on the real line and define $d_p(F, G)$ to be the infimum of $\{E[|X - Y|^p]\}^{1/p}$ over all pairs of random variables X and Y such that X has distribution F and Y has distribution G. It can be shown that the infimum is attained and that d_p is a metric on the space of distributions having a p-th moment. Further, if F has a finite variance $\sigma^2(F)$, then $d_2(F_n, F) \to 0$ is equivalent to F_n converging weakly to F and $\sigma^2(F_n) \to \sigma^2(F)$. Hence, Proposition 1.2.1 may be restated as follows. If F has a finite variance $\sigma^2(F)$ and $d_2(F_n, F) \to 0$, then $J_n(F_n)$ converges weakly to $J(F)$. The metric d_2 is known as the Mallows' metric. For details, see Bickel and Freedman (1981).

Continuing the example of the nonparametric mean, it is of interest to consider the studentized root

$$R_n(x_n, \theta(F)) = n^{1/2}(\bar{X}_n - \theta(F))/\sigma(\hat{F}_n), \qquad (1.13)$$

where $\sigma^2(\hat{F}_n)$ is the usual bootstrap estimate of variance. To obtain consistency of the bootstrap method, we appeal to the following result.

Proposition 1.3.2. *Suppose F is a distribution on the line with finite nonzero variance $\sigma^2(F)$. Let $J_n(F)$ be the distribution of the root (1.13) based on a sample of size n from F. Let $\mathbf{C_F}$ be defined as in Proposition 1.3.1.*

14   1. Bootstrap Sampling Distributions

Then, for any sequence $\{F_n\} \in \mathbf{C_F}$, $J_n(F_n)$ converges weakly to the standard normal distribution.

Before proving this proposition, we first need a weak law of large numbers for a triangular array. The following lemma serves as a suitable version for our purposes.

Lemma 1.3.2. *Suppose $Y_{n,1}, \ldots, Y_{n,n}$ is a triangular array of independent random variables, the n-th row having c.d.f. G_n. Assume G_n converges in distribution to G and*

$$E[|Y_{n,1}|] \to E[|Y|] < \infty$$

as $n \to \infty$, where Y has c.d.f. G.

Then, $n^{-1} \sum_{i=1}^{n} Y_{n,i} \to E(Y)$ as $n \to \infty$.

Proof. Apply (7.5) on p. 235 of Feller (1971) with $s_n = n$, and also make use of the fact that the $\{Y_{n,1}\}$ sequence is uniformly integrable from Remark 1.3.2. ∎

Proof of Proposition 1.3.2. For purposes of the proof, let $X_{n,1}, \cdots, X_{n,n}$ be independent with distribution F_n. By Proposition 1.2.1 and Slutsky's theorem, it is enough to show $\sigma^2(\hat{F}_n) \to \sigma^2(F)$ in probability; that is, for any $\epsilon > 0$,

$$Prob_{F_n} \{ \left| \frac{1}{n} \sum_i (X_{n,i} - \bar{X}_n)^2 - \sigma^2(F) \right| > \epsilon \} \to 0.$$

But, this follows from the previous lemma on the weak law of large numbers for a triangular array.

The consistency of the bootstrap method based on the root (1.13) now follows easily. ∎

It is interesting to consider how the bootstrap behaves when the underlying distribution has an infinite variance (but well-defined mean). The short answer is that the bootstrap procedure considered thus far will fail, in the sense that the convergence in expression (1.7) does not hold. The failure of the bootstrap for the mean in the infinite variance case was first noted by Babu (1984); further elucidation is given in Athreya (1987) and Knight (1989). In fact, the following striking theorem asserts that the simple bootstrap studied thus far will work for the mean if and only if the variance is finite. The theorem, which we state but do not prove, is due to Giné and Zinn (1989).

Theorem 1.3.1. *Let $X_1, \ldots, X_n$ be i.i.d. with c.d.f. F, and let $\hat{F}_n$ be the empirical c.d.f. of $X_1, \ldots, X_n$. The X_i variables are defined on a probability space with outcomes ω. Conditional on $X_1, \ldots, X_n$, let $X_1^*, \ldots, X_n^*$ be i.i.d.*

according to $\hat{F}_n$. Let ρ be any metric metrizing weak convergence. Suppose there exist random variables $\mu_n = \mu_n(X_1, \ldots, X_n)$, and increasing sequence $\{a_n\}$ with $a_n \to \infty$, and a distribution $J(\omega, F)$ (depending possibly on ω or equivalently the infinite sequence $X_1, X_2, \ldots$) such that

$$\rho(\mathcal{L}(\frac{\sum_{i=1}^n X_i^*}{a_n} - \mu_n | X_1, \ldots, X_n), J(\omega, F)) \to 0 \qquad (1.14)$$

with probability one. Here, $\mathcal{L}(\cdot)$ denotes the law of a random variable. Then, $\sigma^2(F) < \infty$, $n/a_n^2 \to c$ for some $c > 0$, $J(\omega, F) = N(0, c\sigma^2(F))$, and μ_n can be taken to be $n\bar{X}_n/a_n$.

Thus, finite variance is an essential assumption for the strong consistency of the bootstrap. Of course, for correct asymptotic level of confidence sets (as in the convergence (1.6)), it is really only necessary to have the convergence (1.14) hold in probability; a related result for the weak consistency of the bootstrap is also provided in Giné and Zinn (1989). The result is that the simple bootstrap studied so far works if and only if the variance is finite. For a nice exposition of these results, see Giné (1997).

Related results for the studentized bootstrap based on approximating the distribution of the root (1.13) were considered by Csörgő and Mason (1989) and Hall (1990). The conclusion is that the bootstrap is strongly or almost surely consistent if and only the variance is finite; the bootstrap is weakly consistent if and only if X_i is in the domain of attraction of the normal distribution.

In fact, it was realized by Athreya (1985) that the bootstrap can be modified so that consistency ensues even with infinite variance. The modification consists in reducing the bootstrap sample size. Indeed, the following is true.

Theorem 1.3.2. Let $X_1, \ldots, X_n$ be i.i.d. with c.d.f. F belonging to the domain of attraction of a α-stable law $J(F)$ for $1 < \alpha < 2$; specifically, assume

$$\rho(\mathcal{L}(\sum_{i=1}^n (X_i - \theta(F))/a_n), J(F)) \to 0$$

for some constants a_n increasing to ∞, where ρ metrizes weak convergence. Suppose $m_n \to \infty$ as $n \to \infty$. Conditional on $X_1, \ldots, X_n$, let $X_1^*, \ldots, X_{m_n}^*$ be i.i.d. according to $\hat{F}_n$, the empirical distribution of $X_1, \ldots, X_n$. Then,

$$\rho(\mathcal{L}(\frac{\sum_{i=1}^{m_n}(X_{n,i}^* - \bar{X}_n)}{a_{m_n}} | X_1, \ldots, X_n), J(F)) \to 0 \text{ in probability}$$

if and only if $m_n/n \to 0$.

The consistency of the bootstrap assuming a reduced sample size m_n with assuming $m_n/n \to 0$ is due to Athreya (1985); the converse is due to Arcones and Giné (1989). A related result on the strong consistency of the bootstrap via reducing the bootstrap sample size is also given in Arcones and Giné (1989). In order to apply these results, one must know the normalizing sequence a_n. However, Arcones and Giné (1991) show that a_n may be estimated so that self-normalizing roots are considered. That is, if $J_n(F)$ denotes the distribution of the root

$$R_n(x_n, \theta(F)) = \frac{\sum_{i=1}^n (X_i - \theta(F))}{(\sum_{i=1}^n X_i^2)^{1/2}}$$

under F. Then, the bootstrap approximation $J_{m_n}(\hat{F}_n)$ satisfies

$$\rho(J_n(F), J_{m_n}(\hat{F}_n)) \to 0 \text{ in probability}$$

if $m_n/n \to 0$ and $m_n \to \infty$.

In fact, these results on the nonparametric mean foreshadow our general results on consistency developed in this treatise. As will be seen in other instances where the simple bootstrap fails, consistency can often be recovered by reducing the bootstrap sample size. The benefit of reducing the bootstrap sample size was recognized first in Bretagnolle (1983). An even more general approach based on subsampling will, of course, be considered.

1.4 Generalizations to Mean-like Statistics

Example 1.4.1 (Linear functional). To generalize Example 1.1.3, let F be a distribution on the line and consider the problem of estimating a linear functional of F, given by

$$\theta(F) = \int \psi(x) dF(x).$$

Let the root $R_n(x_n, \theta(F))$ be defined by

$$R_n(x_n, \theta(F)) = n^{1/2}[\theta(\hat{F}_n) - \theta(F)], \qquad (1.15)$$

where $\hat{F}_n$ is the empirical distribution of the data. Let

$$\tau(F) = \int \psi^2(x) dF(x) - \theta^2(F).$$

In order to verify the conditions in Theorem 1.2.1 for the validity of the bootstrap, the following is needed.

Proposition 1.4.1. *Let F be a distribution on the line such that $\tau(F)$ is finite and nonzero. Let $\mathbf{C_F}$ be the set of sequences of distributions $\{F_n\}$ satisfying $\tau(F_n) \to \tau(F)$, and the distribution of $\psi(Y_n)$ converges weakly to the distribution of $\psi(Y)$ when Y_n has distribution F_n and Y has distribution F. Let $J_n(F)$ be the law of the root (1.15) under F.*

  i. *If $\{F_n\} \in \mathbf{C_F}$, then $J_n(F_n)$ converges weakly to $J(F)$, where $J(F)$ is the normal distribution with mean zero and variance $\tau(F)$.*

1.4 Generalizations to Mean-like Statistics   17

ii. *Hence, the bootstrap is consistent in the sense that*

$$\rho(J_n(F), J_n(\hat{F}_n)) \to 0 \text{ with probability one.}$$

Moreover, the bootstrap confidence interval $B_n(1 - \alpha, x_n)$ given by equation (1.3) is asymptotically valid in the sense of the convergence (1.6).

Proof. The proof of (i) is immediate from Theorem 1.3.1. To prove (ii) of the proposition, appeal to Theorem 1.2.1. Let $\hat{F}_n$ be the empirical distribution of the data. Then, $\{\hat{F}_n\} \in \mathbf{C_F}$ with probability one, and consistency of the bootstrap follows. ∎

Remark 1.4.1. Note that, if ψ is assumed continuous, then $\mathbf{C_F}$ reduces to the set of sequences $\{F_n\}$ such that F_n converges weakly to F and $\tau(F_n) \to \tau(F)$. Without the continuity assumption, this need not imply the distribution of $\psi(Y_n)$ converges weakly to the distribution of $\psi(Y)$. This example illustrates one of many special properties of the empirical distribution function. That is, if Y_n has distribution F_n and Y has distribution F, then Y_n converging in distribution to Y does not imply $\psi(Y_n)$ converges in distribution to $\psi(Y)$ without further assumptions on ψ and/or F. However, suppose $\hat{F}_n$ is the empirical distribution based on a sample $X_1, \cdots, X_n$ from F. Then conditional on $\hat{F}_n$, if Y_n has distribution $\hat{F}_n$, then $\psi(Y_n)$ converges in distribution to $\psi(Y)$ for almost all sample sequences $X_1, X_2, \cdots$, with no assumptions at all on ψ or F (except for measurability assumptions). To see why, conditional on $\hat{F}_n$, $\psi(Y_n)$ has the distribution assigning mass n^{-1} to each of $\psi(X_i)$, $1 \leq i \leq n$. But this is just the empirical distribution of n observations from the distribution of $\psi(Y)$ when Y has distribution F. By Glivenko–Cantelli, it actually follows that the distribution function of $\psi(Y_n)$ (conditional on $\hat{F}_n$) and the distribution function of $\psi(Y)$ are uniformly close as $n \to \infty$ with probability one.

Example 1.4.2 (Multivariate mean). Let $x_n = (X_1, \cdots, X_n)$ be a sample of n observations from F, where X_i takes values in $\mathbb{R}^k$. Let $\theta(F) = E_F(X_i)$ be equal to the mean vector, and let

$$S_n(x_n, \theta(F)) = n^{1/2}(\bar{X}_n - \theta(F)), \tag{1.2.12}$$

where $\bar{X}_n = \sum_i X_i/n$ is the sample mean vector. Let $R_n(x_n, \theta(F)) = \|S_n(x_n, \theta(F))\|$, where $\|\cdot\|$ is any norm on $\mathbb{R}^k$. The validity of the bootstrap method based on the root R_n follows from the following proposition.

Proposition 1.4.2. *Let $L_n(F)$ be the distribution (in $\mathbb{R}^k$) of $S_n(x_n, \theta(F))$ under F, where S_n is defined in (1.2.12). Let $\Sigma(F)$ be the covariance matrix of S_n under F. Let $\mathbf{C_F}$ be the set of sequences $\{F_n\}$ such that F_n converges weakly to F and $\Sigma(F_n) \to \Sigma(F)$, so that each entry of the matrix $\Sigma(F_n)$ converges to the corresponding entry (assumed finite) of $\Sigma(F)$.*

i. *Then, $L_n(F_n)$ converges weakly to $L(F)$, the multivariate normal distribution with mean zero and covariance matrix $\Sigma(F)$.*

18   1. Bootstrap Sampling Distributions

ii. *Assume $\Sigma(F)$ contains at least one nonzero component. Let $\|\cdot\|$ be any norm on $\mathbb{R}^k$ and let $J_n(F)$ be the distribution of $R_n(x_n, \theta(F)) = \|S_n(x_n, \theta(F))\|$ under F. Then, $J_n(F_n)$ converges weakly to $J(F)$, which is the distribution of $\|Z\|$ when Z has distribution $L(F)$.*
iii. *Hence, the bootstrap approximation satisfies*

$$\rho(J_n(F), J_n(\hat{F}_n)) \to 0 \text{ with probability one,}$$

and bootstrap confidence regions based on the root R_n are asymptotically valid in the sense that the convergence (1.6) holds.

Proof. The proof of (i) follows by the Cramer–Wold device and by Proposition 1.3.1.

To prove (ii), note that any norm $\|\cdot\|$ on $\mathbb{R}^k$ is continuous almost everywhere with respect to $L(F)$. A proof can be based on the fact that, for any norm $\|\cdot\|$, the set $\{x \in \mathbb{R}^k : \|x\| = c\}$ has Lebesgue measure zero because it is the boundary of a convex set. (In the case that the support of $L(F)$ is not all of $\mathbb{R}^k$, one can restrict attention to its support and apply the argument since its support is necessarily a linear subspace.) In any case, the continuous mapping theorem applies and so $J_n(F_n)$ converges weakly to $J(F)$.

Part (iii) follows because $\{\hat{F}_n\} \in \mathbf{C_F}$ with probability one, by the Glivenko–Cantelli theorem (on $\mathbb{R}^k$) and the strong law of large numbers. ∎

Note the power of the bootstrap method. Analytical methods based on the root $R_n = \|S_n\|$ would depend heavily on the choice of norm $\|\cdot\|$, but the bootstrap handles them all with equal ease.

Let $\hat{\Sigma}_n = \Sigma(\hat{F})$ be sample covariance matrix. As in the univariate case, one can also bootstrap the root defined by

$$\tilde{R}_n(x_n, \theta(F)) = \|\hat{\Sigma}_n^{-1/2}(\bar{X}_n - \theta(F))\|, \qquad (1.16)$$

provided $\Sigma(F)$ is assumed positive definite.

Example 1.4.3 (Smooth functions of means). Suppose $X_1, \ldots, X_n$ are i.i.d. according to P and take values in some sample S. Suppose $\theta = \theta(P) = (\theta_1, \cdots, \theta_p)$, where $\theta_j = E_P[h_j(X_i)]$ and the h_j are real-valued functions defined on S. Interest focuses on θ or some function f of θ. Let $\hat{\theta}_n = (\hat{\theta}_{n,1}, \cdots, \hat{\theta}_{n,j})$, where $\hat{\theta}_{n,j} = \sum_{i=1}^n h_j(X_i)/n$. Assume moment conditions on the $h_j(X_i)$. Then, by the multivariate mean case, the bootstrap approximation to the distribution of $n^{1/2}(\hat{\theta}_n - \theta)$ is appropriately close in the sense

$$\rho\left(\mathcal{L}_P(n^{1/2}(\hat{\theta}_n - \theta)), \mathcal{L}_{P_n^*}(n^{1/2}(\hat{\theta}_n^* - \hat{\theta}_n))\right) \to 0 \text{ in probability,} \qquad (1.17)$$

where ρ is any metric metrizing weak convergence in $\mathbb{R}^p$. Here, P_n^* refers to the distribution of the data resampled from the empirical distribution conditional on $X_1, \ldots X_n$. Moreover,

$$\rho\left(\mathcal{L}_P(n^{1/2}(\hat{\theta}_n - \theta)), \mathcal{L}(Z)\right) \to 0, \tag{1.18}$$

where Z is multivariate Gaussian with mean zero and covariance matrix Σ having (i,j)-th component

$$Cov(Z_i, Z_j) = Cov[h_i(X_1), h_j(X_1)].$$

To see why, define Y_i to be the vector in $\mathbb{R}^p$ with j-th component $h_j(X_i)$, so that we are exactly back in the multivariate mean case. Now, suppose f is an appropriately smooth function from $\mathbb{R}^p$ to $\mathbb{R}^q$, and interest now focuses on the parameter $\mu = f(\theta)$. Assume $f = (f_1, \cdots, f_q)$, where $f_i(y_1, \cdots, y_p)$ is a real-valued function from $\mathbb{R}^p$ having a nonzero differential at $(y_1, \cdots, y_p) = (\theta_1, \cdots, \theta_p)$. Let D be the $p \times q$ matrix with (i,j) entry $\partial f(y_1, \cdots, y_p)/\partial y_j$ evaluated at $(\theta_1, \cdots, \theta_p)$. Then, the following is true.

Proposition 1.4.3. *Suppose f is a function satisfying the above smoothness assumptions. If $E[h_j^2(X_i)] < \infty$, then equations (1.17) and (1.18) hold. Moreover,*

$$\rho\left(\mathcal{L}_P(n^{1/2}[f(\hat{\theta}_n) - f(\theta)]), \mathcal{L}_{P_n^*}(n^{1/2}[f(\hat{\theta}_n^*) - f(\hat{\theta}_n)])\right) \to 0$$

in probability and

$$\sup_s \left| Prob_P\{\|f(\hat{\theta}_n) - f(\theta)\| \leq s\} - Prob_{P_n^*}\{\|f(\hat{\theta}_n^*) - f(\hat{\theta}_n)\| \leq s\}\right| \to 0$$

in probability.

Proof. The proof follows as equations (1.17) and (1.18) are immediate from the multivariate mean case, and the smoothness assumptions on f imply $n^{1/2}[f(\hat{\theta}_n) - f(\theta)]$ has a limiting multivariate Gaussian distribution with mean 0 and covariance matrix $D\Sigma D'$; see Theorem A of Serfling (1980, p. 122). ∎

1.5 Bootstrapping the Empirical Process

We begin by considering a sample $x_n = (X_1, \ldots, X_n)$ of real-valued observations having a c.d.f F. The empirical c.d.f. $\hat{F}_n$ is defined by

$$\hat{F}_n(t) = n^{-1} \sum_{i=1}^n 1\{X_i \leq t\}.$$

Consider the root

$$S_n(x_n, F) = n^{1/2}[\hat{F}_n(\cdot) - F(\cdot)].$$

Here, we are regarding the root S_n as a random element in a function space $D[-\infty, \infty]$, the space of functions that are right continuous and have left limits on the extended real line. For two such functions F and G, we consider the uniform metric

$$d_\infty(F, G) = \sup_t |F(t) - G(t)|.$$

Finally, $D[-\infty, \infty]$ is endowed with the projection σ-field, so that S_n is indeed a measurable map. S_n is referred to as the empirical process. The distribution of $S_n(x_n, F)$ under F is denoted $L_n(F)$.

With this choice of root, this example deserves special importance. Indeed, if the bootstrap is capable of approximating the distribution of the entire empirical process, then one should be able to conclude that the bootstrap can also approximate the distribution of functionals of the empirical process, such as the distribution of the supremum of the empirical process (which is necessary in Example (1.1.2) for the problem of constructing uniform confidence bands for F). In addition, as will be seen in the next section, estimators that are approximately linear functionals of the empirical process can be studied.

In particular, also consider the root

$$R_n(x_n, \theta(F)) = \sup_t |n^{1/2}[\hat{F}_n(t) - F(t)]|,$$

whose distribution under F is denoted $J_n(F)$. By the continuous mapping theorem, the behavior of $J_n(F)$ can be deduced from the behavior of $L_n(F)$.

As is well known (see Pollard, 1984, Chapter V), $L_n(F)$ converges weakly to $L(F)$, where $L(F)$ is the distribution of a mean zero Gaussian process $Z_F(\cdot)$ whose covariance function is

$$Cov[Z_F(r), Z_F(t)] = F(\min(r, t)) - F(r)F(t). \tag{1.19}$$

This result is usually referred to as Donsker's theorem or the Empirical Central Limit theorem.

In order for the bootstrap approximation $J_n(\hat{F}_n)$ to succeed, we need to understand the behavior of $L_n(F_n)$ for fixed sequences $\{F_n\}$, and then Theorem 1.2.1 can be applied. In fact, the triangular array convergence result in the first part of the following proposition is what is required.

Proposition 1.5.1. *Fix a c.d.f. F on the real line. Let $\{F_n\}$ be a sequence of distributions satisfying $d_\infty(F_n, F) \to 0$ as $n \to \infty$.*

  i. *Then, $L_n(F_n)$ converges weakly to $L(F)$, a mean zero Gaussian process having covariance function given by (1.19).*
  ii. *Hence, the bootstrap satisfies*

$$\rho(J_n(F), J_n(\hat{F}_n)) \to 0 \text{ with probability one,}$$

and uniform confidence bands for F based on the bootstrap are asymptotically valid in the sense of the convergence (1.8).

Proof. For purposes of the proof of (i), let $U_1, U_2, \ldots$ be i.i.d. uniform $(0,1)$ variables. Also, let $B_n(\cdot)$ be the empirical process based on $U_1, \ldots, U_n$;

specifically, let
$$B_n(t) = n^{-1/2} \sum_{i=1}^{n} [1\{U_i \leq t\} - t].$$
Define the quantile function Q_n corresponding to F_n as
$$Q_n(u) = \inf\{t : F_n(t) \geq u\}.$$
Then, construct $X_{n,1}, \ldots, X_{n,n}$ i.i.d. F_n via $X_{n,i} = Q_n(U_i)$, and let $\hat{F}_n$ denote the empirical c.d.f. based on $X_{n,1} \ldots, X_{n,n}$. Note that $F_n(t) \geq u$ if and only if $Q_n(u) \leq t$. Now, the empirical process $S_n(\cdot)$ under F_n satisfies $S_n(t) = B_n(F_n(t))$ because

$$\begin{aligned} S_n(t) &= n^{1/2}[\hat{F}_n(t) - F_n(t)] \\ &= n^{1/2} \sum_{i=1}^{n} [1\{X_{n,i} \leq t\} - F_n(t)] \\ &= n^{1/2} \sum_{i=1}^{n} [1\{X_n(U_i) \leq t\} - F_n(t)] \\ &= n^{1/2} \sum_{i=1}^{n} [1\{U_i \leq F_n(t)\} - F_n(t)] \\ &= B_n(F_n(t)). \end{aligned}$$

Hence, $L_n(F)$, the distribution of $S_n(\cdot)$ under F, is that of $B_n(F_n(\cdot))$. But, by the almost sure representation theorem (see Pollard, 1984, p. 71), there exist versions B_n^+ of B_n such that
$$\sup_s |B_n^+(s) - B(s)| \to 0 \text{ with probability one,}$$
where $B(\cdot)$ is a standard Brownian Bridge process. Hence, $B_n(F_n(\cdot))$ has the same distribution as $B_n^+(F_n(\cdot))$. But,
$$\sup_t |B_n^+(F_n(t)) - B(F(t))|$$
$$\leq \sup_t |B_n^+(F_n(t)) - B(F_n(t))| + \sup_t |B(F_n(t)) - B(F(t))|.$$

The first term on the right-hand side tends to zero with probability one because of the construction of $B_n^+(\cdot)$. The second term tends to zero with probability one because the Brownian Bridge process $B(\cdot)$ has uniformly continuous sample paths with probability one. Thus, the distribution of $B_n(F_n(\cdot))$ converges weakly to that of $B(F(\cdot))$, which is the same limit law of the empirical process under a fixed F. It also follows (by the continuous mapping theorem) that, whenever $d_\infty(F_n, F) \to 0$, $J_n(F_n)$ converges weakly to $J(F)$, which is the law of $\sup_t |B(F(t))|$. Note that, in order to claim the correct asymptotic coverage probability, it must be known that the limit law $J(F)$ is continuous; this follows in this example by

Schmid (1958) or Tsirel'son (1975). Hence, part (ii) follows because $\{\hat{F}_n\}$ converges uniformly to F with probability one. ∎

The bootstrap of the empirical process has been generalized greatly, thanks to the Vapnik–Cervonenkis–Dudley theory of empirical processes. Specifically, let $X_1, \ldots, X_n$ be i.i.d. variables according to a probability law P taking values in a general space S. The empirical measure $\hat{P}_n$ is the law satisfying

$$\hat{P}_n(E) = \frac{1}{n} \sum_{i=1}^{n} 1\{X_i \in E\},$$

for a set E in S. The empirical process S_n indexed by a class of functions **F** is given by

$$S_n(f) = n^{1/2}[\hat{P}_n f - Pf],$$

where $f \in \mathbf{F}$, and the notation Pf means $\int_S f(x) dP(x)$. Clearly, some measurability requirements on the functions f in **F** are required. If we look at the finite dimensional distributions of $S_n(\cdot)$, the multivariate Central Limit theorem applies, so that $(S_n(f_1), \ldots, S_n(f_k))$ converges weakly under P to a multivariate normal distribution with mean zero and covariance that of $Z = (Z_1, \ldots, Z_k)$, where

$$Cov[Z_i, Z_j] = P(f_i f_j) - (Pf_i)(Pf_j);$$

of course, this assumes $Pf_i^2 < \infty$. Hence, by the multivariate mean example, the bootstrap of the empirical process indexed by a finite family of functions is immediate. Needless to say, the technical difficulties appreciate considerably once the family **F** is large. Sufficient conditions for the convergence of the general empirical process can be given in terms of entropy conditions or covering numbers; see Chapter VII of Pollard (1984). In fact, Giné and Zinn (1990) proved a remarkable result that essentially says the following: the empirical process converges to a Gaussian limit if and only if the bootstrap process converges (with probability one) to the same limit; see their paper for a precise statement. The general theory of bootstrapping empirical processes is developed in van der Vaart and Wellner (1996) and in Chapter 2 of Giné (1997).

In the special case when S is k-dimensional Euclidan space, the k-dimensional empirical process was considered in Beran and Millar (1986). Confidence sets for a multivariate distribution based on the bootstrap can then be constructed with asymptotic validation provided by the empirical process theory. Just as in the one-dimensional case, what is required is that the convergence of the empirical process should be the same under a set of sequences $\{P_n\}$ as under P, if $\{P_n\}$ gets close to P. Even in the simplest generalization to the plane, the distribution of supremum of the empirical process varies with P, even if P is assumed continuous (in contrast to the real line). Thus, bootstrapping general empirical processes sometimes allows for the construction of asymptotically valid inferential procedures when no asymptotic approximations are available.

1.6 Differentiability and the Bootstrap

In this section, we consider the consistency of the bootstrap for a certain class of nonlinear estimators. As will be seen, under an assumption that implies a linear approximation to the statistical functional of interest, the bootstrap is asymptotically valid. The differential approach developed here is not the most general, but it is fairly simple and avoids certain technical difficulties encountered in approaches based on weaker assumption. The gist of what we are showing is that the bootstrap is valid for certain nonlinear estimators. Once the ideas of our approach are understood, they can be generalized easily. To date, the weakest form of differentiability which has proved useful in statistical applications is compact or Hadamard differentiability, as initiated in Reeds (1976). We will not consider this form of differentiability; a systematic and general treatment is presented in van der Vaart and Wellner (1996).

As before, consider the problem of constructing a confidence interval for a parameter $\theta(F)$ based on a sample $x_n = (X_1, \cdots, X_n)$ of size n from a distribution F on the real line. Let

$$R_n(x_n, \theta(F)) = n^{1/2}[\theta(\hat{F}_n) - \theta(F)], \qquad (1.20)$$

where $\hat{F}_n$ is the empirical distribution of the X_i. Let $J_n(F)$ be the distribution under F of the root R_n defined by equation (1.20). As we have seen, in order for the bootstrap to be consistent, $J_n(F)$ must be, in some sense, smooth in F. But, smoothness in F can often be traced back to the functional $\theta(\cdot)$. If $\theta(\cdot)$ is an appropriately smooth functional of F, then a generalization of the classical "delta method" allows a linear approximation of the form

$$\theta(\hat{F}_n) - \theta(F) = \int g_F d(\hat{F}_n - F) + o(\|\hat{F}_n - F\|).$$

The term $o(\|\hat{F}_n - F\|)$ should be small (in probability) due to the closeness of $\hat{F}_n$ to F, and the linear term $\int g_F d(\hat{F}_n - F)$ tends in distribution to the normal distribution with mean 0 and variance $Var_F[g_F(X_i)]$. Based on the validity of the bootstrap in the context of estimating F itself, and the validity of the bootstrap for a linear functional, it is plausible that $J_n(\hat{F}_n)$ should also be approximately normal with mean zero and variance $Var_F[g_F(X_i)]$. It is the purpose of this section to investigate the use of differentiability in proving consistency of the bootstrap.

Various notions of differentiability in statistics have been used with varying degrees of success. While the fundamental idea of approximating a nonlinear functional by a linear one is incredibly useful, the goal of striving

24   1. Bootstrap Sampling Distributions

for the weakest notion of differentiability useful for statistics is countered by the difficulty in establishing that differentiability holds. In any case, we begin the development under the assumption of Fréchet differentiability, which is quite strong. However, once some of the issues are understood in this context, the ideas can be appropriately generalized to prove more abstract results.

Let $\theta(\cdot)$ be a real-valued functional defined on a class **F** of distributions on the real line. It will be assumed that **F** contains all distributions with finite support, so that $\theta(\hat{F}_n)$ is well-defined when $\hat{F}_n$ is an empirical distribution. Let d be a semi-metric on **F**. Thus, it will not be assumed that $d(F, G) = 0$ implies $F = G$; the essential requirement is that the triangle inequality hold: $d(F, G) \leq d(F, H) + d(H, G)$. Typically, d will be generated by a semi-norm $\|\cdot\|$. In either case, we will write $\|F - G\|$ for $d(F, G)$. The usual choice for $\|\cdot\|$ is the supremum norm or Kolmogorov distance between distribution functions:

$$\|F - G\|_\infty = d_\infty(F, G) = \sup_x |F(x) - G(x)|.$$

Definition 1.6.1. The functional $\theta(\cdot)$ defined on **F** is Fréchet differentiable at F with respect to $\|\cdot\|$ if there exists a function g_F such that $\int g_F dF = 0$ and for G in **F**,

$$\theta(G) = \theta(F) + \int g_F d(G - F) + o(\|G - F\|) \qquad (1.21)$$

as $\|G - F\| \to 0$. The function g_F is called the Fréchet derivative of $\theta(\cdot)$ at F.

The assumption that $\int g_F dF = 0$ entails no loss of generality. Indeed, without this requirement, if g_F satisfies (1.21), then so does $g_F + c$ for any constant c. For convenience, this lack of uniqueness has been avoided.

It should be noted that different, but similar, definitions of Fréchet differentiability have been given in statistical contexts. In the usual setup, Fréchet differentiability asserts a certain linear approximation of a map from two (usually normed) vector spaces. Typically, **F** can be embedded in a normed vector space, but it seems undesirable to have to do this from the start.

Another difference is that we have assumed the linear approximation has an integral representation (as is done throughout much of the statistical literature), $\int g_F d(G - F)$, whereas Fréchet differentiability typically only asserts the existence of a linear approximation. However, the extra requirement that the approximating linear functional have such an integral representation is generally quite weak, and essentially follows as long as **F** contains all point mass distributions. Indeed, the more standard definition of Fréchet differentiability would replace the term $\int g_F d(G - F)$ by $L_F(G - F)$, where L_F is a linear functional defined on the space of differences of c.d.f.s (or equivalently differences of signed measures). However, we can simply define g_F so that $g_F = L_F(\delta_x - \delta_{x_0})$, where δ_x denotes the distribution which is point mass at x and x_0 is any fixed value. This assumes the space of distributions defining the functional of interest includes point masses, which usually is not much of a restriction for our purpose since we typically consider functionals defined on empirical distributions anyway. It then follows that $L_F(G - F) = \int g_F d(G - F)$ whenever F and G are finite linear combinations of point masses, and then for any limit of such distributions in any topology for which L_F is continuous. A discussion of this point is made in Dudley (1990).

The first major consequence of Fréchet differentiability is the following.

Theorem 1.6.1. *Assume* **F** *contains all distributions F with finite support. Also, assume $\theta(\cdot)$ is Fréchet differentiable at F with respect to $\|\cdot\|$, and whose derivative, g_F, at F satisfies $0 < Var_F[g_F(X)] = \int g_F^2 dF < \infty$. Let $X_1, \cdots, X_n$ be a sample from F, and let $\hat{F}_n$ be the empirical c.d.f. based on $X_1, \cdots, X_n$. Assume*

$$n^{1/2}\|\hat{F}_n - F\| = O_P(1), \tag{1.22}$$

that is, $n^{1/2}\|\hat{F}_n - F\|$ is a tight sequence of random variables.

Then, $n^{1/2}[\theta(\hat{F}_n) - \theta(F)]$ converges in distribution to the normal distribution with mean zero and variance $\tau(F)$, where $\tau(F) = Var_F[g_F(X_i)]$.

Proof. By differentiability,

$$n^{1/2}[\theta(\hat{F}_n) - \theta(F)] = n^{-1/2} \sum_{i=1}^{n} g_F(X_i) + R_n(X_1, \cdots, X_n),$$

where

$$R_n = n^{1/2} o(\|\hat{F}_n - F\|) = n^{1/2} \|\hat{F}_n - F\| o_P(1).$$

But, $n^{1/2}\|\hat{F}_n - F\|$ is tight, implying $R_n \to 0$ in probability. The first term $n^{-1/2} \sum_i g_F(X_i)$ converges in distribution to the normal distribution with mean zero and variance $\tau(F)$. The result now follows by Slutsky's theorem. ∎

Corollary 1.6.1. *Assume* **F** *contains all distributions F with finite support. Also, assume $\theta(\cdot)$ is Fréchet differentiable at F with respect to the supremum norm $\|\cdot\|_\infty$, and whose derivative, g_F, at F satisfies $0 < Var_F[g_F(X)] < \infty$. Let $X_1, \cdots, X_n$ be a sample from F, and let $\hat{F}_n$ be the empirical distribution function of the data.*

Then, $n^{1/2}[\theta(\hat{F}_n) - \theta(F)]$ converges in distribution to the normal distribution with mean zero and variance $Var_F[g_F(X_i)]$.

Proof. Since $n^{1/2}[\hat{F}_n(\cdot) - F(\cdot)]$ converges in distribution in $L_\infty(\mathbb{R})$ to a Gaussian process it follows that $n^{1/2}\|\hat{F}_n - F\|$ actually has a limiting distribution. Thus, the result follows by Theorem 1.6.1. An alternative method to showing the weaker result that $n^{1/2}\|\hat{F}_n - F\|$ is tight is to apply the Dvoretsky, Kiefer, Wolfowitz inequality. ∎

A weaker notion than Fréchet differentiability is Gateaux differentiability.

Definition 1.6.2. The functional $\theta(\cdot)$ defined on **F** is Gateaux differentiable at F in the direction G if $\theta'_F(G - F)$ defined by

$$\theta'_F(G - F) = \lim_{\lambda \to 0^+} [\theta(F + \lambda(G - F)) - \theta(F)]/\lambda \qquad (1.23)$$

is well-defined and exists.

Implicit in the definition is that **F** contains all distributions of the form $F + \lambda(G - F)$ for all sufficiently small λ, which holds when **F** is convex. Assuming that $\theta(\cdot)$ is Gateaux differentiable at F in the direction δ_x, where δ_x is the distribution assigning mass one to x, the influence function (or sometimes called the influence curve) of $\theta(\cdot)$ is defined as

$$g_F(x) = \theta'(\delta_x - F).$$

Then, if for all G in **F**,

$$\theta'(G - F) = \int g_F d(G - F), \qquad (1.24)$$

then $\theta(\cdot)$ is said to have a von-Mises derivative with derivative θ'.

Unfortunately, the notion of Gateaux or von-Mises differentiability is too weak for statistical purposes. Indeed, a functional $\theta(\cdot)$ may have a von-Mises derivative of the form (1.24) such that $\int g_F^2 dF < \infty$, but nevertheless, $n^{1/2}[\theta(\hat{F}_n) - \theta(F)]$ is not asymptotically normal. For an example, see Example 2.2.3 of Fernholz (1983).

The usefulness of the notion of Gateaux differentiability stems from the fact that it is relatively easy to calculate, as it boils down to finding the derivative of a real-valued function of a real variable. Moreover, the next result says that if a functional is Fréchet differentiable it is Gateaux differentiable. Hence, if $\theta(\cdot)$ has an influence function g_F, one immediately knows that, if $\theta(\cdot)$ is Fréchet differentiable as well, then $\theta(\hat{F}_n)$ is asymptotically normal with asymptotic variance $Var_F[g_F(X)]$. In other words, Gateaux differentiability allows one to guess at the asymptotic normality result that could be proved once Fréchet differentiability has been verified. Therefore, a useful approach is to first determine whether a functional is Gateaux differentiable, thus enabling one to assume the form of the Fréchet derivative g_F. Then, one needs to check the validity of the linear approximation by looking directly at the remainder term.

In the following result, $\|\cdot\|$ is actually assumed to be a semi-norm on **F**, that is, if $\lambda > 0$, $\|\lambda(G - F)\| = \lambda\|G - F\|$.

Theorem 1.6.2. *Assume $\theta(\cdot)$ is defined on a convex class **F** of distributions. If $\theta(\cdot)$ has a Fréchet derivative g_F at F with respect to a* semi-norm $\|\cdot\|$, $\theta(\cdot)$ is Gateaux differentiable and

$$\theta'_F(G-F) = \int g_F d(G-F).$$

Proof. Fix G and let $F_\lambda = F + \lambda(G-F)$. Then, $F_\lambda - F = \lambda(G-F)$, so that

$$\|F_\lambda - F\| = \lambda \|G - F\| \to 0$$

as $\lambda \to 0$. So, as $\lambda \to 0$,

$$\theta(F_\lambda) - \theta(F) = \int g_F d(F_\lambda - F) + o(\|F_\lambda - F\|)$$

$$= \lambda \int g_F d(G-F) + \lambda o(1).$$

Thus,

$$\lim_{\lambda \to 0^+} [\theta(F_\lambda) - \theta(F)]/\lambda = \int g_F d(G-F). \quad \blacksquare$$

Example 1.6.1 (Square of mean). Suppose F has a (finite) mean $\mu(F)$; consider $\theta(F) = \mu^2(F)$. Then,

$$\theta(F + \lambda(G-F)) = \{\mu(F) + \lambda[\mu(G) - \mu(F)]\}^2.$$

Taking the derivate of this expression with respect to λ and evaluating it at $\lambda = 0$ yields

$$\theta'_F(G-F) = 2\mu(F)[\mu(G) - \mu(F)]$$

and so the influence function is

$$g_F(x) = \theta'_F(\delta_x - F) = 2\mu(F)[x - \mu(F)].$$

To check Fréchet differentiability, it is easy to see that

$$\theta(G) - \theta(F) - \int g_F d(G-F) = [\mu(G) - \mu(F)]^2.$$

Hence, it is not that case that $\theta(\cdot)$ is Fréchet differentiable with respect to the supremum distance between c.d.f.s $\|\cdot\|_\infty$. Indeed, in order to have Fréchet differentiability with respect to $\|\cdot\|$, it is required that

$$\frac{|\mu(G_n) - \mu(F)|^2}{\|G_n - F\|} \to 0$$

as $\|G_n - F\| \to 0$ for any sequence $\{G_n\} \in \mathbf{F}$. But, taking

$$G_n = (1 - \lambda_n)F + \lambda_n \delta_n$$

28   1. Bootstrap Sampling Distributions

for any sequence λ_n converging to zero, shows this convergence does not hold for $\|\cdot\|_\infty$. However, a simple remedy is to define the semi-norm

$$\|G - F\| = |\mu(G) - \mu(F)|,$$

and Fréchet differentiability trivially holds now. Furthermore, the requirement (1.22) in Theorem 1.6.1 amounts to checking the tightness of $n^{1/2}|\bar{X}_n - \mu(F)|$, which clearly holds if we further restrict $\mathbf{F}$ to c.d.f.s having a finite variance. This example is generalized to functions of means in Dudley (1990), who shows the benefits of changing the norm while retaining Fréchet differentiability.

Example 1.6.2 (Trimmed mean, continuation of Example 1.1.4). Assume F is symmetric about $\theta(F)$, as in Example 1.1.4. The trimmed mean functional is defined as

$$\theta_\alpha(F) = \frac{1}{1-2\alpha} \int_\alpha^{1-\alpha} F^{-1}(p) dp.$$

It can be shown (Serfling, 1980, p. 236) that the influence function is

$$g_F(x) = \begin{cases} \frac{1}{1-2\alpha}[F^{-1}(\alpha) - F^{-1}(\frac{1}{2})], & \text{if } x < F^{-1}(\alpha); \\ \frac{1}{1-2\alpha}[x - F^{-1}(\frac{1}{2})], & \text{if } F^{-1}(\alpha) \le x \le F^{-1}(1-\alpha); \\ \frac{1}{1-2\alpha}[F^{-1}(1-\alpha) - F^{-1}(\frac{1}{2})], & \text{otherwise.} \end{cases}$$

The desired conclusion is that $n^{1/2}[\theta_\alpha(\hat{F}_n) - \theta(F)]$ is asymptotically normal with mean zero and variance $Var_F[g_F(X)]$; one can easily check that $Var_F[g_F(X)]$ agrees with $\sigma^2(\alpha, F)$ given in Equation (1.1). Various approaches to proving asymptotic normality, including the differentiable approach, are presented in Serfling (1980, Chapter 8).

Example 1.6.3 (p-th quantile). Consider the p-th quantile of F defined by

$$\theta(F) = F^{-1}(p) = \inf\{x : F(x) \ge p\}.$$

Assume F has a density f. Then, one can check by simple calculus that the influence function is

$$g_F(x) = \frac{p - 1\{x \le \theta(F)\}}{f(\theta(F))}.$$

Then, the desired conclusion is that $n^{1/2}[\theta(\hat{F}_n) - \theta(F)]$ is asymptotically normal with mean zero and variance

$$Var_F[g_F(X)] = \frac{p(1-p)}{f^2(\theta(F))}.$$

Again, the differentiable approach leads directly to the right answer (though admittedly there are even simpler approaches in this example), and rigorous verification of asymptotic normality is discussed in Serfling (1980, p. 236). Note, however, that the quantile functional is not Fréchet differentiable with respect to the supremum norm. However, weaker notions of differentiability do hold in this example; see van der Vaart and Wellner (1996, Section 3.9).

The following result is the main result of this section. Under a differentiability hypothesis for a nonlinear estimator, the bootstrap is weakly consistent. The theorem assumes Fréchet differentiability with respect to the sup norm, but the proof can be trivially adapted to include other semi-norms as well, such as the one in Example 1.6.1.

Theorem 1.6.3. *Assume $\mathbf{F}$ contains distributions with finite support. Also, assume $\theta(\cdot)$ is Fréchet differentiable at F with respect to the supremum norm $\|\cdot\|_\infty$, and whose derivative, g_F at F satisfies $0 < Var_F[g_F(X)] < \infty$. Let $J_n(F)$ denote the distribution under F of the root*

$$R_n(x_n, \theta(F)) = n^{1/2}[\theta(\hat{F}_n) - \theta(F)].$$

i. *Let $\mathbf{C_F}$ be the set of sequences $\{F_n\}$ satisfying*

$$\limsup_n n^{1/2}\|F_n - F\| < \infty \text{ and } \int g_F^2 dF_n \to \int g_F^2 dF \text{ as } n \to \infty.$$

Then, $J_n(F_n)$ converges to the normal distribution with mean zero and variance $\int g_F^2 dF$.

ii. *Suppose $X_1, \ldots, X_n$ are i.i.d. F with empirical distribution $\hat{F}_n$. Then, the bootstrap approximation $J_n(\hat{F}_n)$ to $J_n(F)$ satisfies*

$$\rho(J_n(\hat{F}_n), J_n(F)) \to 0 \text{ in probability under } F,$$

and bootstrap confidence intervals are valid in the sense of the convergence (1.6).

Proof. To prove (i), simply mimic the proofs of Theorem 1.6.1 and Corollary 1.6.1 with F replaced by F_n. The only new result that is called upon is that, if $\hat{F}_n$ denotes the empirical distribution of n observations which are i.i.d. F_n, then $n^{1/2}\|\hat{F}_n - F_n\|$ is tight (and actually has a limiting distribution), by appealing to Proposition 1.5.1 (i). Indeed, $n^{1/2}[\hat{F}_n(\cdot) - F_n(\cdot)]$ converges weakly in $D[-\infty, \infty]$ equipped with the sup norm. Actually, the tightness holds for any sequence $\{F_n\}$. Then, the differentiability hypothesis yields

$$n^{1/2}[\theta(\hat{F}_n) - \theta(F_n)] = n^{1/2}\int g_F d(\hat{F}_n - F_n) + R_{n,1} + R_{n,2},$$

where $R_{n,1} = n^{1/2}o(\|\hat{F}_n - F\|)$ and $R_{n,2} = n^{1/2}o(\|F_n - F\|)$. Because of $\{F_n\} \in \mathbf{C_F}$, it follows that $R_{n,2} \to 0$ in probability under F_n. The triangle inequality gives

$$R_{n,1} = n^{1/2}o(\|\hat{F}_n - F_n\|) + n^{1/2}o(\|F_n - F\|).$$

The first term on the right side tends to zero in probability by appeal to the above tightness result; the second term tends to zero in probability

30   1. Bootstrap Sampling Distributions

again by definition of $\mathbf{C_F}$. Hence, $J_n(F_n)$, behaves (by Slutsky's theorem) like the distribution of the linear term $n^{1/2}\int g_F d(\hat{F}_n - F_n)$, which is asymptotically normal, mean zero and variance $\int g_F^2 dF$; see Example 1.4.1 and Proposition 1.4.1, part (i). (Note that the assumption $\int g_F^2 dF_n \to \int g_F^2 dF$ is utilized in the linear term.)

To prove (ii), let $\hat{F}_n$ now denote the empirical distribution of n observations from F. Then, we must examine whether $\{\hat{F}_n\}$ falls in $\mathbf{C_F}$. By the strong law of large numbers, $\int g_F^2 d\hat{F}_n \to \int g_F^2 dF$ with probability one. However, it is not true that

$$\limsup_n n^{1/2}\|\hat{F}_n - F\| < \infty \text{ with probability one.}$$

But, we can appeal to the almost sure representation theorem as follows. Letting $S_n(\cdot)$ denote the process $S_n(t) = n^{1/2}[\hat{F}_n(t) - F(t)]$ and utilizing the fact that $S_n(\cdot)$ converges weakly to a Gaussian process having uniformly bounded sample paths (with probability one), there exist versions $S_n^+(\cdot)$ of $S_n(\cdot)$ such that $S_n^+(\cdot)$ has the same distribution as $S_n(\cdot)$ and S_n^+ converges almost surely to a Gaussian process having uniformly bounded sample paths. Then, define $\hat{F}_n^+(\cdot)$ so that $S_n^+(\cdot) = n^{1/2}[F_n^+(\cdot) - F(\cdot)]$ and it follows that $\{F_n^+\} \in \mathbf{C_F}$ with probability one. Hence, $\rho(J_n(F_n^+), J_n(F)) \to 0$ with probability one; thus, the convergence holds in probability if F_n^+ is replaced by $\hat{F}_n$. ■

1.7 Further Examples

Thus far, it has been seen that the bootstrap is asymptotically valid when $J_n(P)$ is somehow smooth as a function of P. Moreover, when the root is appropriately linear and the linear term has a finite variance, the required smoothness holds. However, even in the exact linear case, the bootstrap fails if the assumption of finite variance does not hold. The bootstrap can fail in other ways as well. We defer counterexamples to the bootstrap to the next chapter. We conclude here by outlining some further examples of bootstrap success.

Example 1.7.1 (Median). Suppose $X_1, \ldots, X_n$ are i.i.d. with c.d.f. F having a unique median $\theta(F)$. Assume F has a positive derivative f at $\theta(F)$. As usual, $\hat{F}_n$ denotes the empirical c.d.f., so that $\theta(\hat{F}_n)$ denotes the sample median. Consider the root

$$R_n(x_n, \theta(F)) = n^{1/2}[\theta(\hat{F}_n) - \theta(F)].$$

As discussed in Example 1.6.3, $J_n(F)$ converges weakly to the normal distribution with mean zero and variance $[4f^2((\theta(F))]^{-1}$. Again, Theorem 1.6.3 does not apply because the median functional is not Fréchet differentiable, but as expected and claimed, the bootstrap is still valid. To appreciate why, $J_n(F)$ must be smooth in F, in which case we can appeal to our basic consistency theorem. To this end, let $\mathbf{C}_F$ be the set of sequences $\{F_n\}$ satisfying

$$\lim_{n\to\infty} n^{1/2}[F_n(\theta(F_n) + n^{1/2}x) - \frac{1}{2}] = xf(\theta(F))$$

for every real x. Then, it can be shown that, if $\{F_n\} \in \mathbf{C}_F$, $J_n(F_n)$ converges weakly to $J(F)$. Furthermore, by appealing to the almost sure representation theorem, it can be shown there exist versions of the empirical c.d.f. that fall in $\mathbf{C}_F$ with probability one. Weak consistency of the bootstrap follows. Details are provided in Beran (1984).

Historically, this example was a major triumph for the bootstrap, because the usual delete-1 jackknife estimate of variance of the sample median is inconsistent; moreover, the bootstrap estimate of variance (the variance of $J_n(\hat{F}_n)$) is consistent under a weak moment hypothesis. On the other hand, if F is not differentiable at $\theta(F)$ or if it is but the derivative is zero, the bootstrap fails (see Huang, Sen, and Shao, 1996).

Example 1.7.2 (Linear regression).

Here, we present our first example that strays slightly outside of the i.i.d. setup. Suppose the data x_n is a column vector $(X_1, \ldots, X_n)'$ of observations satisfying the linear regression model

$$x_n = A_n\theta + \epsilon_n,$$

where $A_n = \{a_{i,j,n}\}$ is the $n \times p$ matrix of nonrandom regressors whose rank is p, θ is a $p \times 1$ vector and

$$\epsilon_n = (E_1, \ldots, E_n)'$$

is a column vector of n i.i.d. random variables having c.d.f. F; F is assumed to have mean zero and finite variance $\sigma^2(F)$. Since A_n is fixed, what is unknown about the data generating mechanism is the distribution of the errors F and the regression coefficients θ. The least squares estimator of θ is

$$\hat{\theta}_n = (A'_n A_n)^{-1} A'_n x_n$$

and the usual estimate of $\sigma^2(F)$ is then

$$s_n^2 = (n-p)^{-1} \|x_n - A_n\hat{\theta}_n\|^2.$$

Let $J_n(F)$ denote the sampling distribution of $s_n^{-1}(A'_n A_n)^{1/2}(\hat{\theta}_n - \theta)$ under F; note that this distribution does not depend on θ.

The least squares residuals are $\hat{\epsilon}_n = x_n - A_n\hat{\theta}_n$; the i-th component of $\hat{\epsilon}_n$ is denoted $\hat{E}_{n,i}$. Let F_n^* be the empirical c.d.f. of the centered residuals $\epsilon_n^* = (E_{n,1}^*, \ldots, E_{n,n}^*)$ having i-th component defined by

$$E_{n,i}^* = \hat{E}_{n,i} - n^{-1}\sum_{i=1}^{n} \hat{E}_{n,i}.$$

32   1. Bootstrap Sampling Distributions

The bootstrap estimator of $J_n(F)$ is then $J_n(F_n^*)$.

We now outline consistency of the bootstrap. For the asymptotics, assume p is fixed as $n \to \infty$. Also, assume $\max n^{-1/2}|a_{i,j,n}| \to 0$ and $n^{-1}A_n'A_n \to A$, where A is finite and positive definite. Let $\mathbf{C}_F$ be the set of sequence $\{F_n\}$ satisfying F_n converges weakly to F, the mean of F_n is zero, and $\sigma^2(F_n) \to \sigma^2(F)$. Then, under $\{F_n\} \in \mathbf{C}_F$, the Lindeberg Central Limit theorem gives the distribution of $(A_n'A_n)^{1/2}(\hat{\theta}_n - \theta)$ converges weakly to that of Z, where Z is multivariate normal, mean zero, and covariance matrix I_p, the p by p identity matrix. Also,

$$s_n^2 = \frac{1}{n-p}\|\epsilon_n - B_n\epsilon_n\|^2,$$

where $B_n = A_n(A_n'A_n)^{-1}A_n'$ is a symmetric projection (idempotent) matrix of rank p. So,

$$s_n^2 = \frac{1}{n-p}[\|\epsilon_n\|^2 - \|B_n\epsilon_n\|^2].$$

By the law of large numbers, $\|\epsilon_n\|^2/(n-p) \to \sigma^2(F)$ with probability one. Moreover,

$$E[\frac{1}{n-p}\|B_n\epsilon_n\|^2] = \frac{1}{n-p}tr[B_n\epsilon_n\epsilon_n'B_n] = \frac{1}{n-p}tr[B_n\sigma^2(F_n)I_pB_n]$$

$$= \frac{\sigma^2(F_n)}{n-p}tr(B_n^2) = \frac{\sigma^2(F_n)}{n-p}tr(B_n) = \frac{\sigma^2(F_n)p}{n-p} \to 0.$$

Hence, $s_n^2 \to \sigma^2(F)$ in probability under F_n. Thus, $J_n(F_n)$ converges weakly to the distribution of Z, whenever $\{F_n\} \in \mathbf{C}_F$.

It remains to check $\{F_n^*\}$ falls in $\mathbf{C}_F$ with probability one. By Glivenko–Cantelli, $\hat{F}_n$ converges weakly to F with probability one. By the law of large numbers, the mean of $\hat{F}_n$ converges to 0 with probability one, and $\sigma^2(\hat{F}_n) \to \sigma^2(F)$ with probability one. We also claim that $n^{-1}\|B_n\epsilon_n\|^2 \to 0$ with probability one. The above arguments show this last convergence holds in probability and so it is plausible it holds with probability one as well (see Freedman, 1981, Lemma 2.4). By these convergences, one can conclude that $\{F_n^*\}$ falls in $\mathbf{C}_F$ with probability one (see Beran, 1984). This example was first considered by Freedman (1981), who also points out that the bootstrap fails without the proper centering of the residuals.

Example 1.7.3 (Correlation model). Suppose $X_1, \ldots, X_n$ are i.i.d. random row vectors of dimension $p+1$ with unknown c.d.f. F in $\mathbf{R}^{p+1}$. The first component of X_i is denoted Y_i and T_i is the p-dimensional row vector forming the remaining components of X_i. Assume $E_F(T_i'Y_i)$ and $V(F) = E_F(T_i'T_i)$ exist and are finite, and that $V(F)$ is positive definite. The minimum mean squared error linear predictor of Y_i based on T_i is

$T_i\theta(F)$, where
$$\theta(F) = V^{-1}(F)E_F(T_i'Y_i).$$

Let $\hat{F}_n$ be the empirical c.d.f. of $X_1, \ldots, X_n$. Consider the estimator $\theta(\hat{F}_n)$ of $\theta(F)$. Note that this estimator is the same as the one discussed in the previous linear regression example; however, the different models have a different distribution theory. Let $J_n(F)$ denote the distribution of $n^{1/2}[\theta(\hat{F}_n) - \theta(F)]$ under F. The bootstrap estimator is then $J_n(\hat{F}_n)$.

We outline consistency of the bootstrap (for details, see Beran, 1984). First, define
$$A(F) = E_F[(Y_i - T_i\theta(F))^2 T_i'T_i],$$
which is assumed finite, as is $E_F(T_i'T_i)^2$. Let $\mathbf{C}_F$ be the set of sequence $\{F_n\}$ satisfying F_n converges weakly to F, $\theta(F_n) \to \theta(F)$, $A(F_n) \to A(F)$, and $V(F_n) \to V(F)$. It can be shown by the Lindeberg Central Limit theorem and the law of large numbers that $J_n(F_n)$ converges weakly to $J(F)$, the law of the multivariate normal distribution with mean zero and covariance matrix $V^{-1}A(F)V^{-1}(F)$ whenever $\{F_n\} \in \mathbf{C}_F$. Furthermore, $\{\hat{F}_n\}$ falls in $\mathbf{C}_F$ with probability one, and so the bootstrap is consistent.

1.8 Hypothesis Testing

In this section, we consider the use of the bootstrap for the construction of hypothesis tests. Assume the data $X_1, \ldots, X_n$ are i.i.d. from an unknown law P. The observations are assumed to take values in a sample space S. The null hypothesis H_0 asserts that P belongs to a certain family of distributions $\mathbf{P_0}$, while the alternative hypothesis H_1 asserts that P belongs to a family $\mathbf{P_1}$. Of course, we assume the intersection of $\mathbf{P_0}$ and $\mathbf{P_1}$ is the null set, and the unknown law P belongs to $\mathbf{P}$, the union of $\mathbf{P_0}$ and $\mathbf{P_1}$.

There are several approaches one can take to construct a hypothesis test. First, consider the case when the null hypothesis can be expressed as a hypothesis about a real- or vector-valued parameter $\theta(P)$. Then, one can exploit the familiar duality between confidence regions and hypothesis tests to make inferences about $\theta(P)$. Thus, an asymptotically valid test of the null hypothesis that $\theta(P) = \theta_0$ can be constructed by an asymptotically valid confidence region for $\theta(P)$ by the following rule: accept the null hypothesis if and only if the confidence region includes θ_0. Therefore, all the methods we have thus far discussed for constructing confidence regions may be utilized: methods based on a pivot, an asymptotic pivot, an asymptotic approximation, or the bootstrap. Because of this familiar duality, the details of this approach are left to the reader.

However, not all hypothesis testing problems fit nicely into the framework of testing parameters. For example, consider the problem of testing for normality. For another example, when X_i is vector valued, consider the problem of testing whether X_i has a distribution that is spherically symmetric.

The approach taken in this section is to estimate the null distribution of some test statistic by resampling from a distribution obeying the constraints of the null hypothesis. Indeed, in order to construct a critical value, the distribution of some test statistic must be known, at least under the null hypothesis.

To be explicit, assume the goal is to construct an asymptotically valid test based on a real-valued test statistic $T_n = T_n(X_1, \ldots, X_n)$. Large values of T_n reject the null hypothesis. Thus, having picked a suitable test statistic T_n, our goal is to construct a critical value, say, $c_n(1-\alpha)$, so that the test that rejects if and only if T_n exceeds $c_n(1-\alpha)$ satisfies

$$P\{T_n(X_1, \ldots, X_n) > c_n(1-\alpha)\} \to \alpha \text{ as } n \to \infty$$

when $P \in \mathbf{P_0}$. Furthermore, we require this rejection probability to tend to one when $P \in \mathbf{P_1}$. The question is how to construct a critical value to achieve these goals. Unlike in the classical case, the critical value can and will be constructed to be data dependent. The bootstrap approach offers a solution to the question of how large the critical value should be. To this end, let the true distribution of T_n be denoted by

$$G_n(x, P) = Prob_P\{T_n(X_1, \ldots, X_n) \leq x\}.$$

Note that we have introduced $G_n(\cdot, P)$ instead of utilizing $J_n(\cdot, P)$ to distinguish from the case of confidence intervals where $J_n(\cdot, P)$ represents the distribution of a root which may depend both on the data and on P. In the hypothesis testing context, $G_n(\cdot, P)$ represents the distribution of a statistic under P. Because P is unknown, $G_n(\cdot, P)$ is unknown, Also, let

$$g_n(1-\alpha, P) = \inf\{x : G_n(x, P) \geq 1-\alpha\}.$$

The bootstrap approach is to estimate the null sampling distribution by $G_n(\cdot, \hat{Q}_n)$, where $\hat{Q}_n$ is an estimate of P in $\mathbf{P_0}$ so that $\hat{Q}_n$ satisfies the constraints of the null hypothesis. A bootstrap critical value can then be defined by

$$g_n(1-\alpha, \hat{Q}_n) = \inf\{x : G_n(x, \hat{Q}_n) \geq 1-\alpha\}.$$

Then, the nominal level α test rejects H_0 if and only if $T_n > g_n(1-\alpha, \hat{Q}_n)$.

Notice that we would not want to replace a $\hat{Q}_n$ satisfying the null hypothesis constraints by the empirical distribution function $\hat{P}_n$, the usual resampling mechanism of resampling mechanism of resampling the data with replacement. One might say that the bootstrap is so adept at estimating the distribution of a statistic that $G_n(\cdot, \hat{P}_n)$ is a good estimate of $G_n(\cdot, P)$ whether or not P satisfies the null hypothesis constraints. Hence, the test that rejects when T_n exceeds $g_n(1-\alpha, \hat{P}_n)$ will (under suitable conditions) behave asymptotically like the test that rejects when T_n exceeds $g_n(1-\alpha, P)$, and this test has an asymptotic probability of α of rejecting the null hypothesis. Obviously, when $P \in \mathbf{P_1}$, we want our test to reject with probability that is approaching one.

Thus, the choice of resampling distribution $\hat{Q}_n$ should satisfy the following. If $P \in \mathbf{P_0}$, $\hat{Q}_n$ should be near P so that $G_n(\cdot, P) \approx G_n(\cdot, \hat{Q}_n)$, then $g_n(1-\alpha, P) \approx g_n(1-\alpha, \hat{Q}_n)$ and the asymptotic rejection probability approaches α. If, on the other hand, $P \in \mathbf{P_1}$, $\hat{Q}_n$ should not approach P, but some P_0 in $\mathbf{P_0}$. In this way, the critical value should satisfy

$$g_n(1-\alpha, \hat{Q}_n) \approx g_n(1-\alpha, P_0) \to g(1-\alpha, P_0) < \infty$$

as $n \to \infty$. Then, assuming the test statistic is constructed so that $T_n \to \infty$ under P when $P \in \mathbf{P_1}$, we will have

$$P\{T_n > g_n(1-\alpha, \hat{Q}_n)\} \approx P\{T_n > g(1-\alpha, P_0)\} \to 1$$

as $n \to \infty$, by Slutsky's theorem.

Example 1.8.1 (Testing the mean). Let $X_1, \ldots, X_n$ be real valued with finite mean and variance. The problem is to test whether the mean of the X_i is zero. So, $\mathbf{P_0}$ is the set of distributions with mean zero and finite variance. A first possibility for choice of test statistic is $T_n = n\bar{X}_n^2$, where $\bar{X}_n$ is the sample mean. To construct an appropriate resampling mechanism that obeys the constraints of the null hypothesis, let $\hat{Q}_n$ be the distribution in $\mathbf{P_0}$ closest to the empirical distribution $\hat{P}_n$. One way to describe closeness is the following. For distributions P and Q on the real line, let $\delta_{KL}(P, Q)$ be the (forward) Kullback–Leibler divergence between P and Q defined by

$$\delta_{KL}(P, Q) = \int log(\frac{dP}{dQ}) dP.$$

Note that $\delta_{KL}(P, Q)$ may be ∞, δ_{KL} is not a metric, and it is not even symmetric in its arguments. To construct the resampling distribution, let $\hat{Q}_n$ be the Q that minimizes $\delta_{KL}(\hat{P}_n, Q)$ over Q in $\mathbf{P_0}$. (Another possibility is to minimize the [backward] Kullback-Leibler divergence $\delta_{KL}(Q, \hat{P}_n)$.) This choice for $\hat{Q}_n$ can be shown to be well defined and corresponds to finding the nonparametric maximum likelihood estimator of P assuming P is constrained to have mean zero. The optimization problem can be solved fairly easily. By Efron (1981), $\hat{Q}_n$ assigns mass w_i to X_i, where w_i is proportional to

$$w_i \propto \frac{(1+tX_i)^{-1}}{\sum_{j=1}^n (1+tX_j)^{-1}}$$

and t is chosen so that $\sum_{i=1}^n w_i X_i = 0$.

Now, we address the consistency in level or type I error of the test. As in the construction of confidence intervals, $G_n(\cdot, P)$ must be smooth in P in order for the bootstrap to succeed. In the theorem below, rather than specifying a set of sequences $\mathbf{C_P}$ as was done in Theorem 1.2.1, smoothness is alternatively described in terms of a metric d. For the sake of stating a theorem, either approach could be used; also see Remark 1.2.2. The proof is analogous to the proof of Theorem 1.2.1.

Theorem 1.8.1. *Let $X_1, \ldots, X_n$ be i.i.d. according to $P \in \mathbf{P_0}$. Assume the following triangular array convergence: if $d(P_n, P) \to 0$ and $P \in \mathbf{P_0}$ implies $G_n(\cdot, P_n)$ converges weakly to $G(\cdot, P)$ with $G(\cdot, P)$ a continuous c.d.f. Moreover, assume $\hat{Q}_n$ is an estimator of P based on $X_1, \ldots, X_n$ that satisfies $d(\hat{Q}_n, P) \to 0$ in probability whenever $P \in \mathbf{P_0}$.*

Then, $P\{T_n > g_n(1 - \alpha, \hat{Q}_n)\} \to \alpha$ as $n \to \infty$.

Example 1.8.2 (Goodness of fit). The problem is to test whether the underlying probability distribution P belongs to a parametric family of distributions $\mathbf{P_0} = \{P_\theta, \theta \in \Theta_0\}$, where Θ_0 is an open subset of k-dimensional Euclidean space. Let $\hat{P}_n$ be the empirical measure based on $X_1, \ldots, X_n$. Let $\hat{\theta}_n \in \Theta_0$ be an estimator of θ. Consider the test statistic

$$T_n = n^{1/2} \delta(\hat{P}_n, P_{\hat{\theta}_n}),$$

where δ is some measure (typically a metric) between $\hat{P}_n$ and $P_{\hat{\theta}_n}$. (In fact, δ need not even be symmetric, which is useful sometimes: for example, consider the Cramér–von Mises statistic.) Beran (1986) considers the case where $\hat{\theta}_n$ is a minimum distance estimator, while Romano (1988c) assumes that $\hat{\theta}_n$ is some asymptotically linear estimator (like a typical maximum likelihood estimator). For the resampling mechanism, take $\hat{Q}_n = P_{\hat{\theta}_n}$. Both Beran (1986) and Romano (1988c) give different sets of conditions so that the above theorem is applicable. In fact, much of the necessary tools are already provided in Section 1.5 when the bootstrap of the empirical process was considered. Note, when $P \in \mathbf{P_1}$, $\hat{\theta}_n$ should converge to the value θ_0 satisfying

$$\delta(P, P_{\theta_0}) = \inf_{\theta \in \Theta_0} \delta(P, P_\theta).$$

Of course, this assumes problems of uniqueness and existence are taken into account.

Goodness of fit problems are generalized considerably in Romano (1988c). For testing $\mathbf{P_0}$ versus $\mathbf{P_1}$, suppose $\mathbf{P_0}$ can be characterized by the set of probabilities P satisfying $\tau(P) = P$ for some mapping τ from $\mathbf{P}$ to $\mathbf{P_0}$. Equivalently, if δ is a metric on $\mathbf{P}$, then $\mathbf{P_0}$ is specified as the set of P satisfying $\delta(P, \tau(P)) = 0$. Intuitively, the larger the value of $\delta(P, \tau(P))$, the greater the departure of P from $\mathbf{P_0}$. Actually, we do not need to assume that δ is a metric or even symmetric, but we do assume $\delta(P, \tau(P)) = 0$ if and only if $P \in \mathbf{P_0}$. In essence, the hypothesis testing problem has been reduced to that of testing whether the parameter $\delta(P, \tau(P))$ is zero or greater than zero.

1.8 Hypothesis Testing

In fact, we claim *any* testing problem can be put in this framework. All that is required to construct τ is the following: for $P \in \mathbf{P_0}$, define $\tau(P) = P$; for $P \in \mathbf{P_1}$, define $\tau(P)$ to be any member in $\mathbf{P_0}$. For the bootstrap to succeed, however, $\tau(\cdot)$ should be smooth. One possibility is the following minimum distance construction. Let ρ be a metric (or possibly a pseudometric, and possibly distinct from δ) on the space of probability distributions. If for $P \in \mathbf{P_1}$, $\inf_{Q \in \mathbf{P_0}} \rho(P, Q)$ is uniquely achieved, then define $\tau(P)$ so that

$$\rho(P, \tau(P)) = \inf_{Q \in \mathbf{P_0}} \rho(P, Q).$$

Now, given a particular choice of $\tau(\cdot)$, a test statistic T_n can be constructed by

$$T_n = \tau_n \delta(\hat{P}_n, \tau(\hat{P}_n)),$$

where $\hat{P}_n$ is the empirical measure. Here, τ_n is an appropriate scaling so that the limiting distribution of T_n is nondegenerate (in hopes of satisfying the conditions of the theorem). Notice that when ρ and δ are the same, the test statistic reduces to

$$T_n = \tau_n \inf_{Q \in \mathbf{P_0}} \delta(\hat{P}_n, Q).$$

In this setup, the resampling mechanism is now built into the construction of the test; simply take $\hat{Q}_n = \tau(\hat{P}_n)$.

Example 1.8.1 (continued) Reconsider Example 1.8.1, the problem of testing the mean. In the light of the above discussion, instead of using $n\bar{X}_n^2$ as a choice of test statistic, one can directly use $T_n = n \delta_{KL}(\hat{P}_n, \hat{Q}_n)$, where $\hat{Q}_n$ is the Q minimizing the Kullback–Leibler divergence $\delta_{KL}(\hat{P}_n, Q)$ over Q in $\mathbf{P_0}$. This is equivalent to the test statistic used by Owen (1988) in his construction of empirical likelihood. However, rather than using the asymptotic chi-squared distribution to get critical values, a direct bootstrap approach resamples from $\hat{Q}_n$.

Example 1.8.3 (Testing independence). Let $X_1, \ldots, X_n$ be i.i.d. S-valued random variables and suppose that $X_i = (X_{i,1}, \ldots, X_{i,d})$ is composed of d components. The problem is to test the joint independence of the components. To get started, suppose that the j-th component takes values in a space S_j, and S is the product space $S = \prod_{j=1}^{d} S_j$. If P is a probability on S, let P_j denote the marginal probability on S_j of the j-th component. For any P, define $\tau(P)$ to be the product law $\prod_{j=1}^{d} P_j$. Then, $\tau(P) = P$ if and only if P is a product of its marginals. To construct a test statistic, we need to specify $\delta(P, Q)$. Romano (1988c) takes $\delta(P, Q)$ to be the supremum distance over an appropriate Vapnik–Cervonenkis class of sets, and shows Theorem 1.8.1 is valid. Notice that resampling from $\hat{Q}_n = \tau(\hat{P}_n)$ is equivalent to resampling from the marginals of the empirical distribution independently. Romano (1989) shows that this is quite close to a randomization or permutation test. Other examples, such as testing for spherical symmetry, testing for exchangeability, or testing for a change point are considered as well.

1.9 Conclusions

In this chapter, the basic theory of the bootstrap was presented in the context of independent data. Some fundamental large-sample convergence properties were developed, mainly in the context of statistics that are approximately linear. Attention was focused on the construction of confidence intervals and the construction of hypothesis tests. The goal was to provide the reader with the necessary tools to understand some mathematical properties of the bootstrap, in anticipation of analogous results for subsampling presented in the balance of the book. For more thorough expositions of the bootstrap, the reader is referred to the works of Efron (1982), Beran and Ducharme (1991), Hall (1992), Efron and Tibshirani (1993), Shao and Tu (1995), Giné (1997), and Davison and Hinkley (1997). A glaring omission is the development of a higher-order asymptotic theory (which cannot be achieved in one chapter), and which is presented in Hall (1992). The use of refined percentile methods, such as Efron's BC_a method, is also omitted; a review can be found in DiCiccio and Romano (1988).

2
Subsampling in the I.I.D. Case

2.1 Introduction

In this chapter, a general theory for the construction of confidence intervals or regions is presented. Much of what is presented is extracted from Politis and Romano (1992c, 1994b). The basic idea is to approximate the sampling distribution of a statistic based on the values of the statistic computed over smaller subsets of the data. For example, in the case where the data are n observations that are independent and identically distributed, a statistic is computed based on the entire data set and is recomputed over all $\binom{n}{b}$ data sets of size b. Implicit is the notion of a statistic sequence, so that the statistic is defined for samples of size n and b. These recomputed values of the statistic are suitably normalized to approximate the true sampling distribution.

The approach presented here is perhaps the most general theory for the construction of first-order asymptotically valid confidence regions. That is, it will be seen that, under very weak assumptions on b, the method is valid whenever the original statistic, suitably normalized, has a limit distribution under the true model. Other methods, such as the bootstrap, require that the distribution of the statistic is somehow locally smooth as a function of the unknown model. In fact, many papers have been devoted to showing the convergence of a suitably normalized statistic to its limiting distribution is appropriately uniform as a function of the unknown model in specific situations. In contrast, no such assumption or verification of such smoothness is required in the theory for subsampling. Indeed, the method here is applicable even in the several known situations which represent counterexamples to the bootstrap. To appreciate why subsampling behaves well under such weak assumptions, note that each subset of size b (taken without replacement from the original data) is indeed a sample of size b from the true model. Hence, it should be intuitively clear that one can at least approximate the sampling distribution of the (normalized) statistic based on a sample of size b. But, under the weak convergence hypothesis, the sampling distributions based on samples of size b and n should be close. The bootstrap, on the other hand, is based on recomputing a statistic over a sample of size n from some estimated model that is hopefully close to the true model.

The method has a clear extension to the context of a stationary time series or, more generally, a homogeneous random field. The only difference is that the statistic is computed over a smaller number of subsets of the data that retain the dependence structure of the observations. For example, if $X_1, \ldots, X_n$ represent n observations from some stationary time series, the statistic is recomputed only over the $n - b + 1$ subsets of size b of the form $\{X_i, X_{i+1}, \ldots, X_{i+b-1}\}$. The extensions to time series, homogeneous random fields, and marked point processes will be discussed in later chapters.

The use of subsample values to approximate the variance of a statistic is well known. The Quenouille–Tukey jackknife estimate of bias and variance based on computing a statistic over all subsamples of size $n - 1$ has been well-studied and is closely related to the mean and variance of our estimated sampling distribution with $b = n - 1$. Mahalanobis (1946) suggested the use of subsamples to estimate variability in studying crop yields, though he used the name interpenetrating samples. Half-sampling methods have been well studied in the context of sampling theory (see McCarthy, 1969). Hartigan (1969) has introduced what Efron (1982) calls a random subsampling method, which is based on the computation of a statistic over all $2^n - 1$ nonempty subsets of the data. His method is seen to produce exact confidence limits in the special context of the symmetric location problem. Hartigan (1975) has adapted his finite sample results to a more general context of certain classes of estimators which have asymptotic normal distributions. But, even in this context, his asymptotic results assume the number of subsamples used to recompute the statistic remains fixed as $n \to \infty$, which results in a loss of efficiency.

The jackknife and random subsampling methods are similar in that they both use subsets of the data to approximate standard errors of a statistic, or perhaps even to approximate a sampling distribution. The method presented here retains the conceptual simplicity of these methods and is seen to be applicable under very minimal assumptions.

Efron's (1979) bootstrap, while sharing some similar properties to the aforementioned methods, has corrected some deficiencies in the jackknife, and has tackled the more ambitious goal of approximating an entire sampling distribution. Shao and Wu (1989) have shown that, by basing a jackknife estimate of variance on the statistic computed over subsamples with d observations deleted, many of the deficiencies of the usual $d = 1$ jackknife estimate of variance can be removed. Later, Wu (1990) used these subsample values to approximate an entire sampling distribution by what he calls a jackknife histogram, but only in regular i.i.d. situations where the statistic is appropriately linear so that asymptotic normality ensues. In more broad generality, Sherman and Carlstein (1996) considered the use of subsamples as a diagnostic tool to describe the shape of the sampling distribution of a general statistic, though formal inference procedures, such as the construction of confidence intervals, are not delivered. Here, we show how these subsample values can accurately estimate a sampling distribution without any assumptions of asymptotic normality, by only assuming the existence of a limiting distribution. Moreover, the asymptotic validity of confidence statements follows. In summary, while the method developed in this work is quite related to several well-studied techniques, the simplicity of our arguments has lead to asymptotic justification under the most general conditions.

In Section 2.2, the method is described in the context of i.i.d. observations. The main theorem is presented and several examples are given. Some comparisons with the bootstrap are drawn in Section 2.3. A theorem on stochastic approximation to the subsampling distribution is presented in Section 2.4. In Section 2.5, the theory is extended in a simple way to subsampling studentized roots and to roots designed to construct confidence bands for parameters that are functions. The use of subsampling in the context of hypothesis testing is described in Section 2.6. A general theorem proving consistency of subsampling using random or data-driven choices of the block size is presented in Section 2.7. Subsampling is related to the delete-d jackknife in the context of estimating the variance of a statistic in Section 2.8.

2.2 The Basic Theorem

Throughout this section, $X_1, \ldots, X_n$ is a sample of n independent and identically distributed random variables taking values in an arbitrary sample space S. The common probability measure generating the observations is denoted P. The goal is to construct a confidence region for some parameter $\theta(P)$. For now, assume θ is real valued, but this can and will be considerably generalized to allow for the construction of confidence regions for multivariate parameters or confidence bands for functions.

Let $\hat{\theta}_n = \hat{\theta}_n(X_1, \ldots, X_n)$ be an estimator of $\theta(P)$. It is desired to estimate or approximate the true sampling distribution of $\hat{\theta}_n$ in order to make inferences about $\theta(P)$. Nothing is assumed about the form of the estimator, though it is natural in the i.i.d. context to assume $\hat{\theta}_n$ is symmetric in its arguments (but even this is not necessary).

Define $J_n(P)$ to be the sampling distribution of $\tau_n(\hat{\theta}_n - \theta(P))$ based on a sample of size n from P, where τ_n is a normalizing constant. Also define the corresponding cumulative distribution function:

$$J_n(x, P) = Prob_P\{\tau_n[\hat{\theta}_n(X_1, \ldots, X_n) - \theta(P)] \leq x\}.$$

Essentially, the only assumption that we will need to construct asymptotically valid confidence intervals for $\theta(P)$ is the following.

Assumption 2.2.1. *There exists a limiting law $J(P)$ such that $J_n(P)$ converges weakly to $J(P)$ as $n \to \infty$.*

This assumption will be required to hold for some sequence τ_n. It will be necessary, however, that τ_n is such that the limit law $J(P)$ is nondegenerate.

To describe the method studied in this section, let $Y_1, \ldots, Y_{N_n}$ be equal to the $N_n = \binom{n}{b}$ subsets of size b of $\{X_1, \ldots, X_n\}$, ordered in any fashion. Of course, the Y_i depend on b and n, but this notation has been suppressed. Only a very weak assumption on b will be required. In typical situations, it will be assumed that $b/n \to 0$ and $b \to \infty$ as $n \to \infty$. Now, let $\hat{\theta}_{n,b,i}$ be equal to the statistic $\hat{\theta}_b$ evaluated at the data set Y_i. The approximation to $J_n(x, P)$ we study is defined by

$$L_{n,b}(x) = N_n^{-1} \sum_{i=1}^{N_n} 1\{\tau_b(\hat{\theta}_{n,b,i} - \hat{\theta}_n) \leq x\}. \tag{2.1}$$

The motivation behind the method is the following. For any i, Y_i is a random sample of size b from P. Hence, the *exact* distribution of $\tau_b(\hat{\theta}_{n,b,i} - \theta(P))$ is $J_b(P)$. The empirical distribution of the N_n values of $\tau_b(\hat{\theta}_{n,b,i} - \theta(P))$ should then serve as a good approximation to $J_n(P)$. Of course, $\theta(P)$ is unknown, so we replace $\theta(P)$ by $\hat{\theta}_n$, which is asymptotically permissible because $\tau_b(\hat{\theta}_n - \theta(P))$ is of order $\tau_b/\tau_n \to 0$.

Now, specialize to the case when $\hat{\theta}_n = \bar{X}_n$ is the sample mean of n i.i.d. observations having mean $\theta(P)$. Assuming the variance of X_i is finite, so $\tau_n = n^{1/2}$. One can compute the variance of the distribution $L_{n,b}(\cdot)$, and it is given by $[(n-b)/n] \cdot S_n^2$, where $S_n^2 = \sum_i (X_i - \bar{X}_n)^2/(n-1)$. In this case, this variance depends on b, due to the dependencies of the $\hat{\theta}_{n,b,i}$ with each other and with $\hat{\theta}_n$. So, it seems desirable to then replace τ_b in (2.1) by $\tau_b \cdot [n/(n-b)]^{1/2}$ so that the resulting approximating distribution has variance S_n^2, independent of b. In our theorems, $(n-b)/n \to 1$, so that the first order asymptotic properties are the same in either case. By making this change, however, our estimator more closely resembles that of Wu (1990), and the variance of the approximating distribution corresponds to Shao and Wu's (1989) delete $d = n - b$ jackknife estimate of variance. Henceforth, we do not modify τ_b in (2.1) because the above modification is only justified for linear statistics. Keep in mind, however, that such a modification may improve the approximation in some situations (see Chapter 10).

The above example also helps to explain the lack of consideration of using jackknife pseudo values to approximate a sampling distribution; a notable exception is Wu (1990). In the sample mean case, if $b = n - 1$ and if Y_i is the original sample with X_i deleted, then $(n-1)[\hat{\theta}_{n,b,i} - \hat{\theta}_n] = (\bar{X}_n - X_i)$. The empirical distribution of these values converges uniformly to the distribution of $\theta(P) - X_1$ with probability one. Hence, the use of our technique (with Wu's scaling) results in inconsistency, except in the special case that X_1 is normally distributed. Also, no choice of rescaling can correct the problem. This example points to the failure of the use of the traditional jackknife pseudo values when only one observation is deleted at a time. See Example 2.3.2 for an example where taking b of the same order as n results in inconsistency. In the opposite extreme, if $b = 1$ and $Y_i = X_i$, then $\hat{\theta}_{n,b,i} - \hat{\theta}_n = X_i - \bar{X}_n$, and the empirical distribution of these values converges uniformly to the distribution of $X_1 - \theta(P)$ with probability one; again, inconsistency follows. Aside from these extreme choices of b, our theory is typically applicable when $b/n \to 0$ and $b \to \infty$.

Theorem 2.2.1. *Assume Assumption 2.2.1. Also assume $\tau_b/\tau_n \to 0$, $b \to \infty$, and $b/n \to 0$ as $n \to \infty$.*

i. *If x is a continuity point of $J(\cdot, P)$, then $L_{n,b}(x) \to J(x, P)$ in probability.*

ii. *If $J(\cdot, P)$ is continuous, then*

$$\sup_x |L_{n,b}(x) - J_n(x, P)| \to 0 \text{ in probability.} \quad (2.2)$$

iii. *Let*

$$c_{n,b}(1-\alpha) = \inf\{x : L_{n,b}(x) \geq 1 - \alpha\}.$$

Correspondingly, define

$$c(1-\alpha, P) = \inf\{x : J(x, P) \geq 1 - \alpha\}.$$

If $J(\cdot, P)$ is continuous at $c(1 - \alpha, P)$, then

$$\text{Prob}_P\{\tau_n[\hat{\theta}_n - \theta(P)] \leq c_{n,b}(1-\alpha)\} \to 1 - \alpha \text{ as } n \to \infty. \quad (2.3)$$

Therefore, the asymptotic coverage probability under P of the confidence interval $[\hat{\theta}_n - \tau_n^{-1} c_{n,b}(1-\alpha), \infty)$ is the nominal level $1 - \alpha$.

iv. *Assume $\tau_b(\hat{\theta}_n - \theta(P)) \to 0$ almost surely and, for every $d > 0$,*

$$\sum_n \exp\{-d(n/b)\} < \infty.$$

Then, the convergences in (i) and (ii) hold with probability one.

44   2. Subsampling in the I.I.D. Case

Proof. Let

$$U_n(x) = U_{n,b}(x, P) = N_n^{-1} \sum_{i=1}^{N_n} 1\{\tau_b[\hat{\theta}_{n,b,i} - \theta(P)] \leq x\}. \qquad (2.4)$$

Note that the dependence of $U_n(x)$ on b and P will now be suppressed for notational convenience. To prove (i), it suffices to show $U_n(x)$ converges in probability to $J(x, P)$ for every continuity point x of $J(x, P)$. To see why,

$$L_{n,b}(x) = N_n^{-1} \sum_i 1\{\tau_b[\hat{\theta}_{n,b,i} - \theta(P)] + \tau_b[\theta(P) - \hat{\theta}_n] \leq x\},$$

so that for every $\epsilon > 0$,

$$U_n(x - \epsilon) 1(E_n) \leq L_{n,b}(x) 1(E_n) \leq U_n(x + \epsilon),$$

where $1(E_n)$ is the indicator of the event $E_n \equiv \{\tau_b|\theta(P) - \hat{\theta}_n| \leq \epsilon\}$. But, the event E_n has probability tending to one. So, with probability tending to one,

$$U_n(x - \epsilon) \leq L_{n,b}(x) \leq U_n(x + \epsilon)$$

for any $\epsilon > 0$. Hence, if $x + \epsilon$ and $x - \epsilon$ are continuity points of $J(\cdot, P)$, then $U_n(x \pm \epsilon) \to J(x \pm \epsilon, P)$ in probability implies

$$J(x - \epsilon, P) - \epsilon \leq L_{n,b}(x) \leq J(x + \epsilon, P) + \epsilon$$

with probability tending to one. Now, let $\epsilon \to 0$ so that $x \pm \epsilon$ are continuity points of $J(\cdot, P)$. Therefore, it suffices to show $U_n(x) \to J(x, P)$ in probability for all continuity points x of $J(\cdot, P)$. But, $U_n(x)$ is a U-statistic of degree b. Also, $0 \leq U_n(x) \leq 1$ and $E[U_n(x)] = J_b(x, P)$. By an inequality of Hoeffding (1963) (see Serfling, 1980, Theorem A, p. 201), for any $t > 0$,

$$Prob_P\{U_n(x) - J_b(x, P) \geq t\} \leq \exp\{-2\lfloor n/b \rfloor t^2\}. \qquad (2.5)$$

One can obtain a similar inequality for $t < 0$ by considering the U-statistic $-U_n(x)$. Hence, $U_n(x) - J_b(x, P) \to 0$ in probability. The result (i) follows since $J_b(x, P) \to J(x, P)$.

To prove (ii), given any subsequence $\{n_k\}$, one can extract a further subsequence $\{n_{k_j}\}$ so that $L_{n_{k_j}}(x) \to J(x, P)$ almost surely. Therefore, $L_{n_{k_j}}(x) \to J(x, P)$ almost surely for all x in some countable dense set of the real line. So, $L_{n_{k_j}}$ tends weakly to $J(x, P)$ and this convergence is uniform by Polya's theorem. Hence, the result (ii) holds.

The proof of (iii) is very similar to the proof of Theorem 1 of Beran (1984) given our result (i).

To prove (iv), follow the same argument, using the added assumptions and the Borel–Cantelli Lemma on the inequality equation (2.5). ∎

Remark 2.2.1. The assumptions $b/n \to 0$ and $b \to \infty$ need not imply $\tau_b/\tau_n \to 0$. For example, in the unusual case $\tau_n = \log(n)$, if $b = n^\gamma$ and $\gamma > 0$, the assumption $\tau_b/\tau_n \to 0$ is not satisfied. In regular cases, $\tau_n = $

$n^{1/2}$, and the assumptions on b simplify to $b/n \to 0$ and $b \to \infty$. The further assumption on b in part (iv) of the theorem will then hold, for example, if $b = n^\gamma$ for any $\gamma \in (0,1)$. In fact, it is easy to see that it holds if $b \log(n)/n \to 0$.

Remark 2.2.2. The assumptions on b are as weak as possible under the weak assumptions of the theorem. However, in some cases, the choice $b = O(n)$ yields similar results; this occurs in Wu (1990), where the statistic is approximately linear with an asymptotic Gaussian distribution and $\tau_n = n^{1/2}$. This choice will not work in general (see Example 2.3.2).

Remark 2.2.3. The proof of consistency of the subsampling distribution $L_{n,b}(x)$ boils down to proving consistency of the related U-statistic $U_n(x)$. Rather than using Hoeffding's exponential inequality as done in the proof, it may be instructive to show the variance of $U_n(x)$ tends to zero as follows. Suppose k is the greatest integer less than or equal to n/b. For $j = 1, \ldots, k$, let $R_{n,b,j}$ be equal to the statistic $\hat{\theta}_b$ evaluated at the data set $\hat{\theta}_b(X_{b(j-1)+1}, X_{b(j-1)+2}, \ldots, X_{b(j-1)+b})$ and set

$$\bar{U}_n(x) = k^{-1} \sum_{j=1}^{k} 1\{\tau_b[R_{n,b,j} - \theta(P)] \leq x\}.$$

Clearly, $\bar{U}_n(x)$ and $U_n(x)$ have the same expectation. But, since $\bar{U}_n(x)$ is the average of k i.i.d. variables (each of which is bounded between 0 and 1), it follows that

$$Var[\bar{U}_n(x)] \leq \frac{1}{4k} \to 0$$

as $n \to \infty$. Intuitively, $U_n(x)$ should have a smaller variance than $\bar{U}_n(x)$, because $\bar{U}_n(x)$ uses in the ordering in the sample in an arbitrary way. Indeed, the fact that $U_n(x)$ has a smaller variance than $\bar{U}_n(x)$ can be argued by a sufficiency argument using the Rao–Blackwell theorem. Simply note that we can write

$$U_n(x) = E[\bar{U}_n(x)|\mathbf{X_n}],$$

where $\mathbf{X_n}$ is the collection of the order statistics $\{X_{(1)}, \ldots, X_{(n)}\}$.

Assumption 2.2.1 is satisfied in numerous examples. Next, we offer an interesting example that illustrates the scope of our method, as it falls outside the range of $n^{1/2}$-consistent estimators and normal limits. While methods like the bootstrap are potentially applicable in this example, the validity of the bootstrap is not known. At the very least, the bootstrap method would require a somewhat tedious argument to justify its asymptotic validity.

Example 2.2.1 (Optimal replacement time). Consider the problem of age replacement where replacements of a unit X occur at failure of the unit or at age t, whichever comes first. X is assumed continuous with an increasing failure rate distribution F having density f, finite mean, and $F(0) = 0$. Suppose a cost c_1 is incurred for each failed unit that is replaced and a cost $c_2 < c_1$ is suffered for each nonfailed unit that is exchanged. It is easy to see that the average cost per time unit, over an infinite time horizon, based on the strategy of preventively replacing the unit at time t is given by

$$A(t, F) = \frac{c_1 F(t) + c_2[1 - F(t)]}{\int_0^t [1 - F(x)] dx}.$$

The problem is to find $\theta(F)$, the value that minimizes $A(t, F)$ over t. If $r(x) = f(x)/[1 - F(x)]$ is the failure rate of F, and if $r(x)$ is assumed continuous and increasing to ∞, then $\theta(F)$ is well defined. The optimal minimum cost is then

$$\beta(F) = (c_1 - c_2) r(\theta(F)).$$

In practice, F is unknown, so our problem is to construct a confidence interval for $\theta(F)$ based on a random sample $X_1, \ldots, X_n$ from F. Let $\hat{F}_n$ denote the empirical distribution of the data, and let $\hat{\theta}_n$ be a value of t minimizing $A(t, \hat{F}_n)$; that is, $\hat{\theta}_n = \theta(\hat{F}_n)$. For the purposes of the discussion here, do not worry about problems of existence or uniqueness; see Arunkumar (1972) for a careful description. Arunkumar (1972) has shown that $n^{1/3}[\hat{\theta}_n - \theta(F)]$ has a nondegenerate limiting distribution, so our Assumption 2.2.1 is verified with $\tau_n = n^{1/3}$. The asymptotic distribution is the distribution of $c(F)$ times the value of t which minimizes $[W(t) - t^2]$, where $W(t)$ is a two-sided Wiener–Lévy process and the constant $c(F)$ depends on intricate properties of F such as $f(\theta(F))$. Hence, the asymptotic distribution is of little use toward the construction of confidence intervals for $\theta(F)$. Léger and Cléroux (1990) have constructed bootstrap confidence intervals for $\beta(F)$. The subsampling approach may be used for this problem as well because $n^{1/2}[\beta(\hat{F}_n) - \beta(F)]$ has a limiting normal distribution.

Remark 2.2.4. In fact, one can remove the assumption that $\tau_b/\tau_n \to 0$ if the goal is to construct an asymptotically valid confidence interval for $\theta(P)$, but at the small expense of bypassing consistent estimation of $J_n(\cdot, P)$. To see how, let

$$u_{n,b}(1 - \alpha, P) = \inf\{x : U_{n,b}(x, P) \geq 1 - \alpha\},$$

where $U_{n,b}(\cdot, P)$ is defined in (2.4). Under continuity assumptions on $J(\cdot, P)$, the proof of Theorem 2.2.1 shows $U_{n,b}(x, P)$ converges in probability to $J(x, P)$; it follows that $u_{n,b}(1 - \alpha, P)$ converges in probability (under P) to $c(1 - \alpha, P)$. Moreover, the assumption $\tau_b/\tau_n \to 0$ is not used. Note, however, that $u_{n,b}(1 - \alpha, P)$ is not an estimator since it depends on P. Nevertheless, with P fixed, the event

$$\{\tau_n(\hat{\theta}_n - \theta(P)) \leq u_{n,b}(1 - \alpha, P)\} \qquad (2.6)$$

has an asymptotic probability of $1 - \alpha$ under P (assuming $J(\cdot, P)$ is continuous at $c(1 - \alpha, P)$). But,

$$u_{n,b}(1 - \alpha, P) = c_{n,b}(1 - \alpha) + \tau_b(\hat{\theta}_n - \theta(P)). \tag{2.7}$$

Hence, the event (2.6) is exactly the same as the event

$$\{\tau_n(\hat{\theta}_n - \theta(P)) \le c_{n,b}(1 - \alpha) + \tau_b(\hat{\theta}_n - \theta(P))\}, \tag{2.8}$$

or equivalently,

$$\{(\tau_n - \tau_b)(\hat{\theta}_n - \theta(P) \le c_{n,b}(1 - \alpha)\} \tag{2.9}$$

By solving for $\theta(P)$, the following nominal level $1 - \alpha$ confidence interval is obtained:

$$[\hat{\theta}_n - (\tau_n - \tau_b)^{-1} c_{n,b}(1 - \alpha), \infty). \tag{2.10}$$

This interval (2.10) can be computed without knowledge of P, it has asymptotic coverage probability under P of $1 - \alpha$, and the assumption $\tau_b/\tau_n \to 0$ was not needed. Clearly, the only difference between this interval (2.10) and the interval presented in Theorem 2.2.1 is the factor $(\tau_n - \tau_b)^{-1}$ here replaces the factor τ_n^{-1} there. This discussion leads to the following corollary.

Corollary 2.2.1. *Assume Assumption 2.2.1. Also, assume $b \to \infty$ and $b/n \to 0$ as $n \to \infty$. If $J(\cdot, P)$ is continuous at $c(1 - \alpha, P)$, then the interval (2.10) contains $\theta(P)$ with asymptotic probability $1 - \alpha$ under P.*

In fact, the interval (2.10) can be viewed in the following way. Rather than approximate $J_n(x, P)$ by $L_{n,b}(x, P)$, consider the distribution

$$\tilde{L}_{n,b}(x) = L_{n,b}(\frac{\tau_n - \tau_b}{\tau_n} \cdot x), \tag{2.11}$$

so that the correction factor $(\tau_n - \tau_b)/\tau_n$ is employed. Then, the $1 - \alpha$ quantile of $\tilde{L}_{n,b}(\cdot)$ is just $\tau_n \cdot c_{n,b}(1 - \alpha)/(\tau_n - \tau_b)$. Hence, solving for $\theta(P)$ in the inequality

$$\{\tau_n(\hat{\theta}_n - \theta(P)) \le \tilde{L}_{n,b}^{-1}(1 - \alpha)\}$$

leads to the interval (2.10). A similar, but different, correction factor will be studied in Chapter 10 with the goal of improved higher-order accuracy.

2.3 Comparison with the Bootstrap

The usual bootstrap approximation to $J_n(x, P)$ is $J_n(x, \hat{Q}_n)$, where $\hat{Q}_n$ is some estimate of P. In many (but not all) i.i.d. situations, $\hat{Q}_n$ is taken to be the empirical distribution of the sample $X_1, \ldots, X_n$. The analogous results to (2.2) and (2.3) with $L_{n,b}(x)$ replaced by $J_n(x,\hat{Q}_n)$ have been proved in many situations (see Bickel and Freedman, 1981, and Beran, 1984). In fact, dozens of other papers exist whose sole purpose is to prove such results in very specific situations. Our theorem immediately applies very generally with no further work.

To elaborate a little further, analogous bootstrap limit results are typically proved in the following manner. For some choice of metric (or pseudo-metric) d on the space of probability measures, it must be known that $d(P_n, P) \to 0$ implies $J_n(P_n)$ converges weakly to $J(P)$. That is, Assumption 2.2.1 must be strengthened so that the convergence of $J_n(P)$ to $J(P)$ is suitably locally uniform in P. In addition, the estimator $\hat{Q}_n$ must then be known to satisfy $d(\hat{Q}_n, P) \to 0$ almost surely or in probability under P. In contrast, no such strengthening of Assumption 2.2.1 is required in Theorem 2.2.1. In the known counterexamples to the bootstrap, it is precisely a certain lack of uniformity in convergence that leads to failure of the bootstrap.

In some special cases, it has been realized that a sample size trick can often remedy the inconsistency of the bootstrap. To describe how, focus on the case where $\hat{Q}_n$ is the empirical measure, denoted by $\hat{P}_n$. Rather than approximating $J_n(P)$ by $J_n(\hat{P}_n)$, the suggestion is to approximate $J_n(P)$ by $J_b(\hat{P}_n)$ for some b that usually satisfies $b/n \to 0$ and $b \to \infty$. The resulting estimator $J_b(x, \hat{P}_n)$ is obviously quite similar to our $L_{n,b}(x)$ given in (2.1). In words, $J_b(x, \hat{P}_n)$ is the bootstrap approximation defined by the distribution (conditional on the data) of $\tau_b[\hat{\theta}_b(X_1^*,\ldots,X_b^*) - \hat{\theta}_n]$, where $X_1^*,\ldots,X_b^*$ are chosen with replacement from $X_1,\cdots,X_n$. In contrast, $L_{n,b}(x)$ is the distribution (conditional on the data) of $\tau_b[\hat{\theta}_b(Y_1^*,\ldots,Y_b^*) - \hat{\theta}_n)]$, where $Y_1^*,\ldots,Y_b^*$ are chosen *without* replacement from $X_1,\ldots,X_n$. Clearly, these two approaches must be similar if b is so small that sampling with and without replacement are essentially the same. Indeed, if one resamples b numbers (or indices) from the set $\{1,\ldots,n\}$, then the chance that none of the indices is duplicated is $\Pi_{i=1}^{b-1}(1 - \frac{i}{n})$. This probability tends to 0 if $b^2/n \to 0$. (To see why, take logs and do a Taylor expansion analysis.) Hence, the following is true.

Corollary 2.3.1. *Under the further assumption that $b^2/n \to 0$, parts (i–iii) of Theorem 2.2.1 remain valid if $L_{n,b}(x)$ is replaced by the bootstrap approximation $J_b(x, \hat{P}_n)$.*

The bootstrap approximation with smaller resample size, $J_b(\hat{P}_n)$, is further studied in Bickel, Götze, and van Zwet (1997). In spite of the corollary, we point out that $L_{n,b}$ is more generally valid. Indeed, without the assumption $b^2/n \to 0$, $J_b(x, \hat{P}_n)$ can be inconsistent. To see why, let P be any distribution on the real line with a density (with respect to Lebesgue measure). Consider any statistic $\hat{\theta}_n$, τ_n, and $\theta(P)$ satisfying Assumption 2.2.1. Even the sample mean will work here. Now, modify $\hat{\theta}_n$ to $\tilde{\theta}_n$ so that the statistic $\tilde{\theta}_n(X_1,\ldots,X_n)$ completely misbehaves if any pair of the observations $X_1,\ldots,X_n$ are identical. The bootstrap approximation to the distribution of $\hat{\theta}_n$ must then misbehave as well unless $b^2/n \to 0$, while the consistency of $L_{n,b}$ remains intact.

The above example, though artificial, was designed to illustrate a point. Below, some known counterexamples to the bootstrap are reviewed.

Example 2.3.1 (U-statistics of degree 2). Let $X_1,\ldots,X_n$ be i.i.d. on the line with c.d.f. F. Denote by $\hat{F}_n$ the empirical distribution of the data. Let

$$\theta(F) = \int\int \omega(x,y)dF(x)dF(y)$$

and assume $\omega(x,y) = \omega(y,x)$. Assume $\int \omega^2(x,y)dF(x)dF(y) < \infty$. Set $\tau_n = n^{1/2}$ and $\hat{\theta}_n = \theta(\hat{F}_n)$. Then, it is well known that $J_n(F)$ converges weakly to $J(F)$, the normal distribution with mean 0 and variance given by

$$v^2(F) = 4[\int \{\int \omega(x,y)dF(y)\}^2 dF(x) - \theta^2(F)].$$

Hence, our condition 2.2.1 is satisfied. However, in order for the bootstrap to succeed, the additional condition $\int \omega^2(x,x)dF(x) < \infty$ is required. Bickel and Freedman (1981) give a counterexample to show the inconsistency of the bootstrap without this additional condition.

Interestingly, the bootstrap may fail even if $\int \omega^2(x,x)dF(x) < \infty$, stemming from the possibility that $v^2(F) = 0$. (Otherwise, Bickel and Freedman's argument justifies the bootstrap.) As an example, let $w(x,y) = xy$. In this case, $\theta(\hat{F}_n) = \bar{X}_n - S_n^2/n$, where S_n^2 is the usual unbiased sample variance. If $\theta(F) = 0$, then $v(F) = 0$. Then, $n[\theta(\hat{F}_n) - \theta(F)]$ converges weakly to $\sigma^2(F)(Z^2 - 1)$, where Z denotes a standard normal random variable and $\sigma^2(F)$ denotes the variance of F. However, it is easy to see that the bootstrap approximation to the distribution of $n[\theta(\hat{F}_n) - \theta(F)]$ has a representation $\sigma^2(F)Z^2 + 2Z\sigma(F)n^{1/2}\bar{X}_n$. Thus, failure of the bootstrap follows.

In the context of U-statistics, the possibility of using a reduced sample size in the resampling has been considered in Bretagnolle (1983); an alternative correction is given by Arcones (1991).

Example 2.3.2 (Extreme order statistic). The following counterexample is provided by Bickel and Freedman (1981). If $X_1,\ldots,X_n$ are i.i.d. according to a uniform distribution on $(0,\theta)$, then $n[\max(X_1,\ldots,X_n) - \theta]$ has a limit distribution given by the distribution of $-\theta X$, where X is exponential with mean one. Hence, Assumption 2.2.1 is satisfied here. However, the usual bootstrap fails. Note in Theorem 2.2.1 that the conditions on b (with $\tau_n = n$) reduce to $b/n \to 0$ and $b \to \infty$. In this example, at least, it is clear that we cannot assume $b/n \to c$, where $c > 0$. Indeed, $L_{n,b}(x)$

50   2. Subsampling in the I.I.D. Case

places mass b/n at 0. Thus, while it is sometimes true that, under further conditions such as Wu (1990) assumes, we can take b to be of the same order as n, this example makes it clear that we cannot in general weaken our assumptions on b without imposing further structure.

Example 2.3.3 (Mean in infinite variance case). Consider the case where $X_1, \ldots, X_n$ are i.i.d. real-valued random variables with c.d.f. F and mean $\theta(F)$. If the variance of F is not finite, the bootstrap is known to fail (see Babu, 1984, Athreya, 1987, Knight, 1989, and Kinateder, 1992). For further discussion of this point, see the end of Section 1.3. In this example, it has been realized that taking a smaller bootstrap sample size can result in consistency of the bootstrap (see Arcones, 1991, who attributes the idea to an unpublished report of Athreya; also see Wu, Carlstein, and Cambanis, 1989, and Arcones and Giné, 1989). For our method, Theorem 2.2.1 is generally applicable if F is in the domain of attraction of a stable law of index greater than one. Note, however, that the proper choice of the normalizing sequence τ_n will depend on the value of the index; hence, the basic theorem of Section 2 is not directly applicable because the rate of convergence must be known or at least estimated. In fact, the theory presented in this chapter will later be extended to cases like this where the rate of convergence is unknown, and subsampling offers a viable and general approach to dealing with such difficult situations. For a more detailed discussion, see Chapter 11.

Example 2.3.4 (Density estimation). Assume $X_1, \ldots, X_n$ are i.i.d. real-valued random variables with density f. Suppose f is smooth and unimodal, with mode denoted by $\theta(f)$. Let $\hat{f}_{n,h}(t) = (nh)^{-1} \sum_i K[(t - X_i)/h]$, and let $\hat{\theta}_{n,h}$ denote a mode of $\hat{f}_{n,h}$. Here, $K(\cdot)$ is a kernel function (typically an even probability density function), and h is the bandwidth. Under regularity conditions, $(nh^3)^{1/2}(\hat{\theta}_{n,h} - \theta(f))$ has a limiting normal distribution, so Theorem 2.2.1 is applicable. The optimal choice for h in this case (in the sense of minimizing mean squared error) is to take h proportional to $n^{-1/7}$. For such a choice of h, bootstrap methods based on resampling from the empirical distribution are not consistent. Indeed, bootstrap methods based on resampling from the empirical will work if the choice of bandwidth h for estimating the mode is suboptimal. In the optimal choice of bandwidth case, the bootstrap can be fixed by resampling from $\hat{f}_{n,g}$, where g is large compared with h. This phenomenon of having to resample from an appropriately oversmooth density estimate occurs in many other contexts involving functionals of a density. In short, the results depend upon delicate choices of smoothing parameters (for details, see Romano, 1988a, 1988b).

Example 2.3.5 (Superefficient estimator). Consider Hodges' famous example of a superefficient estimator. In this example, $X_1, \ldots, X_n$ are i.i.d. according to the normal distribution with mean $\theta(P)$ and variance one. Let

$\hat{\theta}_n = c\bar{X}_n$ if $|\bar{X}_n| \leq n^{-1/4}$ and $\hat{\theta}_n = \bar{X}_n$ otherwise. Here, $c > 0$. As is well known, $n^{1/2}(\hat{\theta}_n - \theta(P))$ has a limit distribution for every θ, so the conditions for our Theorem 2.2.1 remain applicable. However, Beran (1984) showed that the sampling distribution of $n^{1/2}(\hat{\theta}_n - \theta(P))$ cannot be bootstrapped, even if one is willing to apply a parametric bootstrap!

We have claimed that subsampling is superior to the bootstrap in a first-order asymptotic sense, since it is more generally valid. However, in many typical situations, the bootstrap is far superior and has some compelling second-order asymptotic properties. These are well-studied in Hall (1992). In nice situations, such as when the statistic or root is a smooth function of sample means, a bootstrap approach is often very satisfactory. In other situations, especially those where it is not known that the bootstrap works even in a first-order asymptotic sense, subsampling is preferable. Still, in other situations (such as the mean in the infinite variance case), the bootstrap may work, but only with a reduced sample size. The issue becomes whether to sample with or without replacement (as well as the choice of resample size). Although this question is not yet answered unequivocally, some preliminary evidence in Bickel, Götze, and van Zwet (1997) suggests that the bootstrap approximation $J_b(x, \hat{P}_n)$ might be more accurate; more details on the issue of higher-order accuracy of the subsampling approximation $L_{n,b}(x)$ are given in Chapter 10. Hartigan's subsampling method of recomputing the statistic over all subsamples, half-sample methods, and bootstrap methods are compared in Babu (1992).

2.4 Stochastic Approximation

Because $\binom{n}{b}$ can be large, $L_{n,b}$ may be difficult to compute. Instead, a stochastic approximation may be employed. For example, let $I_1, \ldots I_B$ be chosen randomly with or without replacement from $\{1, 2, \ldots, N_n\}$. Then, $L_{n,b}(x)$ may be approximated by

$$\hat{L}_{n,b}(x) = \frac{1}{B} \sum_{i=1}^{B} 1\{\tau_b(\hat{\theta}_{n,b,I_i} - \hat{\theta}_n) \leq x\}. \tag{2.12}$$

Corollary 2.4.1. *Under the assumptions of Theorem 2.2.1 and the assumption $B \to \infty$ as $n \to \infty$, the results of Theorem 2.2.1 are valid if $L_{n,b}(x)$ is replaced by $\hat{L}_{n,b}(x)$.*

Proof. If the I_i are sampled with replacement, $\sup_x |\hat{L}_{n,b}(x) - L_{n,b}(x)| \to 0$ almost surely by the Dvoretzky, Kiefer, Wolfowitz inequality (see Serfling, 1980, p. 59). This result is also true in the case the I_i are sampled without replacement by a similar inequality (see Romano, 1989). ∎

52   2. Subsampling in the I.I.D. Case

An alternative approach, which also requires fewer computations, is the following. Rather than employing all $\binom{n}{b}$ subsamples of size b from $X_1, \ldots, X_n$, just use the $n-b+1$ subsamples of size b of the form $\{X_i, X_{i+1}, \ldots, X_{i+b-1}\}$. Notice that the ordering of the data is fixed and retained in the subsamples. Indeed, this is the approach that is applied in the next chapter for time series data. Even when the i.i.d. assumption seems reasonable, this approach may be desirable to ensure robustness against possible serial correlation. Most inferential procedures based on i.i.d. models are simply not valid (i.e., not even first order correct) if the independence assumption is violated, so it seems worthwhile to account for possible dependencies in the data if we do not sacrifice too much in efficiency.

2.5 General Parameters and Other Choices of Root

In general, it may be desirable to approximate the sampling distribution of other roots. Below, two common choices are considered. The first generalization concerns the approximation of studentized root. The second generalization applies to general parameters that need not be real valued. In particular, the setup is designed to handle confidence bands for functional parameters. We leave it to the reader to consider further generalizations whose asymptotic validity may be justified by mimicking the proof of Theorem 2.2.1.

2.5.1 Studentized Roots

Here, the goal is to approximate the distribution of $\tau_n[\hat{\theta}_n - \theta(P)]/\hat{\sigma}_n$, where $\hat{\sigma}_n$ is some estimate of scale. Let $\hat{\sigma}_{n,b,i}$ be equal to the estimate of scale based on the i-th subsample of size b from the original data. Analogous to (2.1), define

$$L^*_{n,b}(x) = N_n^{-1} \sum_{i=1}^{N_n} 1\{\tau_b(\hat{\theta}_{n,b,i} - \hat{\theta}_n)/\hat{\sigma}_{n,b,i} \leq x\}. \qquad (2.13)$$

Then, the following theorem holds. The proof is similar to that of Theorem 2.2.1 and so it is omitted.

Theorem 2.5.1. *Assume Assumption 2.2.1. Also assume $\tau_b/\tau_n \to 0$, $b \to \infty$, and $b/n \to 0$ as $n \to \infty$. Assume $\hat{\sigma}_n \to \sigma$ in probability, where $\sigma = \sigma(P)$ is a positive constant. Let $x \cdot \sigma(P)$ be a continuity point of $J(\cdot, P)$.*

  i. *If $x \cdot \sigma(P)$ is a continuity point of $J(\cdot, P)$, then $L^*_{n,b}(x) \to J(x \cdot \sigma(P), P)$ in probability.*
  ii. *If $J(\cdot, P)$ is continuous, then $\sup_x |L^*_{n,b}(x) - J_n(x \cdot \sigma(P), P)| \to 0$ in probability.*

iii. Let
$$c^*_{n,b}(1-\alpha) = \inf\{x : J^*_{n,b}(x) \geq 1-\alpha\}.$$
If $J(\cdot, P)$ is continuous at $c(1-\alpha, P) \cdot \sigma(P)$, then
$$Prob_P\{\tau_n[\hat{\theta}_n - \theta(P)]/\hat{\sigma}_n \leq c_{n,b}(1-\alpha)\} \to 1-\alpha, \text{ as } n \to \infty.$$
Thus, the asymptotic coverage probability under P of the interval $[\hat{\theta}_n - \hat{\sigma}_n \tau_n^{-1} c^*_{n,b}(1-\alpha), \infty)$ is the nominal level $1-\alpha$.

iv. Assume $\tau_b(\hat{\theta}_n - \theta(P)) \to 0$ almost surely, $\hat{\sigma}_n \to \sigma(P)$ almost surely, and, for every $d > 0$, $\sum_n \exp\{-d(n/b)\} < \infty$. Then, the convergences in (i) and (ii) hold with probability one.

2.5.2 General Parameter Space

It is often desirable to construct confidence regions for multivariate parameters, or for parameters taking values in a function space. For example, consider the problem of constructing confidence bands for the density or distribution function, which may form the basis of a goodness of fit test. Assume $\theta(P)$ takes values in a normed linear space Θ, with norm denoted by $\|\cdot\|$. Let $\hat{\theta}_n$ be an estimate of $\theta(P)$. Assume Assumption 2.2.1, with the interpretation that $\tau_n[\hat{\theta}_n - \theta(P)]$ has a distribution in θ. Here, Θ is endowed with an appropriate σ-field so that $\tau_n[\hat{\theta}_n - \theta(P)]$ is measurable and an appropriate weak convergence theory ensues, though we omit such measurability issues here. Let $J_{n,\|\cdot\|}(P)$ denote the distribution of $\tau_n\|\hat{\theta}_n - \theta(P)\|$ under P, with corresponding c.d.f $J_{n,\|\cdot\|}(\cdot, P)$. Assumption 2.2.1 implies that $J_{n,\|\cdot\|}(P)$ converges weakly to $J_{\|\cdot\|}(P)$, the distribution of $\|Z\|$, where Z has distribution $J(P)$. The corresponding c.d.f. of $J_{\|\cdot\|}(P)$ is denoted $J_{\|\cdot\|}(\cdot, P)$. The approximation to $J_{n,\|\cdot\|}(P)$ we study is defined analogously to (2.1):
$$L_{n,b,\|\cdot\|}(x) = N_n^{-1} \sum_{i=1}^{N_n} 1\{\tau_b\|\hat{\theta}_{n,b,i} - \hat{\theta}_n\| \leq x\}.$$

Theorem 2.5.2. *Assume Assumption 2.2.1. Also assume $\tau_b/\tau_n \to 0$, $b \to \infty$, and $b/n \to 0$ as $n \to \infty$.*

i. *If x is a continuity point of $J_{\|\cdot\|}(\cdot, P)$, then $L_{n,b,\|\cdot\|}(x) \to J_{\|\cdot\|}(x, P)$ in probability.*
ii. *If $J_{\|\cdot\|}(\cdot, P)$ is continuous, then $\sup_x |L_{n,b,\|\cdot\|}(x) - J_{\|\cdot\|}(x, P)| \to 0$ in probability.*
iii. *Let*
$$c_{n,b,\|\cdot\|}(1-\alpha) = \inf\{x : L_{n,b,\|\cdot\|}(x) \geq 1-\alpha\}.$$
Correspondingly, define
$$c_{\|\cdot\|}(1-\alpha, P) = \inf\{x : J_{\|\cdot\|}(x, P) \geq 1-\alpha\}.$$

If $J_{\|\cdot\|}(\cdot, P)$ is continuous at $c_{\|\cdot\|}(1-\alpha, P)$, then

$$Prob_P\{\tau_n\|\hat{\theta}_n - \theta(P)\| \leq c_{n,b,\|\cdot\|}(1-\alpha)\} \to 1-\alpha \text{ as } n \to \infty.$$

Thus, the asymptotic coverage probability under P of the set $\{\theta \in \Theta : \tau_n\|\hat{\theta}_n - \theta\| \leq c_{n,b,\|\cdot\|}(1-\alpha)\}$ is the nominal level $1-\alpha$.

iv. *Assume $\tau_b(\hat{\theta}_n - \theta(P)) \to 0$ almost surely and, for every $d > 0$, $\sum_n \exp\{-d[n/b]\} < \infty$. Then, the convergences in (i) and (ii) hold with probability one.*

The proof of the above theorem is similar to that of Theorem 2.2.1, and is omitted. Immediate applications of the theorem result in uniform confidence bands for a cumulative distribution function F, based on i.i.d. observations from F or in the case where observations are censored. The theory is also applicable to biased sampling models, including stratified sampling, enriched stratified sampling, choice-based sampling, and case-control studies; these models are developed in Gill, Vardi, and Wellner (1988), where they show Assumption 2.2.1 is satisfied under weak assumptions. Since distributional theory is quite hard in these models, our method offers simplicity in implementation and mathematical justification.

2.6 Hypothesis Testing

In this section, we consider the use of subsampling for the construction of hypothesis tests. As before, $X_1, \ldots, X_n$ is a sample of n independent and identically distributed observations taking values in a sample space S. The common unknown distribution generating the data is denoted by P. This unknown law P is assumed to belong to a certain class of laws $\mathbf{P}$. The null hypothesis H_0 asserts $P \in \mathbf{P_0}$, and the alternative hypothesis H_1 is $P \in \mathbf{P_1}$, where $\mathbf{P_i} \subset \mathbf{P}$ and $\mathbf{P_0} \bigcup \mathbf{P_1} = \mathbf{P}$.

There are several general approaches one can take for the construction of asymptotically valid tests, depending on the nature of the problem. In the special (but usual) case where the null hypothesis translates into a null hypothesis about a real- or vector-valued parameter $\theta(P)$, one can construct a confidence region for $\theta(P)$—by subsampling, bootstrapping, asymptotic approximations, or other methods—and then exploit the usual duality between the construction of confidence regions for parameters and the construction of hypothesis tests about those parameters. This is the approach taken in Politis and Romano (1996b), and the details are left to the reader.

Of course, not all hypothesis testing problems fit nicely into the aforementioned framework. An alternative bootstrap approach summarized in Section 1.8 of Chapter 1 that was based on bootstrapping from a distribution obeying the constraints of the null hypothesis; such an approach applies directly to many problems in testing for goodness of fit. None of the above approaches easily handles the following example, taken from Bickel and Ren (1997), but we will see that an appropriate simple subsampling scheme applies here as well. The approach is similar to that of Bickel and Ren (1997), who consider the related bootstrap with smaller resample size.

Example 2.6.1 (Goodness of fit for censored data). Suppose that $U_1, \ldots, U_n$ are i.i.d. random variables with cumulative distribution function F. The null hypothesis H_0 asserts $F = F_0$, where F_0 is some specified distribution. In this problem, however, we do not necessarily observe the full data $U_1, \ldots, U_n$ because the observations U_i are left and right censored. Specifically, assume (Y_i, Z_i) are independent and identically distribution pairs with $Z_i < Y_i$ (with probability one), and the (Y_i, Z_i) pairs are independent of $U_1, \ldots U_n$. Define

$$V_i = \begin{cases} U_i, & \text{if } Z_i < U_i \leq Y_i; \\ Y_i, & \text{if } U_i > Y_i; \\ Z_i, & \text{if } U_i \leq Z_i \end{cases}$$

and

$$\delta_i = \begin{cases} 1, & \text{if } Z_i < U_i \leq Y_i; \\ 2, & \text{if } U_i > Y_i; \\ 3, & \text{if } U_i \leq Z_i. \end{cases}$$

The actual observations available are $X_i = (V_i, \delta_i)$. Let $\hat{F}_n$ be the nonparametric maximum likelihood estimator of F based on $X_1, \ldots X_n$; this can be computed numerically by the algorithms described in Mykland and Ren (1996). Now, consider the Cramér–von Mises test statistic given by

$$T_n = n \int_{-\infty}^{\infty} [\hat{F}_n(x) - F_0(x)]^2 dF_0(x).$$

Under suitable conditions and when F is the true distribution for U_i, $n^{1/2}[\hat{F}_n(\cdot) - F(\cdot)]$, viewed as a process on $D[-\infty, \infty]$, converges weakly to a mean zero Gaussian process with covariance depending on the joint distribution of (Z_i, Y_i) (see Giné and Zinn, 1990, and Bickel and Ren, 1996). Hence, T_n possesses a limiting distribution as well, both under the null hypothesis and against a sequence of contiguous alternatives; the notion of contiguity is presented in Bickel et al. (1993, Section A.9). The difficulty that the bootstrap has in trying to approximate this limiting distribution is that Y_i and Z_i are never observed together for any i, so that any information on the joint distribution is not available. Note, however, in the right-censoring case (with $Z_i = -\infty$), $\hat{F}_n$ is the Kaplan–Meier estimator, and the distribution of the censoring variables can be estimated and the bootstrap offers a viable approach.

We now return to the general setup of testing the null hypothesis H_0 that $P \in \mathbf{P_0}$ versus the alternative hypothesis H_1 that $P \in \mathbf{P_1}$. The goal is to construct an asymptotically valid test based on a given test statistic,

$$T_n = \tau_n t_n(X_1, \ldots, X_n),$$

where, as before, τ_n is a fixed nonrandom normalizing sequence (though even this assumption can and will be weakened in a later chapter). Let

$$G_n(x, P) = Prob_P\{\tau_n t_n(X_1, \ldots, X_n) \leq x\}.$$

At this point, not too much is assumed about T_n, though it is certainly natural in the i.i.d. case presented here that $t_n(X_1, \ldots, X_n)$ is symmetric in its arguments. As before, we will be assuming that $G_n(\cdot, P)$ converges in distribution, at least for $P \in \mathbf{P_0}$. Of course, this would imply (as long as $\tau_n \to \infty$) that $t_n(X_1, \ldots, X_n) \to 0$ in probability for $P \in \mathbf{P_0}$. Naturally, t_n should somehow be designed to distinguish between the competing hypotheses. The theorem we will present will assume t_n is constructed to satisfy the following: $t_n(X_1, \ldots, X_n) \to t(P)$ in probability, where $t(P)$ is a constant which satisfies $t(P) = 0$ if $P \in \mathbf{P_0}$ and $t(P) > 0$ if $P \in \mathbf{P_1}$. This assumption easily holds in every conceivable example.

To describe the test construction, let $Y_1, \ldots, Y_{N_n}$ be equal to the $N_n = \binom{n}{b}$ subsets of $\{X_1, \ldots, X_n\}$, ordered in any fashion. Let $t_{n,b,i}$ be equal to the statistic t_b evaluated at the data set Y_i. The sampling distribution of T_n is then approximated by

$$\hat{G}_{n,b}(x) = N_n^{-1} \sum_{i=1}^{N_n} 1\{\tau_b t_{n,b,i} \leq x\}. \tag{2.14}$$

Using this estimated sampling distribution, the critical value for the test is obtained as the $1 - \alpha$ quantile of $\hat{G}_{n,b}(\cdot)$; specifically, define

$$g_{n,b}(1 - \alpha) = \inf\{x : \hat{G}_{n,b}(x) \geq 1 - \alpha\}. \tag{2.15}$$

Finally, the nominal level α test rejects H_0 if and only if $T_n > g_{n,b}(1 - \alpha)$.

The following theorem gives the consistency of this procedure, under the null hypothesis, the alternative hypothesis, and a sequence of contiguous alternatives.

Theorem 2.6.1.

i. *Assume, for $P \in \mathbf{P_0}$, $G_n(P)$ converges weakly to a continuous limit law $G(P)$, whose corresponding cumulative distribution function is $G(\cdot, P)$ and whose $1 - \alpha$ quantile is $g(1 - \alpha, P)$. Assume $b/n \to 0$ and $b \to \infty$ as $n \to \infty$. If $G(\cdot, P)$ is continuous at $g(1 - \alpha, P)$ and $P \in \mathbf{P_0}$, then*

$$g_{n,b}(1 - \alpha) \to g(1 - \alpha, P) \text{ in probability}$$

and

$$Prob_P\{T_n > g_{n,b}(1 - \alpha)\} \to \alpha \text{ as } n \to \infty.$$

2.6 Hypothesis Testing   57

ii. *Assume the test statistic is constructed so that* $t_n(X_1, \ldots, X_n) \to t(P)$ *in probability, where* $t(P)$ *is a constant which satisfies* $t(P) = 0$ *if* $P \in \mathbf{P_0}$ *and* $t(P) > 0$ *if* $P \in \mathbf{P_1}$. *Assume* $b/n \to 0$, $b \to \infty$, *and* $\liminf_n (\tau_n/\tau_b) > 1$. *Then, if* $P \in \mathbf{P_1}$, *the rejection probability satisfies*

$$Prob_P\{T_n > g_{n,b}(1-\alpha)\} \to 1 \text{ as } n \to \infty.$$

iii. *Suppose* P_n *is a sequence of alternatives such that, for some* $P_0 \in \mathbf{P_0}$, $\{P_n^n\}$ *is contiguous to* $\{P_0^n\}$. *Assume* $b/n \to 0$ *and* $b \to \infty$ *as* $n \to \infty$. *Then,*

$$g_{n,b}(1-\alpha) \to g(1-\alpha, P_0) \text{ in } P_n^n\text{-probability.}$$

Hence, if T_n *converges in distribution to* T *under* P_n *and* $G(\cdot, P_0)$ *is continuous at* $g(1-\alpha, P_0)$, *then*

$$P_n^n\{T_n > g_{n,b}(1-\alpha)\} \to Prob\{T > g(1-\alpha, P_0)\}.$$

Proof. To prove (i), note again that $\hat{G}_{n,b}(x)$ is a U-statistic of degree b, with expectation under P equal to $G_b(x, P)$. An argument analogous to the one used in the proof of Theorem 2.2.1 (but easier because there is no centering) shows that $\hat{G}_{n,b}(x) \to G(x, P)$ in probability. Indeed, the variance of the U-statistic tends to zero by the same exponential inequality. It follows that $g_{n,b}(1-\alpha) \to g(1-\alpha, P)$ in probability. Thus, by Slutsky's theorem, the asymptotic rejection probability of the event $T_n > g_{n,b}(1-\alpha)$ is exactly α.

To prove (ii), rather than considering $\hat{G}_{n,b}(x)$, just look at the empirical distribution of the values of $t_{n,b,i}$ (not scaled by τ_b); so define

$$\hat{G}_{n,b}^0(x) = N_n^{-1} \sum_{i=1}^{N_n} 1\{t_{n,b,i} \leq x\} = \hat{G}_{n,b}(\tau_b x).$$

But, by a now familiar argument, $\hat{G}_{n,b}^0$ is a U-statistic with expectation

$$E_P[\hat{G}_{n,b}^0(x)] = Prob_P\{t_b(X_1, \ldots, X_b) \leq x\},$$

and so $\hat{G}_{n,b}^0(\cdot)$ converges in distribution to a point mass at $t(P)$. It also follows that a $1-\alpha$ quantile, say $g_{n,b}^0(1-\alpha)$, of $\hat{G}_{n,b}^0(\cdot)$ converges in probability to $t(P)$. But, our test rejects when $(\tau_n/\tau_b) \cdot t_n(X_1, \ldots, X_n)$ exceeds $g_{n,b}^0(1-\alpha)$. Since $\liminf_n (\tau_n/\tau_b) > 1$ and $t_n(X_1, \ldots, X_n) \to t(P)$ in probability (with $t(P) > 0$), it follows by Slutsky's theorem that the asymptotic rejection probability is one.

Finally, to prove (iii), we know that $g_{n,b}(1-\alpha) \to g(1-\alpha, P_0)$ in probability under P_0; contiguity forces $g_{n,b}(1-\alpha) \to g(1-\alpha, P_0)$ in probability under P_n. ∎

Remark 2.6.1. Consider the special case of testing a real-valued parameter. Specifically, suppose $\theta(\cdot)$ is a real-valued function from $\mathbf{P}$ to the real line. The null hypothesis is specified by $\mathbf{P_0} = \{P : \theta(P) = \theta_0\}$. Assume the alternative hypothesis is one-sided and specified by $\{P : \theta(P) > \theta_0\}$. Suppose we simply take

$$t_n(X_1,\ldots,X_n) = \hat{\theta}_n(X_1,\ldots,X_n) - \theta_0.$$

Then, it can be checked that the test construction accepts the null hypothesis if and only if the confidence interval (2.10) contains the value θ_0. Thus, in this special case, the test construction presented in this section has an exact duality with the interval presented in (2.10). This is not surprising, because the argument leading up to (2.10) was based on the relationship (2.7) and the asymptotic coverage probability of the event (2.6). Moreover, in the testing context, $\theta(P) = \theta_0$ is fixed and known under the null hypothesis, in which case $u_{n,b}(\alpha, P)$ in (2.7) can be computed, at least under the null hypothesis.

In addition, if $\hat{\theta}_n$ is a consistent estimator of $\theta(P)$, then the hypothesis on t_n in part (ii) of the theorem is satisfied (just take the absolute value of t_n for a two-sided alternative). Thus, the hypothesis on t_n in part (ii) of the theorem boils down to verifying a consistency property and is rather weak, though this assumption can in fact be weakened further. The convergence hypothesis of part (i) is satisfied by typical test statistics; in regular situations, $\tau_n = n^{1/2}$.

Remark 2.6.2. In Example 2.6.1, simply take

$$t_n = \int_{-\infty}^{\infty} [\hat{F}_n(x) - F_0(x)]^2 dF_0(x).$$

Then, t_n (under reasonable conditions) will converge to

$$t(F) = \int_{-\infty}^{\infty} [F(x) - F_0(x)]^2 dF_0(x),$$

if F is the distribution of U_i. Clearly, $t(F) = 0$ if and only if the null hypothesis is true.

Remark 2.6.3. The interpretation of part (iii) of the theorem is the following. Suppose, instead of using the subsampling construction, one could use the test that rejects when $T_n > g_n(1 - \alpha, P)$, where $g_n(1 - \alpha, P)$ is the exact $1 - \alpha$ quantile of the true sampling distribution $G_n(\cdot, P)$. Of course, this test is not available in general because P is unknown and so is $g_n(1 - \alpha, P)$. Then, the asymptotic power of the subsampling test against a sequence of contiguous alternatives $\{P_n\}$ to P with P in $\mathbf{P_0}$ is the same as the asymptotic power of this fictitious test against the same sequence of alternatives. Hence, to the order considered, there is no loss in efficiency in terms of power.

2.7 Data-Dependent Choice of Block Size

The basic theorems we have presented so far prove that subsampling works under weak conditions. In particular, the conditions on the choice of block size b are quite weak. Inevitably, the choice of block size will be data driven (as in Chapter 9) and higher order asymptotic considerations will come into play (as described in Chapter 10). At this point, we are not concerned with an optimality result (and it seems doubtful there will ever exist a universal prescription for choice of block size anyway). Rather, we present a result which shows subsampling works quite generally even with a data-driven choice of block size.

Theorem 2.7.1. *Assume Assumption 2.2.1. Let $1 \leq j_n \leq k_n \leq n$ be integers satisfying $j_n \to \infty$, $k_n/n \to 0$, $\tau_{k_n}/\tau_n \to 0$, and, for every $d > 0$, $k_n \exp(-d\lfloor \frac{n}{k_n} \rfloor) \to 0$ as $n \to \infty$. Also, assume $\{\tau_n\}$ is nondecreasing in n.*

i. *If x is a continuity point of $J(\cdot, P)$, then*
$$\sup_{j_n \leq b \leq k_n} |L_{n,b}(x) - J(x, P)| \to 0 \text{ in probability.}$$

ii. *Hence, if $\{\hat{b}_n\}$ is a data-dependent sequence (that is, a measurable function of $X_1, \ldots, X_n$), and*
$$Prob_P\{j_n \leq \hat{b}_n \leq k_n\} \to 1,$$
then
$$L_{n,\hat{b}_n}(x) \to J(x, P) \text{ in probability.}$$

iii. *If $J(\cdot, P)$ is continuous, then*
$$\sup_x |L_{n,\hat{b}_n}(x) - J(x, P)| \to 0 \text{ in probability.}$$
In fact,
$$\sup_{j_n \leq b \leq k_n} \sup_x |L_{n,b}(x) - J(x, P)| \to 0 \text{ in probability.}$$

iv. *Let*
$$c_{n,\hat{b}_n}(1-\alpha) = \inf\{x : L_{n,\hat{b}_n}(x) \geq 1 - \alpha\}.$$
Then, if $J(\cdot, P)$ is continuous,
$$Prob_P\{\tau_n[\hat{\theta}_n - \theta(P)] \leq c_{n,\hat{b}_n}(1-\alpha)\} \to 1 - \alpha$$
as $n \to \infty$. Therefore, the asymptotic coverage probability under P of the confidence interval $[\hat{\theta}_n - \tau_n^{-1} c_{n,\hat{b}_n}(1-\alpha), \infty)$ is the nominal level $1-\alpha$.

Proof. Let $\hat{\theta}_{n,b,i}$ be equal to the statistic $\hat{\theta}_b$ evaluated at the i-th of the $\binom{n}{b}$ data sets of size b; any ordering of these $\binom{n}{b}$ values will do. As in the proof of Theorem 2.2.1, define

$$U_{n,b}(x) = \binom{n}{b}^{-1} \sum_{i=1}^{\binom{n}{b}} 1\{\tau_b[\hat{\theta}_{n,b,i} - \theta(P)] \le x\}.$$

Here, the notation $U_{n,b}(x)$ clearly includes the dependence on b since, unlike Theorem 2.2.1, we are considering simultaneously a range of b values. First, we claim that, for each continuity point x of $J(\cdot, P)$,

$$\sup_{j_n \le b \le k_n} |U_{n,b}(x) - J(x,P)| \to 0 \text{ in probability.} \quad (2.16)$$

But,

$$\sup_{j_n \le b \le k_n} |J_b(x,P) - J(x,P)| \to 0,$$

because, if this convergence failed, there would exist $\{b_n\}$ with $b_n \in [j_n, k_n]$ such that $J_{b_n}(x, P)$ does not converge to $J(x, P)$, which is a contradiction since $b_n \ge j_n \to \infty$. So, to show the convergence (2.16), it suffices to show

$$\sup_{j_n \le b \le k_n} |U_{n,b}(x) - J_b(x,P)| \to 0 \text{ in probability.} \quad (2.17)$$

But, for any $t > 0$,

$$Prob_P\{\sup_{j_n \le b \le k_n} |U_{n,b}(x) - J_b(x,P)| \ge t\}$$

$$\le \sum_{b=j_n}^{k_n} Prob_P\{|U_{n,b}(x) - J_b(x,P)| \ge t\}$$

$$\le k_n \sup_{j_n \le b \le k_n} Prob_P\{|U_{n,b}(x) - J_b(x,P)| \ge t\}$$

$$\le 2k_n \sup_{j_n \le b \le k_n} \exp\{-2\lfloor \frac{n}{b} \rfloor t^2\},$$

making use of Hoeffding's inequality as in the inequality (2.5). But this last expression is bounded above by $2k_n \exp\{-2\lfloor \frac{n}{k_n} \rfloor t^2\}$, which tends to zero by assumption on $\{k_n\}$. Thus, the convergence (2.17) holds, as does (2.16). Now, note that

$$L_{n,b}(x) = \binom{n}{b}^{-1} \sum_{i=1}^{\binom{n}{b}} 1\{\tau_b[\hat{\theta}_{n,b,i} - \theta(P)] + \tau_b[\theta(P) - \hat{\theta}_n] \le x\}.$$

Fix any $\epsilon > 0$ so that $x \pm \epsilon$ are continuity points of $J(\cdot, P)$. Then,

$$U_{n,b}(x - \epsilon)1(E_{n,b}) \le L_{n,b}1(E_{n,b}) \le U_{n,b}(x + \epsilon), \quad (2.18)$$

where $1(E_{n,b})$ is the indicator of the event $E_{n,b} \equiv \{\tau_b|\theta(P) - \hat{\theta}_n| \leq \epsilon\}$. By the monotonicity of $\{\tau_n\}$,

$$1(E_{n,k_n}) \leq 1(E_{n,b}) \leq 1(E_{n,j_n})$$

and $\tau_{k_n}/\tau_n \to 0$ implies $Prob_P(E_{n,k_n}) \to 1$. So,

$$L_{n,b}(x)1(E_{n,k_n}) \leq U_{n,b}(x + \epsilon).$$

Thus, on the set E_{n,k_n},

$$\sup_{j_n \leq b \leq k_n} L_{n,b}(x) - J(x, P) \leq \sup_{j_n \leq b \leq k_n} U_{n,b}(x + \epsilon) - J(x, P)$$

$$\leq \sup_{j_n \leq b \leq k_n} |U_{n,b}(x + \epsilon) - J(x + \epsilon, P)| + J(x + \epsilon, P) - J(x, P).$$

But, by (2.16), it follows that, for every $\delta > 0$,

$$\sup_{j_n \leq b \leq k_n} L_{n,b}(x) - J(x, P) \leq \delta + J(x + \epsilon, P) - J(x, P)$$

with probability tending to one. Similarly, replacing $x + \epsilon$ by $x - \epsilon$ and using the first inequality in (2.18), we get, for every $\eta > 0$,

$$\sup_{j_n \leq b \leq k_n} |L_{n,b}(x) - J(x, P)| \leq \eta$$

with probability tending to one, which is equivalent to statement (i) of the theorem. Part (ii) is obvious. The rest of the theorem is proved as in the proof of Theorem 2.2.1. ∎

Remark 2.7.1. In some cases, one finds that an optimal choice of $b = b_n$ should satisfy

$$b_n/n^p \to \xi(P),$$

for some $p \in (0, 1)$, where $\xi(P)$ is a constant typically depending on the unknown probability mechanism P. For example, see equation (9.4). In an ad hoc way, one can sometimes estimate $\xi(P)$ consistently by $\hat{\xi}_n$ (say, by a plug-in approach), which leads to the choice of block size

$$\hat{b}_n = \lfloor \hat{\xi}_n n^p \rfloor.$$

Such a construction for $\hat{b}_n$ will easily satisfy the conditions of the theorem. Simply take $j_n = \lfloor \epsilon n^p \rfloor$ and $k_n = \lfloor n^p/\epsilon \rfloor$ for small enough ϵ. Moreover, the condition $\tau_{k_n}/\tau_n \to 0$ will be satisfied in the typical case τ_n is proportional to n^β for some $\beta \in (0, 1)$. In practice, the parameter $\xi(P)$ may be difficult to estimate, and even if consistent estimation is possible, the resulting estimator may have poor finite-sample performance. The point of this section is to show subsampling has some asymptotic validity across a broad range of choices for the subsample size.

Remark 2.7.2. The monotonicity assumption on $\{\tau_n\}$ can be replaced by the condition

$$\sup_{j_n \leq b \leq k_n} [\tau_b/\tau_n] \to 0,$$

as the proof essentially shows. Actually, the assumption can be removed altogether by the modification leading to Corollary 2.2.1.

Remark 2.7.3. The convergence in probability statements in the theorem can be strengthened to be almost sure convergences, provided $\tau_{k_n}[\hat{\theta}_n - \theta(P)] \to 0$ almost surely, and for every $d > 0$,

$$\sum_{n=1}^{\infty} k_n \exp(-d \lfloor \frac{n}{k_n} \rfloor) < \infty.$$

The last condition holds whenever k_n can be taken to be $O(n^p)$ with $p < 1$.

2.8 Variance Estimation: The Delete-d Jackknife

Consider the case where the univariate parameter of interest $\theta(F)$ is a real-valued function of the cumulative distribution function F, and the objective is to consistently estimate the variance of the estimator $\hat{\theta}_n = \theta(\hat{F}_n)$. Here, $\hat{F}_n$ denotes the empirical distribution function of the sample $X_1, \ldots, X_n$. For example, it may be the case that the limiting distribution of $\hat{\theta}_n$ is the mean-zero normal with unknown variance. This is the setup of the original paper/abstract by Tukey (1958) on the jackknife that pioneered the notion of computer-intensive statistical methods such as subsampling and resampling.

As mentioned earlier, Tukey's jackknife is closely related to subsampling with block size $b = n - 1$. As a matter of fact, assuming that the rate satisfies $\tau_n = \sqrt{n}$, the simple jackknife estimator of the variance $Var(\sqrt{n}\hat{\theta}_n)$ is equal to the variance of a random variable with distribution equal to the subsampling distribution $L_{n,b}(x) = N_n^{-1} \sum_{i=1}^{N_n} 1\{\tau_b(\hat{\theta}_{n,b,i} - \hat{\theta}_n) \leq x\}$, with $b = n - 1$.

Although the simple jackknife was shown to be successful in estimating the variance of some important statistics, there are many cases where the simple jackknife variance estimator is inconsistent. The most prominent such example is given by $\hat{\theta}_n$ being the sample median (for example, see Efron, 1982).

To remove these deficiencies, Wu (1986), and later Shao and Wu (1989), introduced and studied the so-called "delete-d jackknife." The delete-d jackknife estimator of the variance $Var(\sqrt{n}\hat{\theta}_n)$ is (approximately) equal to the variance of a random variable with distribution given by $L_{n,b}(\cdot)$ with $b = n - d$; here d is a positive integer less than n.

2.8 Variance Estimation: The Delete-d Jackknife

To give the exact form of the delete-d jackknife variance estimator, we now focus on the case where $\tau_n = \sqrt{n}$, and the statistic $\hat{\theta}_n = \theta(\hat{F}_n)$ happens to be Fréchet differentiable, that is, almost linear (see Section 1.6). Now recall that subsampling entails sampling *without* replacement from the *finite* population $(X_1, \ldots, X_n)$. Therefore, it is not surprising that the finite population correction factor $f = b/n$ may be useful. Hence, we employ the finite population correction to define

$$\tilde{L}_{n,b}(x) = \frac{1}{N_n} \sum_{i=1}^{N_n} 1\{\tau_r(\hat{\theta}_{n,b,i} - \hat{\theta}_n) \leq x\},$$

where $r = b/(1-f)$; see Section 10.3.1 for more details on this construction. Now, the delete-d jackknife estimator of the variance $Var(\sqrt{n}\hat{\theta}_n)$ is denoted by $\widetilde{Var}_{n,b}(\sqrt{n}\hat{\theta}_n)$, and is defined as the variance of a random variable with distribution given by $\tilde{L}_{n,b}(\cdot)$ above with $b = n - d$. The following lemma addresses the issue of consistency of the delete-d jackknife estimator $\widetilde{Var}_{n,b}(\sqrt{n}\hat{\theta}_n)$.

Lemma 2.8.1. *Assume that the statistic $\hat{\theta}_n$ admits the representation*

$$\hat{\theta}_n = \theta(F) + \frac{1}{n}\sum_{i=1}^{n} \phi_F(X_i) + o_P(1/\sqrt{n}), \tag{2.19}$$

for some (measurable) function $\phi_F(\cdot)$ satisfying $Var(\phi_F(X_1)) = \sigma_F^2 > 0$ and $E\phi_F(X_1) = 0$. Let $b \to \infty$, as $n \to \infty$, but with $b/n < 1 - \varepsilon$, for some constant $\varepsilon \in (0,1)$.

Then,

$$\widetilde{Var}_{n,b}(\sqrt{n}\hat{\theta}_n) \xrightarrow{P} \sigma_F^2.$$

Proof. See Theorem 2.11 in Shao and Tu (1995). ∎

Note that, at the expense of assuming the special form (2.19) for $\hat{\theta}_n$ (that is just slightly weaker than Fréchet differentiability), the requirement $b/n \to 0$ was replaced by the weaker $b/n < 1 - \varepsilon$ in Lemma 2.8.1; for instance, b can be chosen to be asymptotically proportional to n and still the delete-d jackknife variance estimator will be consistent in the case at hand where the data are i.i.d.

In Section 3.8, we will discuss the subsampling variance estimator for general statistics that are not necessarily Fréchet differentiable in the case of time series data. Since i.i.d. data can be thought of as the special case of a time series with no dependence, Lemma 3.8.1 in Section 3.8 constitutes an extension of Lemma 2.8.1 above. Nevertheless, as discussed in Section 3.8, the requirement $b/n \to 0$ is unavoidable in a time series setup.

2.9 Conclusions

In this chapter, the basic notions of subsampling in the context of i.i.d. data were presented. Some general consistency theorems were stated and proved, validating our statements that subsampling presents a viable approach to inference under very weak conditions. Subsampling was compared to the bootstrap as well, so that it is clear that subsampling works in a first-order asymptotic sense under weaker conditions than the bootstrap. Subsampling was seen to be related to the delete-d jackknife. In the context of hypothesis testing, two general approaches were presented in the chapter. The first approach consists of constructing a confidence interval for a parameter and exploiting the duality between test procedures and confidence intervals. The second approach, detailed in Section 2.6, avoids having to relate the hypotheses to a parameter and merely requires recomputing a test statistic over subsamples. In Section 2.7, a general theorem asserts that subsampling is consistent, even accounting for a data-dependent choice of subsampling size.

3
Subsampling for Stationary Time Series

3.1 Introduction

It is well known that inference methods for i.i.d. data or, more generally, independent data are simply not consistent when the underlying sequence is dependent. Therefore, the resampling and subsampling methods discussed in the previous chapters need to be modified to be applicable with time series data.

Historically, the first extension was to apply Efron's (1979) bootstrap to an approximate i.i.d. setup by focusing on residuals of some general regression model. In that case, the residuals are resampled, not the original observations. Examples where this idea is successful include linear regression (e.g., Freedman, 1981, 1984; Wu, 1986; Liu, 1988) autoregressive time series (e.g., Efron and Tibshirani, 1986; Bose, 1988), and nonparametric regression and nonparametric kernel spectral estimation (e.g., Franke and Härdle, 1992). However, this approach is restricted to situations where a general regression model can be relied upon.

A more general approach is to apply resampling to the original data sequence by considering blocks of data rather than single data points as in the i.i.d. setup. The motivation is that within each block the dependence structure of the underlying model is preserved and if the block size is allowed to tend to infinity with the sample size, asymptotically correct inference can ensue. A pioneering work was provided by Carlstein (1986), who used a blocking scheme to approximate the variance of a general statistic. His idea was to divide the original sequence in (nonoverlapping) blocks of size

$b < n$, recompute the statistic of interest on these blocks, and use the sample variance of the block statistics, after some suitable normalization. Later, Künsch (1989) and Liu and Singh (1992) independently introduced the "moving blocks" bootstrap, which besides variance estimation can also be used to estimate the sampling distribution of a statistic so that confidence intervals or regions for unknown parameters can be constructed. The method is analogous to the Efron's i.i.d. bootstrap in that it constructs pseudo-data sequences whose (known) data-generating mechanism mimicks the (unknown) data-generating mechanism that gave rise to the observed sequence. The key difference is that blocks of size $b < n$ of observations, resampled with replacement from the data, are concatenated to form such a pseudo sequence rather than single data points. Note that in contrast to Carlstein's (1986) approach, the moving blocks bootstrap uses overlapping blocks, a scheme that is generally more efficient. Politis and Romano (1992a,b, 1993a, 1994a,c) proposed variants of the moving blocks bootstrap that possess interesting features such as random block sizes and a circular, periodic extension of the original data sequence.

It can be easily seen that Efron's bootstrap is just a special case of the moving blocks bootstrap, namely, if blocks of size $b = 1$ are used. Therefore, one of the main problems with the i.i.d. bootstrap carries over to the block bootstrap. Its consistency also has to be proved by a case-by-case analysis, depending on the statistic of interest. For complicated statistics, such an analysis can be tedious or even impossible.

The goal of this chapter is then to extend the subsampling methodology for the i.i.d. setup to the case of dependent data. We aim for the same simplicity of application and generality of applicability. The basis for the material presented was laid in Politis and Romano (1992c, 1994b). As in the case of independent data, the main condition will be that the statistic of interest, properly normalized, has a nondegenerate limiting distribution. However, the crucial difference is that only blocks of consecutive data points, of size $b < n$, will be allowed as proper subsamples, rather than any collection of b points sampled without replacement from the original sequence. Also, some bound on the dependence of the underlying sequence will be required to ensure that, with the sample size, the available information in the data tend to infinity in an appropriate way. We first focus on the case of stationary time series. A generalization of the theory to nonstationary time series will be provided in Chapter 4.

In Section 3.2, the method is described for the general univariate parameter case after some intuition is provided by focusing on the sample mean. A general theorem is stated, allowing for asymptotically valid inference under very weak conditions. An extension to the general multivariate parameter case is given in Section 3.3. The usefulness of the general theory is demonstrated by discussing a large number of examples in Section 3.4. In many hypothesis testing problems the null hypothesis does not translate into a null hypothesis on a parameter. Section 3.5 shows how subsampling can be used for testing purposes in those situations. The issues of bias reduction and variance estimation are discussed in Sections 3.7 and 3.8, respectively. Finally, Section 3.9 provides a brief comparison with the moving blocks bootstrap.

Note that the theory for general parameters, such as function-valued parameters, in the context of time series data is developed in Chapter 7.

3.2 Univariate Parameter Case

3.2.1 Some Motivation: The Simplest Example

To fix ideas, suppose the sample $X_1, \ldots, X_n$ is known or suspected to exhibit serial dependence, and that it can generally be modeled as a stationary time series. Assume that the parameter of interest $\theta(P)$ is the common mean EX_1, and the statistic $\hat{\theta}_n$ is the sample mean $\bar{X}_n = n^{-1}\sum_{t=1}^{n} X_t$. If the serial dependence is weak enough such that $\sum |R(k)| < \infty$, where $R(k) = Cov(X_1, X_{1+k})$, then (under regularity conditions; cf. Brockwell and Davis, 1991) $\bar{X}_n$ is asymptotically normal, that is, $\sqrt{n}(\bar{X}_n - \theta(P))$ has the limiting normal $N(0, \sigma_\infty^2)$ distribution, where

$$\sigma_\infty^2 = Var(X_1) + 2\sum_{k=1}^{\infty} R(k).$$

Note that $\sigma_\infty^2 = \lim_{n \to \infty} \sigma_n^2$, where

$$\sigma_n^2 = Var(\sqrt{n}\bar{X}_n) = Var(X_1) + 2\sum_{k=1}^{n}(1 - |k|/n)R(k).$$

To construct confidence intervals for $\theta(P)$ using the previously mentioned Central Limit theorem, a consistent estimator of σ_∞^2 is required. A straightforward approach would be to estimate the covariances $R(k)$, $k = 1, \ldots, n$, and plug them in the formula for σ_n^2, since $\sigma_n^2 \to \sigma_\infty^2$. However, this naive procedure is not consistent, because the estimates of $R(k)$ for k close to n are highly inaccurate; indeed, since σ_∞^2 is just a constant multiple of the spectral density of the time series evaluated at point zero, this is a well-known difficulty in the literature concerning spectral estimation (see, for example, Priestley, 1981). Apparently, based on a sample of size n we could only accurately estimate $R(k)$, for $k = 1, \ldots, b$, where $b \ll n$. It then follows that we can only hope to estimate well σ_b^2, and not σ_n^2, where $\sigma_b^2 = Var(X_1) + 2\sum_{t=1}^{b}(1 - |t|/b)R(k)$. But there is a natural way to estimate σ_b^2 from $X_1, \ldots, X_n$, namely, to look at the sample variability of $\frac{1}{\sqrt{b}}\sum_{a=t}^{t+b-1} X_a$, for $t = 1, \ldots, n-b+1$. This is equivalent to considering the "sample variance" estimator

$$\hat{\sigma}_b^2 = \frac{1}{n-b+1} \sum_{t=1}^{n-b+1} (\frac{1}{\sqrt{b}} \sum_{a=t}^{t+b-1} X_a - \sqrt{b}\bar{X}_n)^2$$

that was studied in Carlstein (1986) and in Politis and Romano (1993b). This proposed estimator is consistent, under some moment and mixing conditions, essentially because both σ_b^2 and σ_n^2 converge asymptotically to σ_∞^2, if both b and n are assumed to tend to infinity (see Section 3.8 for more details).

In the same light, one could look at the more general problem of estimating the distribution of $\sqrt{n}(\bar{X}_n - \theta(P))$. In principle, this could be done by looking at the sample variability of $\frac{1}{\sqrt{b}} \sum_{a=t}^{t+b-1}(X_a - \theta(P))$, for $t = 1, \ldots, n-b+1$. The problem is that $\theta(P)$ is unknown. However, a feasible solution consists of replacing $\theta(P)$ by its estimate $\bar{X}_n$ and looking at the sample variability of $\frac{1}{\sqrt{b}} \sum_{a=t}^{t+b-1}(X_a - \bar{X}_n)$, for $t = 1, \ldots, n-b+1$. We then can define the corresponding "empirical" distribution

$$L_{n,b}(x) = \frac{1}{n-b+1} \sum_{t=1}^{n-b+1} 1\{\sqrt{b}(b^{-1} \sum_{a=t}^{t+b-1} X_a - \bar{X}_n) \leq x\}$$

as an approximation to the sampling distribution of $\sqrt{n}(\bar{X}_n - \theta(P))$. Here, the underlying principle is that both $\sqrt{b}(\bar{X}_b - \theta(P))$ and $\sqrt{n}(\bar{X}_n - \theta(P))$ have the same asymptotic distribution—which just happens to be the normal $N(0, \sigma_\infty^2)$ distribution—where of course $\bar{X}_b = b^{-1} \sum_{a=1}^{b} X_a$.

Although variance estimation is intimately linked with the assumption of asymptotic normality, this more general idea of directly approximating the sampling distribution will work in a variety of different situations, including cases where asymptotic normality does not hold, where the rate of convergence is not $\sqrt{n}$, or where variance estimation is not consistent.

3.2.2 Theory and Methods for the General Univariate Parameter Case

To describe the case of a general parameter, suppose $\{\ldots, X_{-1}, X_0, X_1, \ldots\}$ is a sequence of random variables taking values in an arbitrary sample space S, and defined on a common probability space. Denote the joint probability law governing the infinite sequence by P. By stationarity, all finite-dimensional marginal distributions are shift-invariant; that is, for any integer m the joint distribution of $X_k, X_{k+1}, \ldots, X_{k+m}$ does not depend on k. The goal is to construct a confidence interval for some parameter $\theta(P)$, on the basis of observing $X_1, \ldots, X_n$. The sequence $\{X_t\}$ will be assumed to satisfy a certain weak dependence condition, namely, the condition of α-mixing (see Definition A.0.1 in Appendix A). Let $\hat{\theta}_n = \hat{\theta}_n(X_1, \ldots, X_n)$ be an estimator of $\theta(P) \in \mathbb{R}$, the parameter of interest.

3.2 Univariate Parameter Case

The crux of the subsampling method is to approximate the sampling distribution of a statistic by recomputing it on subsamples of smaller size of the observed data. In the context of independent data, subsamples of size $b < n$ are generated by sampling b observations without replacement from the original data sequence of size n (see Chapter 2). Since this approach does not take the order of the original sequence into account, it generally fails for time series data. The key, therefore, is to only use *blocks* of size b of consecutive observations as legitimate subsamples, the first one being $\{X_1, X_2, \ldots, X_b\}$, and the last one being $\{X_{n-b+1}, X_{n-b+2}, \ldots, X_n\}$. Note that there are $q = n - b + 1$ such blocks. Obviously, $n - b + 1 \ll \binom{n}{b}$, the number of available subsamples in the independent case. For this reason, no stochastic approximations as in Section 2.4 are necessary for time series data, unless the sample size n is huge.

Define $\hat{\theta}_{n,b,t} = \hat{\theta}_b(X_t, \ldots, X_{t+b-1})$, the estimator of $\theta(P)$ based on the subsample $\{X_t, \ldots, X_{t+b-1}\}$. Let $J_b(P)$ be the sampling distribution of

$$\tau_b(\hat{\theta}_{n,b,1} - \theta(P)),$$

where τ_b is an appropriate normalizing constant. Also define the corresponding cumulative distribution function:

$$J_b(x, P) = \mathrm{Prob}_P\{\tau_b(\hat{\theta}_{n,b,1} - \theta(P)) \leq x\}. \tag{3.1}$$

For convenience, denote $J_n(P) = J_{n,1}(P)$, the sampling distribution of $\tau_n(\hat{\theta}_n - \theta(P))$.

Essentially, the only assumption that will be needed to consistently estimate $J_n(P)$ is the following.

Assumption 3.2.1. *There exists a limiting law $J(P)$ such that $J_n(P)$ converges weakly to $J(P)$ as $n \to \infty$.*

This means that the estimator, properly centered and normalized, has a limiting distribution. It is hard to conceive of any asymptotic theory free of such a requirement.

In order to describe our method, let Y_t be the block of size b of the consecutive data $\{X_t, \ldots, X_{t+b-1}\}$. Only a very weak assumption on b will be required. Typically, $b/n \to 0$ and $b \to \infty$ as $n \to \infty$. The approximation to $J_n(x, P)$ we study is the analogue of (2.1) for the i.i.d. case and defined by

$$L_{n,b}(x) = \frac{1}{n-b+1} \sum_{t=1}^{n-b+1} 1\{\tau_b(\hat{\theta}_{n,b,t} - \hat{\theta}_n) \leq x\}. \tag{3.2}$$

The motivation behind the method is the following. For any t, Y_t is a true subsample of size b from the true model P. Hence, the *exact* distribution of $\tau_b(\hat{\theta}_{n,b,t} - \theta(P))$ is $J_b(P)$. By stationarity, the empirical distribution of the $n-b+1$ values of $\tau_b(\hat{\theta}_{n,b,t} - \theta(P))$ should serve as good approximation to $J_n(P)$, at least for large n. Replacing $\theta(P)$ by $\hat{\theta}_n$ is permissible because

$\tau_b(\hat{\theta}_n - \theta(P))$ is of order τ_b/τ_n in probability and we will assume that $\tau_b/\tau_n \to 0$.

The following theorem could be coined "a general asymptotic validity result under minimal conditions." It states sufficient conditions under which the subsampling method will yield asymptotically valid results for very general statistics. No equivalent theorem is available for resampling methods, such as the moving blocks bootstrap or the stationary bootstrap. Instead, such methods require a much more difficult case by case analysis. We denote the α-mixing sequence corresponding to $\{X_t\}$ by $\alpha_X(\cdot)$.

Theorem 3.2.1. *Assume Assumption 3.2.1 and that $\tau_b/\tau_n \to 0, b/n \to 0$, and $b \to \infty$ as $n \to \infty$. Also, assume that $\alpha_X(m) \to 0$ as $m \to \infty$.*

i. *If x is a continuity point of $J(\cdot, P)$, then $L_{n,b}(x) \to J(x, P)$ in probability.*
ii. *If $J(\cdot, P)$ is continuous, then $\sup_x |L_{n,b}(x) - J(x, P)| \to 0$ in probability.*
iii. *For $\alpha \in (0, 1)$, let*

$$c_{n,b}(1-\alpha) = \inf\{x : L_{n,b}(x) \geq 1-\alpha\}.$$

Correspondingly, define

$$c(1-\alpha, P) = \inf\{x : J(x, P) \geq 1-\alpha\}.$$

If $J(\cdot, P)$ is continuous at $c(1-\alpha, P)$, then

$$Prob_P\{\tau_n[\hat{\theta}_n - \theta(P)] \leq c_{n,b}(1-\alpha)\} \to 1-\alpha \text{ as } n \to \infty.$$

Thus, the asymptotic coverage probability under P of the interval $I_1 = [\hat{\theta}_n - \tau_n^{-1}c_{n,b}(1-\alpha), \infty)$ is the nominal level $1-\alpha$.

Remark 3.2.1. In most examples, the rate of convergence satisfies $\tau_n = n^\gamma$, for some $\gamma > 0$, and the assumptions on b simplify to $b/n \to 0$ and $b \to \infty$. In fact, the condition $\tau_b/\tau_n \to 0$ can be removed if the goal is confidence interval construction, by the same argument as in the i.i.d. case; see the development leading up to Corollary 2.2.1.

Remark 3.2.2. For reasons analogous to those stated in Section 2.2, the conditions on the block size b are in general not only sufficient, but also necessary. For the scenario of dependent observations, it is even more clear that keeping the block size fixed will result in failure of the method. On the other hand, for the case of the sample mean, Lahiri (1998) showed explicitly how subsampling (and block bootstrap methods) fail when the block size grows at the same rate as the sample size. In the case $b/n \to \lambda \in (0, 1)$, the subsampling approximation has a random limit on the space of all probability measures on the real line. In the case $b/n \to 1$, the approximation collapses to a point mass at zero. By linearization, his results carry over to statistics that can be approximated by smooth functions of means.

Remark 3.2.3. Note that, besides the mixing condition, the main difficulty in applying the theorem lies in checking whether the properly standardized estimator has a nondegenerate limiting distribution, whose shape, however, does not have to be known. Much more work is typically necessary to demonstrate the validity of bootstrap methods.

Proof of Theorem 3.2.1. Without loss of generality, we may think of b as a function of n. Therefore, the notational burden can be reduced by omitting the b-subscripts. For example, $L_n(\cdot) \equiv L_{n,b}(\cdot)$, $c_n(\alpha) \equiv c_{n,b}(\alpha)$, etc. To simplify the notation further, introduce $q \equiv q_n \equiv n - b + 1$. Let

$$U_n(x) = \frac{1}{q} \sum_{t=1}^{q} 1\{\tau_b[\hat{\theta}_{n,b,t} - \theta(P)] \leq x\}.$$

To prove (i), it suffices to show that $U_n(x)$ converges in probability to $J(x, P)$ for every continuity point x of $J(\cdot, P)$. This can be seen by noting that

$$L_n(x) = \frac{1}{q} \sum_{t=1}^{q} 1\{\tau_b[\hat{\theta}_{n,b,t} - \theta(P)] + \tau_b[\theta(P) - \hat{\theta}_n] \leq x\},$$

so that for every $\epsilon > 0$,

$$U_n(x - \epsilon)1\{E_n\} \leq L_n(x)1\{E_n\} \leq U_n(x + \epsilon),$$

where $1\{E_n\}$ is the indicator of the event $E_n = \{\tau_b|\theta(P) - \hat{\theta}_n| \leq \epsilon\}$. But the event E_n has probability tending to one. So, with probability tending to one,

$$U_n(x - \epsilon) \leq L_n(x) \leq U_n(x + \epsilon).$$

Thus, if $x + \epsilon$ and $x - \epsilon$ are continuity points of $J(\cdot, P)$, then $U_n(x \pm \epsilon)$ converging to $J(x \pm \epsilon, P)$ in probability implies

$$J(x - \epsilon, P) - \epsilon \leq L_n(x) \leq J(x + \epsilon, P) + \epsilon$$

with probability tending to one. Now, let $\epsilon \to 0$ such that $x \pm \epsilon$ are continuity points of $J(\cdot, P)$ to conclude that $L_n(x) \to J(x, P)$ in probability as well. Therefore, we may restrict our attention to $U_n(x)$.

Since $E(U_n(x)) = J_b(x, P)$, the proof of (i) reduces by Assumption 3.2.1 to showing that $Var(U_n(x))$ tends to zero (as n tends to infinity). Define

$$I_{b,t} = 1\{\tau_b[\hat{\theta}_{n,b,t} - \theta(P)] \leq x\}, \quad t = 1, \ldots, q,$$

$$s_{q,h} = \frac{1}{q} \sum_{t=1}^{q-h} Cov(I_{b,t}, I_{b,t+h}).$$

Then,

$$Var(U_n(x)) = \frac{1}{q}(s_{q,0} + 2\sum_{h=1}^{q-1} s_{q,h})$$

$$= \frac{1}{q}(s_{q,0} + 2\sum_{h=1}^{b-1} s_{q,h} + 2\sum_{h=b}^{q-1} s_{q,h})$$
$$\equiv A^* + A,$$

where $A^* = \frac{1}{q}(s_{q,0} + 2\sum_{h=1}^{b-1} s_{q,h})$ and $A = \frac{2}{q}\sum_{h=b}^{q-1} s_{q,h}$.

It is readily seen that $|A^*| = O(b/q)$. To handle A, we apply Lemma A.0.2. For $h \geq b$

$$|Cov(I_{b,t}, I_{b,t+h})| \leq 4\alpha_X(h - b + 1)$$

and therefore,

$$|A| \leq \frac{8}{q}\sum_{h=1}^{q-b} \alpha_X(h).$$

By the monotonicity of mixing coefficients, $\alpha_X(m) \to 0$ (as $m \to \infty$) and therefore $M^{-1}\sum_{m=1}^{M} \alpha_X(m) \to 0$ (as $M \to \infty$), which implies that A converges to zero. Thus, both A and A^* converge to zero, which completes the proof of (i).

To prove (ii), given any subsequence $\{n_k\}$, one can extract a further subsequence $\{n_{k_j}\}$ such that $L_{n_{k_j}}(x) \to J(x, P)$ for all x in some countable dense set of the real line, almost surely. It then follows that, on a set of probability one, $L_{n_{k_j}}(x)$ tends weakly to $J(x, P)$. By the continuity of $J(\cdot, P)$, this convergence is uniform by Polya's theorem.

The proof of (iii) is very similar to the proof of Theorem 1 of Beran (1984) given our result (i). ∎

The interval I_1 defined in Theorem 3.2.1 corresponds to a one-sided hybrid percentile interval in the bootstrap literature (e.g., Hall, 1992). A two-sided *equal-tailed* confidence interval can be obtained by forming the intersection of two one-sided intervals. The two-sided analogue of I_1 is

$$I_2 = [\hat{\theta}_n - \tau_n^{-1} c_{n,b}(1 - \alpha/2), \ \hat{\theta}_n - \tau_n^{-1} c_{n,b}(\alpha/2)].$$

I_2 is called equal-tailed because it has approximately equal probability in each tail:

$$Prob_P\{\theta(P) < \hat{\theta}_n - \tau_n^{-1} c_{n,b}(1 - \alpha/2)\} \simeq \alpha$$

and

$$Prob_P\{\theta(P) > \hat{\theta}_n - \tau_n^{-1} c_{n,b}(\alpha/2)\} \simeq \alpha/2.$$

As an alternative approach, two-sided *symmetric* confidence intervals can be constructed. A two-sided symmetric confidence interval is given by $[\hat{\theta}_n - \hat{c}, \hat{\theta}_n + \hat{c}]$, where $\hat{c}$ is chosen so that $Prob_P\{|\hat{\theta}_n - \theta(P)| > \hat{c}\} \simeq \alpha$. Hall (1988), in the context of independent observations, showed that symmetric bootstrap confidence intervals may enjoy enhanced coverage and, even in asymmetric circumstances, can be shorter than equal-tailed confidence intervals. To construct two-sided symmetric subsampling intervals in practice, we follow the traditional approach and estimate the two-sided distribution function

$$J_{n,|\cdot|}(x, P) = Prob_P\{\tau_n \left|\hat{\theta}_n - \theta(P)\right| \leq x\}. \tag{3.3}$$

The subsampling approximation to $J_{n,|\cdot|}(x, P)$ is defined by

$$L_{n,b,|\cdot|}(x) = \frac{1}{n-b+1} \sum_{t=1}^{n-b+1} 1\{\tau_b \left|\hat{\theta}_{n,b,t} - \hat{\theta}_n\right|\} \leq x\}. \tag{3.4}$$

By Theorem 3.2.1 the asymptotic validity of two-sided symmetric subsampling intervals easily follows.

Corollary 3.2.1. *Make the same assumptions as in Theorem 3.2.1. Denote by $J_{|\cdot|}(P)$ the law of $|Z|$, where Z is a random variable with distribution $J(P)$.*

i. *If x is a continuity point of $J_{|\cdot|}(\cdot, P)$, then $L_{n,b,|\cdot|}(x) \to J_{|\cdot|}(x, P)$ in probability.*
ii. *If $J_{|\cdot|}(\cdot, P)$ is continuous, then $\sup_x |L_{n,|\cdot|}(x) - J_{|\cdot|}(x, P)| \to 0$ in probability.*
iii. *For $\alpha \in (0,1)$, let*

$$c_{n,b,|\cdot|}(1-\alpha) = \inf\{x : L_{n,b,|\cdot|}(x) \geq 1-\alpha\}.$$

Correspondingly, define

$$c_{|\cdot|}(1-\alpha, P) = \inf\{x : J_{|\cdot|}(x, P) \geq 1-\alpha\}.$$

If $J_{|\cdot|}(\cdot, P)$ is continuous at $c_{|\cdot|}(1-\alpha, P)$, then

$$Prob_P\{\tau_n \left|\hat{\theta}_n - \theta(P)\right| \leq c_{n,b,|\cdot|}(1-\alpha)\} \to 1-\alpha \text{ as } n \to \infty.$$

Thus, the asymptotic coverage probability under P of the interval $I_{SYM} = [\hat{\theta}_n - \tau_n^{-1}c_{n,b,|\cdot|}(1-\alpha), \hat{\theta}_n + \tau_n^{-1}c_{n,b,|\cdot|}(1-\alpha)]$ is the nominal level $1-\alpha$.

Proof. The proof follows immediately from Theorem 3.2.1 and the continuous mapping theorem. ∎

Remark 3.2.4. Symmetric confidence intervals are not necessarily a superior choice. Results about good performance are only asymptotic in character and need not hold for all sample sizes. Furthermore, it is not always desirable to constrain a confidence interval to be symmetric. The asymmetry of an equal-tailed confidence interval may contain some valuable information about the location of the true parameter and the skewness of the sampling distribution of the estimator. We will compare the finite sample performance of symmetric and equal-tailed intervals in some simulation studies in later chapters.

3.2.3 Studentized Roots

A result pertaining to studentized statistics, analogous to Theorem 2.5.1 for the i.i.d. case, could easily be stated and proven. We will leave the details to the reader.

Alternatively, such a result immediately follows as a corollary of Theorem 12.2.2, which is more general by allowing for estimators of scale that may converge in distribution to a nondegenerate law rather than in probability to a constant.

3.3 Multivariate Parameter Case

It is often necessary to construct confidence regions for multivariate parameters $\theta(P) \in \mathbb{R}^k$, with $k > 1$. For example, this problem arises when we want to draw simultaneous inference on the regression coefficients in a multivariate regression problem. Another example would be making joint inference for the first k autocorrelation coefficients of a stationary time series. Fortunately, the subsampling method can be quite easily extended to higher dimensions.

Again, let $\hat{\theta}_{n,b,t} = \hat{\theta}_b(X_t, \ldots, X_{t+b-1})$ be an estimator of $\theta(P)$ based on the subsample $\{X_t, \ldots, X_{t+b-1}\}$. Define $J_b(P)$ to be the sampling distribution of $\tau_b(\hat{\theta}_{n,b,1} - \theta(P))$. Rather than working with multivariate distribution functions, we will look at indicators of Borel sets. For any Borel set $A \in \mathbb{R}^k$, define

$$J_b(A, P) = Prob_P\{\tau_b(\hat{\theta}_{n,b,1} - \theta(P)) \in A\}. \tag{3.5}$$

Assumption 3.2.1 then needs to be modified as follows.

Assumption 3.3.1. *There exists a limiting law $J(P)$ such that $J_n(P)$ converges weakly to $J(P)$ as $n \to \infty$. This means, for any Borel set A whose boundary has mass zero under $J(P)$, we have $J_n(A, P) \to J(A, P)$ as $n \to \infty$.*

The method is analogous to the univariate case: Let Y_t be the block of size b of the consecutive data $\{X_t, \ldots, X_{t+b-1}\}$. Only a very weak assumption on b will be required. Typically, $b/n \to 0$ and $b \to \infty$ as $n \to \infty$. Now, let $\hat{\theta}_{n,b,t}$ be equal to the statistic $\hat{\theta}_b$ evaluated at the data set Y_t. The approximation to $J_n(A, P)$ we study is defined by

$$L_{n,b}(A) = \frac{1}{n-b+1} \sum_{t=1}^{n-b+1} 1\{\tau_b(\hat{\theta}_{n,b,t} - \hat{\theta}_n) \in A\}. \tag{3.6}$$

The following theorem states sufficient conditions for this approximation to be consistent, and also describes how the approximation can be used to construct asymptotically valid confidence regions for $\theta(P)$.

Theorem 3.3.1. *Assume Assumption 3.3.1 and that $\tau_b/\tau_n \to 0$, $b/n \to 0$, and $b \to \infty$ as $n \to \infty$. Also, assume that $\alpha_X(m) \to 0$ as $m \to \infty$.*

i. $L_{n,b}(A) \to J(A, P)$ *in probability, for each Borel set A whose boundary has mass zero under $J(P)$.*

ii. $\rho_k(L_{n,b}, J(P)) \to 0$ *in probability, for every metric ρ_k that metrizes weak convergence on $\mathbb{R}^k$.*

iii. *Let $\{Z_n\}$ and Z be random vectors with $\mathcal{L}(Z_n) = L_{n,b}$ and $\mathcal{L}(Z) = J(P)$, where $\mathcal{L}(\cdot)$ denotes the law of a random variable. Then, for any almost everywhere $J(P)$ continuous real function f on $\mathbb{R}^k$ and any metric ρ_1 which metrizes weak convergence on $\mathbb{R}$,*

$$\rho_1(\mathcal{L}(f(Z_n)), \mathcal{L}(f(Z))) \to 0 \text{ in probability.}$$

In particular, for any norm $\|\cdot\|$ on $\mathbb{R}^k$,

$$\rho_1(\mathcal{L}(\|Z_n\|), \mathcal{L}(\|Z\|)) \to 0 \text{ in probability.}$$

iv. *Let Z be a random vector with $\mathcal{L}(Z) = J(P)$. For a norm $\|\cdot\|$ on $\mathbb{R}^k$, define univariate distributions $L_{n,\|\cdot\|}$ and $J_{\|\cdot\|}(P)$ in the following way:*

$$L_{n,b,\|\cdot\|}(x) = \frac{1}{n-b+1} \sum_{t=1}^{n-b+1} \mathbf{1}\{\|\tau_b(\hat{\theta}_{n,b,t} - \hat{\theta}_n)\| \leq x\},$$

$$J_{\|\cdot\|}(x, P) = \text{Prob}_P\{\|Z\| \leq x\}.$$

For $\alpha \in (0, 1)$, let

$$c_{n,b,\|\cdot\|}(1-\alpha) = \inf\{x : L_{n,b,\|\cdot\|}(x) \geq 1 - \alpha\}.$$

Correspondingly, define

$$c_{\|\cdot\|}(1-\alpha, P) = \inf\{x : J_{\|\cdot\|}(x, P) \geq 1 - \alpha\}.$$

If $J_{\|\cdot\|}(\cdot, P)$ is continuous at $c_{\|\cdot\|}(1-\alpha, P)$, then

$$\text{Prob}_P\{\|\tau_n(\hat{\theta}_n - \theta(P))\| \leq c_{n,b,\|\cdot\|}(1-\alpha)\} \to 1 - \alpha \text{ as } n \to \infty.$$

Thus, the asymptotic coverage probability under P of the region $\{\theta : \|\tau_n(\theta - \hat{\theta}_n)\| \leq c_{n,b,\|\cdot\|}(1-\alpha)\}$ is the nominal level $1 - \alpha$.

Remark 3.3.1. Remark 3.2.3 applies here in spirit as well.

Proof of Theorem 3.3.1. Without loss of generality, we may think of b as a function of n. Therefore we can reduce the notational burden by omitting the b-subscripts. For example, $L_n(\cdot) \equiv L_{n,b}(\cdot)$, $c_n(\alpha) \equiv c_{n,b}(\alpha)$, etc. We introduce the following notation before setting out to prove the theorem. For a set A, denote its interior, closure, and boundary by A^o, A^-, and $\partial(A)$, respectively. For a set A and a positive constant ϵ, define sets $A_{-\epsilon} \subseteq A \subseteq A_{+\epsilon}$ in the following way. Set $M_{A,\epsilon} \equiv \bigcup_{x \in \partial(A)} B(x, \epsilon)$, where $B(x, \epsilon)$ denotes the (closed) ball with center x and radius ϵ. Then $A_{+\epsilon} \equiv A \cup M_{A,\epsilon}$

and $A_{-\epsilon} \equiv A \setminus M_{A,\epsilon}$. For instance, if A is the ball with center y and radius Δ, then $A_{-\epsilon}$ is the open ball with center y and radius $\Delta - \epsilon$ (where a ball with negative radius is defined as the empty set) and $A_{+\epsilon}$ is the closed ball with center y and radius $\Delta + \epsilon$. As before, define $q \equiv q_n \equiv n - b + 1$. Now let

$$U_n(A) = \frac{1}{q} \sum_{t=1}^{q} 1\{\tau_b(\hat{\theta}_{n,b,t} - \theta(P)) \in A\}.$$

To prove (i), it suffices to show that $U_n(A)$ converges in probability to $J(A, P)$ for every Borel set A whose boundary has measure zero under $J(P)$. This can be seen by noting that

$$L_n(A) = \frac{1}{q} \sum_{t=1}^{q} 1\{\tau_b(\hat{\theta}_{n,b,t} - \theta(P)) + \tau_b(\theta(P) - \hat{\theta}_n) \in A\},$$

so that for every $\epsilon > 0$,

$$U_n(A_{-\epsilon})1\{E_n\} \leq L_n(A)1\{E_n\} \leq U_n(A_{+\epsilon}),$$

where $1\{E_n\}$ is the indicator of the event $E_n \equiv \{\tau_b \|\theta(P) - \hat{\theta}_n\| \leq \epsilon\}$. But, the event E_n has probability tending to one. So, with probability tending to one,

$$U_n(A_{-\epsilon}) \leq L_n(A) \leq U_n(A_{+\epsilon}).$$

Thus, if $A_{+\epsilon}$ and $A_{-\epsilon}$ are Borel sets whose boundaries have mass zero under $J(P)$, then $U_n(A_{\pm\epsilon}) \to J(A_{\pm\epsilon}, P)$ in probability implies

$$J(A_{-\epsilon}, P) - \epsilon \leq L_n(A) \leq J(A_{+\epsilon}, P) + \epsilon$$

with probability tending to one. Now, let $\epsilon \to 0$ such that $A_{\pm\epsilon}$ are Borel sets whose boundaries have mass zero under $J(P)$. Therefore, we may restrict our attention to $U_n(A)$.

Since $E(U_n(A)) = J_b(A, P)$, the proof of (i) reduces by Assumption 3.3.1 to showing that $Var(U_n(A)) \to 0$ as $n \to \infty$. This is accomplished completely analogously to the proof of (i) in Theorem 3.2.1 and thus is omitted.

In order to prove (ii), we need the following result from Billingsley (1968).

Lemma 3.3.1. *Let $\{Q_n\}$ and Q be probability measures on $\mathbb{R}^k$. Also, let $\mathcal{U}$ be a subclass of $\mathcal{B}^k$ such that*

　　a. *$\mathcal{U}$ is closed under the formation of finite intersections, and*
　　b. *for every x in $\mathbb{R}^k$ and every positive ϵ, there is an A in $\mathcal{U}$ with $x \in A^\circ \subseteq A \subseteq B^\circ(x, \epsilon)$.*

If $Q_n(A) \to Q(A)$ for every A in $\mathcal{U}$, then $Q_n \stackrel{\mathcal{L}}{\Longrightarrow} Q$, or, equivalently, $\rho_k(Q_n, Q) \to 0$, for any metric ρ_k that metrizes weak convergence on $\mathbb{R}^k$.

Define a class of sets $\mathcal{U}_{J(P)}$ (the subscript indicates that the class may depend on $J(P)$) in the following way. Let D be a dense countable subset in the set of all points in $\mathbb{R}^k$ that have mass zero under $J(P)$. For each $x \in D$, let E_x be a dense countable subset containing positive real numbers ϵ for which $\partial(B^o(x,\epsilon))$ has mass zero under $J(P)$. Now, set $\mathcal{V}_{J(P)} \equiv \bigcup_{x \in D, \epsilon_x \in E_x} S(x, \epsilon_x)$. Since $\mathcal{V}_{J(P)}$ is a countable union of countable sets, it is countable itself. Finally, define $\mathcal{U}_{J(P)}$ to contain all the finite intersections of elements of $\mathcal{V}_{J(P)}$. We see immediately that $\mathcal{U}_{J(P)}$ is countable again and meets the conditions (a) and (b) of Lemma 3.3.1 Furthermore, each set $A \in \mathcal{U}_{J(P)}$ has a boundary of mass zero under $J(P)$.

Let $\{n_j\}$ be a subsequence of $\{n\}$. For each $A \in \mathcal{U}_{J(P)}$, we can then find a further subsequence $\{n_{j_l}\}$ such that $L_{n_{j_l}}(A) \to J(A, P)$ almost surely (by the fact that $L_n(A) \to J(A, P)$ in probability). Since $\mathcal{U}_{J(P)}$ is countable, there is a common subsequence $\{n_{j_m}\}$ such that, on a set of probability one, $L_{n_{j_m}}(A) \to J(A, P)$ for all $A \in \mathcal{U}_{J(P)}$. By Lemma 3.3.1 then, $\rho_k(L_{n_{j_m}}, J(P)) \to 0$ almost surely and this shows that $\rho_k(L_n, J(P)) \to 0$ in probability.

The proof of (iii) is obvious, once we have (ii). Finally, (iv) follows easily from (iii) and Theorem 3.2.1. ∎

Remark 3.3.2. The distinction between equal-tailed and symmetrical two-sided confidence intervals for univariate parameters does not carry over to higher dimensions. By default, we will typically estimate the equivalent of the two-sided distribution function, which corresponds to symmetric confidence intervals.

3.4 Examples

In this section, we describe how the subsampling method can be applied to some specific examples and show how the general theorems from Sections 3.2 and 3.3 can be used to demonstrate the validity of the method. The generic starting point is the example of the univariate mean. Thereafter, we will discuss smooth functions of the (multivariate) mean, multivariate linear regression, autocorrelations, variance ratios, and robust statistics from time series.

In the previous sections, it was demonstrated that, except for a mixing condition on the underlying sequence, the main condition for subsampling to work is that the properly normalized estimator has a nondegenerate limiting distribution. Hence, in the discussion of our examples, we will focus on sufficient conditions that ensure the existence of such a distribution. Note that to show the validity of bootstrap methods, given that they apply, typically, additional conditions would have to be added; a notable exception is provided by Radulović (1996) for the mean case.

Example 3.4.1 (Univariate mean). Suppose $\{X_t\}$ is a stationary sequence of univariate random variables with mean $\theta(P)$. The goal is to construct a confidence interval for $\theta(P)$, on the basis of observing $X_1, \ldots, X_n$.

Let $\hat{\theta}_{n,b,t} \equiv\equiv b^{-1} \sum_{a=t}^{t+b-1} X_a \equiv \bar{X}_{n,b,t}$ be the estimator of $\theta(P)$ based on the block of size b of the consecutive data $\{X_t, \ldots, X_{t+b-1}\}$. Define $J_b(P)$ to be the sampling distribution of $b^{\frac{1}{2}}(\bar{X}_{n,b,t} - \theta(P))$. Also, define the corresponding cumulative distribution function:

$$J_b(x, P) = Prob_P\{b^{\frac{1}{2}}(\bar{X}_{n,b,1} - \theta(P)) \le x\}. \tag{3.7}$$

The approximation to $J_n(x, P)$ we study is defined by

$$L_{n,b}(x) = \frac{1}{n-b+1} \sum_{t=1}^{n-b+1} 1\{b^{\frac{1}{2}}(\bar{X}_{n,b,t} - \bar{X}_n) \le x\}. \tag{3.8}$$

The following central limit theorem states sufficient moment and mixing conditions for which the subsampling technique allows us to draw a first-order correct inference about $\theta(P)$.

Theorem 3.4.1. *Let $\{X_t\}$ be a stationary sequence of random variables defined on a common probability space. Denote the corresponding mixing coefficients by $\alpha_X(\cdot)$. Define*

$$T_{k,t} \equiv k^{-\frac{1}{2}} \sum_{a=t}^{t+k-1} X_a \quad \text{and} \quad \sigma_k^2 \equiv Var(T_{k,1}).$$

Assume the following conditions. For some $\delta > 0$:

- $E|X_1|^{2+2\delta} \le \Delta,$ (3.9)
- $\sigma_n^2 \to \sigma^2 > 0,$ (3.10)
- $C(4, \delta) \equiv \sum_{k=1}^{\infty} (k+1)^2 \alpha_X^{\frac{\delta}{4+\delta}}(k) \le K.$ (3.11)

Furthermore, assume that $b/n \to 0$ and $b \to \infty$ as $n \to \infty$, and let $J(P) = N(0, \sigma^2)$.

Then, conclusions (i–iii) of Theorem 3.2.1 are true.

Proof. For the proof we will assume, without loss of generality, that true mean is equal to zero, that is, $\theta(P) = 0$. By Theorem 3.2.1 it is sufficient to verify our Assumption 3.2.1, as the mixing condition $\alpha_X(k) \to 0$ is implied by assumption (3.11). But, Assumption 3.2.1 follows immediately from Theorem B.0.1 of Appendix B. ∎

Example 3.4.2 (Smooth function of the mean). Natural extensions of the univariate mean case are the multivariate mean and smooth functions of the univariate mean. It turns out that we easily can cover both extensions by stating a theorem for smooth functions of the multivariate mean.

Suppose $\{X_t\}$ is a sequence of multivariate random variables with common mean $\zeta \in \mathbb{R}^k$, and $k \ge 1$. Denote the joint probability law governing the sequence by P. Assume that on the basis of observing $\{X_1,\ldots,X_n\}$, we are interested in finding a confidence region for $\theta(P) \equiv f(\zeta)$, where $f(\cdot)$ is a smooth function from $\mathbb{R}^k$ to $\mathbb{R}^p$.

Let $\hat{\theta}_{n,b,t} = \hat{\theta}_b(X_t,\ldots,X_{t+b-1}) \equiv f(b^{-1}\sum_{a=t}^{t+b-1} X_a) \equiv f(\bar{X}_{n,b,t})$ be the estimator of $\theta(P)$ based on the block of size b of the consecutive data $\{X_t,\ldots,X_{t+b-1}\}$. Define $J_b(P)$ to be the sampling distribution of $b^{\frac{1}{2}}(f(\bar{X}_{n,b,1}) - \theta(P))$. Also define the corresponding analogue of a distribution function. For any Borel set $A \in \mathbb{R}^p$, let

$$J_b(x,P) = Prob_P\{b^{\frac{1}{2}}(f(\bar{X}_{n,b,1}) - \theta(P)) \in A\}. \tag{3.12}$$

The approximation to $J_n(A,P)$ we study is defined by

$$L_{n,b}(A) = \frac{1}{n-b+1} \sum_{t=1}^{n-b+1} 1\{b^{\frac{1}{2}}(f(\bar{X}_{n,b,t}) - f(\bar{X}_n)) \in A\}. \tag{3.13}$$

Theorem 3.4.2. *Let $\{X_t\}$ be a sequence of stationary random vectors with common mean $\zeta \in \mathbb{R}^k$. Denote the corresponding mixing coefficients by $\alpha_X(\cdot)$. Define*

$$T_{l,t} \equiv l^{-\frac{1}{2}} \sum_{a=t}^{t+l-1} X_a \quad \text{and} \quad \Sigma_l \equiv Cov(T_{l,1}).$$

Assume the following conditions hold. For some $\delta > 0$:

- $E|X_{t,j}|^{2+\delta} \leq \Delta$ *for all t and all $1 \leq j \leq k$,* (3.14)
- $\Sigma_n \to \Sigma > 0$, (3.15)
- $C(4,\delta) \equiv \sum_{k=1}^{\infty}(k+1)^2 \alpha_X^{\frac{\delta}{4+\delta}}(k) \leq K.$ (3.16)
- $f: \mathbb{R}^k \to \mathbb{R}^p$ *is differentiable in a neighborhood of ζ.* (3.17)

Furthermore, assume that $b/n \to 0$ and $b \to \infty$ as $n \to \infty$, and let $J(P) = N(0, J'\Sigma J)$, where J is the $k \times p$ Jacobian matrix of f evaluated at ζ.

Then, conclusions (i–iv) of Theorem 3.3.1 are true.

Remark 3.4.1. In the case $p = 1$, $\theta(P)$ is a univariate parameter and the results of Section 3.2 apply alternatively. In particular, we can construct both equal-tailed and symmetric two-sided confidence intervals for $\theta(P)$.

Proof of Theorem 3.4.2. For the proof we will assume, without loss of generality, that the true mean is equal to zero, that is, $\zeta = 0$. The mixing condition in Theorem 3.3.1 is clearly implied by our assumption (3.16). Therefore, it will be sufficient to show that Assumption 3.3.1 is satisfied.

To do so, let A be a Borel set whose boundary has mass zero under $J(P)$. We will accomplish the proof in two steps.

Step 1: $T_{n,1} \overset{\mathcal{L}}{\Longrightarrow} N(0, \Sigma)$.

Proof: By the Cramér–Wold device, it will suffice to look at linear combinations of the type $\lambda' \Sigma^{-\frac{1}{2}} T_{n,1}$, where λ is a vector in $\mathbb{R}^p$. Without loss of generality, we may assume that λ has unit length. We observe that

i. $E(\lambda' \Sigma^{-\frac{1}{2}} X_t) = 0$ for $1 \leq t \leq n$,
ii. by assumption (3.14) and Minkowski, $E|\lambda' \Sigma^{-\frac{1}{2}} X_t|^{2+2\delta} \leq \Delta^*$, where Δ^* is a constant that depends only on Δ and Σ (since $\|\lambda\| = 1$),
iii.
$$\sigma_n^2 \equiv Var(\lambda' \Sigma^{-\frac{1}{2}} T_{n,1})$$
$$= \lambda' \Sigma^{-\frac{1}{2}} Cov(T_{n,1}) \Sigma^{-\frac{1}{2}} \lambda$$
$$= \lambda' \Sigma^{-\frac{1}{2}} \Sigma_n \Sigma^{-\frac{1}{2}} \lambda \to 1.$$

Apply Theorem B.0.1 to deduce that $\lambda' \Sigma^{-\frac{1}{2}} n^{-\frac{1}{2}} \sum_{t=1}^{n} X_t \xrightarrow{\mathcal{L}} N(0,1)$ (to do so let $X_{n,t} = \lambda' \Sigma^{-\frac{1}{2}} X_t$ there). Since we did not specify λ, it follows from the Cramér–Wold device that $\Sigma^{-\frac{1}{2}} T_{n,1} \xrightarrow{\mathcal{L}} N(0,I)$ or, equivalently, that $T_{n,1} \xrightarrow{\mathcal{L}} N(0,\Sigma)$.

Step 2: $n^{\frac{1}{2}}(f(\bar{X}_n) - f(\zeta)) \xrightarrow{\mathcal{L}} N(0, J'\Sigma J)$.
Proof: This follows immediately from Step 1, Condition (3.17), and the multivariate δ-method. ∎

Example 3.4.3 (Linear regression). Linear models are one of the mainstays of statistics. In many situations it is reasonable to assume independent errors. One can then use ordinary least squares (OLS) or, if heteroskedasticity is suspected, turn to generalized least squares (GLS) or robust methods such as the one of White (1980). However, none of those approaches will generally work when the errors exhibit serial correlation. For example, this is the case in most situations where data are collected over time. OLS still yields consistent estimates of the regression parameters, but the corresponding standard errors, that is, the corresponding estimated standard deviations, are incorrect. In the field of econometrics, many ways have been discussed for getting the standard errors right, at least asymptotically. If one is interested in making inference for the regression parameters, one could base it on their limiting normal distribution (under some regularity conditions), in conjunction with the estimated variance-covariance matrix. We present here the subsampling method as an alternative way of making inference (for example, see Politis, Romano, and Wolf, 1997, or Sherman, 1997). In what follows, we will prove that subsampling gives asymptotically valid results under very weak conditions.

Consider the linear model

$$y = X\beta + \epsilon, \tag{3.18}$$

where y and ϵ are $n \times 1$ vectors, β is a $p \times 1$ vector and X a $n \times p$ matrix. Here, X may be stochastic. The goal is to construct a confidence region for

β. In order to be able to apply the subsampling method, we need to define subvectors and submatrices:

$$y_{b,t} \equiv (y_t, \ldots, y_{t+b-1})', \quad \epsilon_{b,t} \equiv (\epsilon_t, \ldots, \epsilon_{t+b-1})' \quad \text{and}$$

$$X_{b,t} \equiv \begin{pmatrix} x'_t \\ \vdots \\ x'_{t+b-1} \end{pmatrix}, \quad \text{where } X = \begin{pmatrix} x'_1 \\ \vdots \\ x'_n \end{pmatrix}.$$

The estimator of β based on $X_{b,t}$ and $y_{b,t}$ is given by

$$\hat{\beta}_{n,b,t} \equiv (X'_{b,t} X_{b,t})^{-1} X'_{b,t} y_{b,t}. \tag{3.19}$$

Define $J_b(P)$ to be the sampling distribution of $b^{\frac{1}{2}}(\hat{\beta}_{n,b,1} - \beta)$. Also, for any Borel set $A \in \mathbb{R}^p$, let

$$J_b(A, P) = Prob_P\{b^{\frac{1}{2}}(\hat{\beta}_{n,b,1} - \beta) \in A\}. \tag{3.20}$$

The approximation to $J_n(A, P)$ we study is defined by

$$L_{n,b}(A) = \frac{1}{n-b+1} \sum_{t=1}^{n-b+1} 1\{b^{\frac{1}{2}}(\hat{\beta}_{n,b,t} - \hat{\beta}_n) \in A\}. \tag{3.21}$$

Theorem 3.4.3. *Let $\{(X_t, \epsilon_t)\}$ be a sequence of stationary random vectors defined on a common probability space. Denote the mixing coefficients for the $\{(X_t, \epsilon_t)\}$ sequence by $\alpha(\cdot)$. Define*

$$T_{k,t} \equiv k^{-\frac{1}{2}} \sum_{a=t}^{t+k-1} X_a \epsilon_a, \quad V_k \equiv Cov(T_{k,1}), \quad \text{and} \quad M_k \equiv E(X'_{k,1} X_{k,1}/k).$$

Assume the following conditions hold. For some $\delta > 0$,

- $E(x_t \epsilon_t) = 0 \quad$ *for all* t, \hfill (3.22)
- $E|x_{t,j} \epsilon_t|^{2+2\delta} \leq \Delta_1 \quad$ *for all t and all $1 \leq j \leq p$,* \hfill (3.23)
- $E|x_{t,j}|^{4+2\delta} \leq \Delta_2 \quad$ *for all t and all $1 \leq j \leq p$,* \hfill (3.24)
- $V_k \to V > 0$, \hfill (3.25)
- $M_k \to M > 0$, \hfill (3.26)
- $C(4) \equiv \sum_{k=1}^{\infty} (k+1)^2 \alpha^{\frac{\delta}{4+\delta}}(k) \leq K.$ \hfill (3.27)

Furthermore, assume that $b/n \to 0$ and $b \to \infty$ as $n \to \infty$, and let $J(P) = N(0, M^{-1}VM^{-1})$.

Then, conclusions (i–iv) of Theorem 3.3.1 are true, provided that we replace $\theta(P)$ by β, $\hat{\theta}_n$ by $\hat{\beta}_n$, and $\hat{\theta}_{n,b,t}$ by $\hat{\beta}_{n,b,t}$.

Proof. The mixing condition in Theorem 3.3.1 is clearly implied by our assumption (3.27). Therefore, it will be sufficient to show that Assumption 3.3.1 is satisfied. We will accomplish the proof in two steps.

Step 1: $n^{-\frac{1}{2}} X' \epsilon \xrightarrow{\mathcal{L}} N(0, V)$.

Proof: By the Cramér–Wold device, it will suffice to look at linear combinations of the type $\lambda' V^{-\frac{1}{2}} n^{-\frac{1}{2}} X' \epsilon$, where λ is a vector in $\mathbb{R}^p$. Without loss of generality, we may assume that λ has unit length. We observe that

i. $E(\lambda' V^{-\frac{1}{2}} X_t \epsilon_t) = 0$ for $1 \leq t \leq n$,
ii. by assumption (3.23) and Minkowski, $E|\lambda' V^{-\frac{1}{2}} X_t \epsilon_t|^{2+2\delta} \leq \Delta_1^*$, where Δ_1^* is a constant that depends only on Δ_1 and V (since $\|\lambda\| = 1$),
iii.

$$\sigma_n^2 \equiv Var(\lambda' V^{-\frac{1}{2}} n^{-\frac{1}{2}} \sum_{t=1}^n X_t \epsilon_t)$$

$$= \lambda' V^{-\frac{1}{2}} Cov(n^{-\frac{1}{2}} \sum_{t=1}^n X_t \epsilon_t) V^{-\frac{1}{2}} \lambda$$

$$= \lambda' V^{-\frac{1}{2}} V_n V^{-\frac{1}{2}} \lambda \to 1.$$

Apply Theorem B.0.1 to deduce that $\lambda' V^{-\frac{1}{2}} n^{-\frac{1}{2}} \sum_{t=1}^n X_t \epsilon_t \xrightarrow{\mathcal{L}} N(0, 1)$ (to do so let $X_{n,t} = \lambda' V^{-\frac{1}{2}} X_t \epsilon_t$ there). Since we did not specify λ, it follows by the Cramér-Wold device that $V^{-\frac{1}{2}} n^{-\frac{1}{2}} X_T \epsilon \xrightarrow{\mathcal{L}} N(0, I)$ or, equivalently, that $n^{-\frac{1}{2}} X_T \epsilon \xrightarrow{\mathcal{L}} N(0, V)$.

Step 2: $n^{\frac{1}{2}} (\hat{\beta}_n - \beta) \xrightarrow{\mathcal{L}} N(0, M^{-1} V M^{-1})$.

Proof: $n^{\frac{1}{2}} (\hat{\beta}_n - \beta) = (X'X/n)^{-1} n^{-\frac{1}{2}} X' \epsilon$. By the imposed mixing and moment conditions on the sequence $\{X_t\}$, we have $(X'X/n) - M_n \to 0$ in probability. In fact, denote the (l, j)-th entry of $[(X'X/n) - M_n]$ by $(D_n)_{l,j}$. Then,

$$(D_n)_{l,j} = \frac{1}{n} \sum_{k=1}^n x_{k,l} x_{k,j} - E(x_{k,l} x_{k,j}).$$

By definition, $E((D_n)_{l,j}) = 0$. Therefore, it will be sufficient to check that $Var((D_n)_{l,j}) \to 0$. Define

$$s_{n,d} = \frac{1}{n} \sum_{a=1}^{n-d} Cov(x_{a,l} x_{a,j}, x_{a+d,l} x_{a+d,j}).$$

Then,

$$Var((D)_{l,j}) = \frac{1}{n}(s_{n,0} + 2 \sum_{d=1}^{n-1} s_{n,d}).$$

By Lemma A.0.1,

$$Cov(x_{a,l} x_{a,j}, x_{a+d,l} x_{a+d,j}) \leq 8 \|x_{a,l} x_{a,j}\|_{2+\delta} \|x_{a+d,l} x_{a+d,j}\|_{2+\delta} \alpha(d)^{\frac{\delta}{2+\delta}}$$

$$\leq 8 \Delta_2^{\frac{2}{2+\delta}} \alpha(d)^{\frac{\delta}{2+\delta}} \quad \text{(by assumption (3.25))}.$$

Therefore, $s_{n,d} \leq 8\Delta^{\frac{2}{2+\delta}}\alpha(d)^{\frac{\delta}{2+\delta}}$ and that implies

$$Var((D)_{t,j}) \leq \frac{16}{n}\Delta^{\frac{2}{2+\delta}}\sum_{d=0}^{n-1}\alpha(d)^{\frac{\delta}{2+\delta}} \to 0.$$

We thus have $((X'X/n) - M_n) \xrightarrow{P} 0$. Together with $M_n \to M$ this implies $(X'X/n) \xrightarrow{P} M$. By Step 1, $n^{-\frac{1}{2}}X'\epsilon$ is $O_P(1)$ and therefore

$$[(X'X/n)^{-1} - M^{-1}]n^{-\frac{1}{2}}X'\epsilon \to 0 \quad \text{in probability}.$$

But, $M^{-1}b^{-\frac{1}{2}}X'\epsilon \xRightarrow{\mathcal{L}} N(0, M^{-1}VM^{-1})$ by Step 1 again and this completes the proof of Assumption 3.3.1. ∎

Remark 3.4.2. Obviously, a theorem analogous to Theorem 3.4.2 allows for the construction of confidence regions for a smooth function of β. The details are left to the reader.

Example 3.4.4 (Autocorrelations). A frequent question in time series analysis is whether the first k autocorrelation coefficients of a stationary time series are equal to zero. Here, the time series may an observed one or one consisting of fitted residuals after a suitable modeling process. The usual starting point is to consider the first k sample autocorrelations $\hat{\rho}(j)$, $1 \leq j \leq k$, defined by

$$\hat{\rho}_n(j) = \hat{R}_n(j)/\hat{R}_n(0) \quad \text{and} \quad \hat{R}_n(j) = \frac{1}{n}\sum_{t=1}^{n-j}(X_t - \bar{X}_n)(X_{t+k} - \bar{X}_n).$$

Here, $\hat{R}_n(j)$ is the usual estimator for $R(j) = Cov(X_0, X_{0+j})$. It is well known, and widely used, that if $\{X_t\}$ is an i.i.d. series, then under some regularity conditions, the first k sample autocorrelations are asymptotically normal with mean $[\rho(1), \ldots, \rho(n)]$ and covariance matrix equal to $n^{-1}I_n$. For example, a common diagnostic check on a fitted model is to apply this result to the estimated autocorrelation function of the residuals. Typically, $\pm 1.96 n^{-1/2}$ limits are superimposed on a graph of $\hat{\rho}_n(j)$ versus j.

The point of this example is that such "checks" can be very misleading if the underlying sequence is uncorrelated only, rather than independent. In particular, there exist uncorrelated processes for which the asymptotic variance of $n^{1/2}(\hat{\rho}_n(j) - \rho(j))$ can be any positive value. The following lemma gives a precise formula for computing the asymptotic variance of the first sample autocorrelation.

Lemma 3.4.1. *Let $\{X_t\}$ be a stationary sequence of random variables with positive second moment and zero first autocorrelation. Denote the corresponding mixing coefficients by $\alpha_X(\cdot)$. Assume the following conditions. For some $\delta > 0$,*

- $\|X_1\|_{4+2\delta} \leq \Delta,$ (3.28)

84   3. Subsampling for Stationary Time Series

- $\sum_{k=1}^{\infty} \alpha_X^{\frac{\delta}{2+\delta}}(k) \leq K.$ (3.29)

Then, $n^{1/2}\hat{\rho}_n(1)$ is asymptotically normal with mean zero and variance

$$\tau^2 = Var^{-2}(X_1)[Var(X_1X_2) + 2\sum_{t=1}^{\infty} Cov(X_1X_2, X_{t+1}X_{t+2})].$$

Proof. The proof can be found in Romano and Thombs (1996). ∎

This result shows that for uncorrelated sequences the limiting variance can be any positive value, depending on the covariance structure of the sequence; see Romano and Thombs (1996) for some explicit examples. Therefore, any inference based on a limiting variance of one can be arbitrarily misleading.

On the other hand, we can use the subsampling method to make an asymptotically correct joint inference for the first k autocorrelations under weak assumptions. The following result will provide the basis for our method. Note that $\kappa(r, s, v)$ denotes the fourth joint cumulant of the distribution of $(X_j, X_{j+r}, X_{j+s}, X_{j+v})$; for example, see (5.3.19) of Priestley (1981).

Lemma 3.4.2. *Let $\{X_t\}$ be a stationary sequence of mean zero random variables with positive second moment. Denote the corresponding mixing coefficients by $\alpha_X(\cdot)$. Assume the following conditions. For some $\delta > 0$,*

- $\|X_1\|_{4+2\delta} \leq \Delta,$ (3.30)

- $\sum_{k=1}^{\infty} \alpha_X^{\frac{\delta}{2+\delta}}(k) \leq K.$ (3.31)

Then, for any fixed nonnegative integer k,

$$n^{1/2}[\hat{\rho}_n(1) - \rho(1), \ldots, \hat{\rho}_n(k) - \rho(k)]$$

is asymptotically normal with mean zero and covariance matrix **T**, *where* **T** *is a $k \times k$ matrix with the (i, j)-th entry given by*

$$\tau_{i,j} \equiv \lim_{n \to \infty} \{n \, Cov[\hat{\rho}_n(i), \hat{\rho}_n(j)]\}$$
$$= \lim_{n \to \infty} nR^{-2}(0) Cov[\hat{R}_n(i) - \rho(i)\hat{R}_n(0), \hat{R}_n(j) - \rho(j)\hat{R}_n(0)]$$
$$= R^{-2}(0)[c_{t+1,j+1} - \rho(i)c_{1,j+1} - \rho(j)c_{1,i+1} + \rho(i)\rho(j)c_{1,1}] \quad (3.32)$$

Here, $c_{t,j}$ is defined as follows.

$$c_{t+1,j+1} \equiv \lim_{n \to \infty} \{n \, Cov[\hat{R}_n(i), \hat{R}_n(j)]\}$$
$$= \sum_{d=-\infty}^{\infty} [R(d)R(d+j-i) + R(d+j)R(d-i) + \kappa(d, i, j-i)]$$

$$= \sum_{d=-\infty}^{\infty} Cov(X_0 X_t, X_d X_{d+j}). \quad (3.33)$$

Proof. The proof can be found in Romano and Thombs (1996). ∎

Under the conditions of Lemma 3.4.2, it is then clear that the subsampling method allows us to make joint inference for $\rho(1), \ldots, \rho(n)$, without falsely appealing to a limiting identity covariance matrix. The asymptotic justification of the subsampling approach follows from Theorem 3.3.1, since Assumption 3.3.1 and the strong mixing condition of the $\{X_t\}$ sequence are implied by Lemma 3.4.2.

Example 3.4.5 (Spectral density function). Consider a stationary sequence $\{X_t\}$ with mean zero and autocovariance $R(k) = E(X_0 X_k)$. As a weak dependence condition, we may assume that $\sum |R(k)| < \infty$, and define the spectral density function f by $f(w) = \frac{1}{2\pi} \sum_{k=-\infty}^{\infty} R(k) e^{-ikw}$. Fix a point $w \in [-\pi, \pi]$, and consider a kernel smoothed estimator of $f(w)$ given by $\hat{f}(w) = \frac{1}{2\pi} \sum_{k=-n}^{n} B_n(k) \hat{R}(k) e^{-ikw}$, where $\hat{R}(k) = \frac{1}{n} \sum_{t=1}^{n-k} X_t X_{t+k}$ is the usual sample autocovariance, and $B_n(k)$ is the 'lag–window'. Under regularity conditions (cf. Priestley, 1981), there is a sequence τ_n, corresponding to a particular choice of a sequence of lag–windows $B_n(\cdot)$, such that $\tau_n(\hat{f}(w) - f(w))$ has an asymptotic normal distribution.

To fix ideas, suppose $B_n(\cdot)$ is the Parzen window (cf. Priestley, 1981, p. 443); here m_n is a sequence of design parameters that should be chosen appropriately depending on n. Then, under moment and weak dependence conditions (the latter having a correspondence to conditions on the smoothness of the spectral density; cf. Ibragimov and Rozanov, 1978), it can be calculated (see Priestley, 1981, p. 462) that asymptotically

$$Bias(\hat{f}(w)) = E\hat{f}(w) - f(w) \sim \frac{b_w}{m_n^2},$$

$$Var(\hat{f}(w)) \sim \frac{\sigma_w^2 m_n}{n},$$

and that $\tau_n(\hat{f}(w) - E\hat{f}(w))$ is asymptotically normal $N(0, \sigma_w^2)$, where $\tau_n = \sqrt{n/m_n}$, and b_w and σ_w^2 are constants depending on w and on f.

By Slutsky's theorem, it therefore follows that for a choice of m_n satisfying $n^{1/5}/m_n \to 0$, as $n \to \infty$, $\sqrt{n/m_n}(\hat{f}(w) - f(w))$ is also asymptotically normal $N(0, \sigma_w^2)$. Thus, Assumption 3.2.1 is satisfied, and the subsampling methodology can be used to set confidence intervals for $f(w)$. The same ideas are directly applicable in the case of homogeneous random fields in d dimensions ($d > 1$); kernel smoothed estimators of $f(w)$ for $w \in [-\pi, \pi]^d$ are formed in analogous manner, and are shown to be asymptotically normally distributed under regularity conditions (cf. Rosenblatt, 1985).

Note however that in order to have a most accurate (from the point of view of asymptotic mean squared error) estimator $\hat{f}(w)$, we should choose $m_n \sim (4b_w^2/\sigma_w^2)^{1/5} n^{1/5}$. In this case, the $Bias(\hat{f}(w))$ is significant and is of the same order as $\sqrt{Var(\hat{f}(w))}$; the asymptotic distribution of $\sqrt{n/m_n}(\hat{f}(w) - f(w))$ is now normal $N(\pm \frac{1}{2}\sigma_w, \sigma_w^2)$, where the $\pm$ sign corresponds to the sign of b_w. Since b_w and σ_w^2 are generally unknown, they could either be estimated, or the choice $m_n \sim A n^{1/5}$ can be made, for some constant $A > 0$; this choice would imply that the asymptotic distribution of $\sqrt{n/m_n}(\hat{f}(w) - f(w))$ is normal $N(const., \sigma_w^2)$. In other words, if m_n is of the optimal order of magnitude $n^{1/5}$, there exists an asymptotic distribution for $\tau_n(\hat{f}(w) - f(w))$, but it generally has nonzero mean. Therefore, a bias correction in the spirit of our Equation (3.39) in Section 3.7 is useful here. To ensure that we have a nonnegative estimate of $f(w)$, the positive part of the bias corrected estimator may also be taken which is also justified by asymptotic considerations (for example, see Politis and Romano, 1995).

Example 3.4.6 (Variance ratio). Whether or not an economic time series follows a random walk has been of great interest to researchers and practitioners for a long time. The classical random walk hypothesis states that a time series evolves as the sum of i.i.d. random variables. However, this has been recognized as somewhat to strict for many interesting time series. A weaker assumption is that the increments are uncorrelated rather than i.i.d. There are several ways of testing the random walk hypothesis. Widely used are variance ratio tests, exploiting the fact that the variance of the sum of uncorrelated random variables is equal to the sum of the individual variances. Such tests date back to the Hausman (1978) specification test. Except for unrealistic scenarios, the sampling distributions of corresponding test statistics cannot be derived exactly, and so one has to rely on asymptotic approximations. The most general asymptotic theory for variance ratio tests so far has been developed in Lo and MacKinley (1988). Corresponding simulation studies are reported in Lo and MacKinley (1989).

Suppose $\{X_t\}$ is a time series defined by the following relationship:

$$X_t = X_{t-1} + \mu + \epsilon_t, \quad E(\epsilon_t) = 0, \quad \text{for all } t,$$

or

$$Z_t = \mu + \epsilon_t, \quad Z_t \equiv X_t - X_{t-1},$$

where the constant μ is an arbitrary drift parameter. The classical "strong" random walk hypothesis states that the ϵ_t are i.i.d. according to a mean zero distribution. Since it is well known that this assumption is violated for many interesting economical time series, focus is on a weaker random walk hypothesis which allows uncorrelated increments. Of course, in order for variance ratio tests to be applicable, the existence of a nonzero finite second moment is assumed.

3.4 Examples

We are interested in the null hypothesis of the ϵ_t being serially uncorrelated; that is,

$$H_0: E(\epsilon_t \epsilon_{t-\tau}) = 0 \text{ for all } \tau \neq 0. \qquad (3.34)$$

Suppose the sample consists of $mq+1$ observations $X_0, \ldots, X_{mq}$, where m and q are some positive integers. Consistent estimators of $\mu = E(Z_t)$ and $\sigma^2 = Var(Z_t)$ are given by

$$\hat{\mu} = \frac{1}{mq} \sum_{t=1}^{mq} Z_t$$

$$\hat{\sigma}_1^2 = \frac{1}{mq-1} \sum_{t=1}^{mq} (Z_t - \hat{\mu})^2.$$

Under H_0, the following is also a consistent estimator of σ^2,

$$\hat{\sigma}_q^2 = \frac{1}{l} \sum_{t=q}^{mq} (\sum_{k=1}^{q} Z_{t+1-k} - q\hat{\mu})^2,$$

where $l = q(mq - q + 1)(1 - q/(mq))$. Actually, it should be pointed out that $\hat{\sigma}_q^2$ is nothing else but the Carlstein subsampling estimator of variance of $n^{1/2} \bar{Z}_n$ based on subsamples of size q (see Section 3.8). However, if H_0 is not true and the increments ϵ_t are serially correlated, the convergence $\hat{\sigma}_q^2 \to \sigma^2$ no longer holds. Under positive serial correlation, $\hat{\sigma}_q^2$ will tend to be greater than σ^2; under negative serial correlation, it will tend to be smaller than σ^2. A q-period variance ratio test statistic can therefore be defined by

$$\bar{M}_r(q) = \frac{\hat{\sigma}_q^2}{\hat{\sigma}_1^2}.$$

It is easily seen to be a consistent (although biased) estimator of the general q-period variance ratio

$$VR(q) = 1 + 2 \sum_{k=1}^{q-1} (1 - \frac{k}{q}) \rho(k),$$

where $\rho(k)$ is the k-th-order autocorrelation coefficient of Z_t. Indeed, $\bar{M}_r(q)$ is asymptotically equivalent to the natural estimator $\widehat{VR}(q)$ of $VR(q)$:

$$\bar{M}_r(q) \simeq \widehat{VR}(q) \equiv 1 + 2 \sum_{k=1}^{q-1} (1 - \frac{k}{q}) \hat{\rho}_n(k). \qquad (3.35)$$

Note that $\widehat{VR}(q)$ is just the Bartlett estimator of the (normalized) spectral density of the $\{Z_t\}$ process evaluated at zero (for example, see Politis and Romano, 1995). In equation (3.35), the $\hat{\rho}_n(k)$ are the sample autocorrelations of $\{Z_t\}$ defined in the usual way,

$$\hat{\rho}_n(k) = \hat{R}_n(k)/\hat{R}_n(0), \quad \hat{R}_n(k) = \frac{1}{n}\sum_{t=1}^{n-k}(Z_t - \bar{Z})(Z_{t+k} - \bar{Z}).$$

The statistic $\bar{M}_r(q)$ was introduced by Lo and MacKinley (1988). It is obvious that under H_0 it should be close to unity. If the increments returns ϵ_t exhibit positive serial correlation, it will tend to be greater then unity, and vice versa.

Under the null hypothesis, $\bar{M}_r(q)$ will have a limiting normal distribution with mean one, which is even robust against some heteroskedasticity. However, the limiting variance very much depends on the the dependence structure of the $\{\epsilon_t\}$ process and is nontrivial to estimate. Lo and MacKinley (1988) derived the limiting distribution under H_0 explicitly, making use of the following two additional assumptions. Note that α-mixing with coefficients $\alpha(m)$ of size $r/(r-1)$, $r > 1$, means that $\alpha(m) = O(m^{-\lambda})$ for $\lambda > r/(r-1)$.

(A2) $\{\epsilon_t\}$ is α-mixing with coefficients $\alpha(m)$ of size $r/(r-1)$, $r > 1$, such that for any $\tau \geq 0$, there exists some $\delta > 0$ for which

$$E|\epsilon_t \epsilon_{t-\tau}|^{2(r+\delta)} < \Delta < \infty.$$

(A4) $E(\epsilon_t \epsilon_{t-j} \epsilon_t \epsilon_{t-k}) = 0$, for any non-zero j, k where $j \neq k$.

It can then be shown that $\bar{M}_r(q) \overset{\mathcal{L}}{\Rightarrow} N(1, V(q))$ (see Lo and MacKinley, 1988). The limiting covariance matrix and a consistent estimator are given by the following equations. Here, $\delta(k)$ denotes the asymptotic variance of the sample autocorrelation $\hat{\rho}_n(k)$.

$$V(q) = 4\sum_{k=1}^{q-1}(1-\frac{k}{q})^2 \delta(k), \quad \hat{V}(q) = 4\sum_{k=1}^{q-1}(1-\frac{k}{q})^2 \hat{\delta}(j), \qquad (3.36)$$

$$\hat{\delta}(j) = \frac{\sum_{t=j+1}^{nq}(Z_t - \hat{\mu})^2(Z_{t-j} - \hat{\mu})^2}{[\sum_{t=1}^{nq}(Z_t - \hat{\mu})^2]^2}.$$

The critical assumption is (A4). It implies that the sample autocorrelations of Z_t are asymptotically uncorrelated. Under H_0 the $\hat{\delta}_j$ are consistent estimators of the asymptotic variance $\delta(k)$ of the autocorrelations $\hat{\rho}_n(k)$. Lo and MacKinley (1988, 1989) mention in a footnote that assumption (A4) is mainly needed for computational simplicity. In case it is violated, the limiting covariance $V(q)$ also depends on the asymptotic covariances of the $\hat{\delta}_j$, as is evident from equation (3.35). Indeed, in this general case the limiting variance is given by

$$V(q) = 4\sum_{k=1}^{q-1}(1-\frac{k}{q})^2 \delta(k) + 8\sum_{1\leq k<j\leq q-1}(1-\frac{k}{q})(1-\frac{j}{q})\gamma(k,j),$$

where $\gamma(k,j)$ are the asymptotic covariances of the autocorrelations $\hat{\rho}_n(k)$ and $\hat{\rho}_n(j)$. Therefore, if the autocorrelations are not asymptotically uncorrelated, the Lo and MacKinley test gives misleading answers.

On the other hand, the subsampling method will be robust against asymptotic correlations of the sample autocorrelations. Its application is based on the fact that we can implicitly carry out the test by constructing a subsample confidence interval for q-period variance ratio $\theta(P) = VR(q)$, based on the estimator $\hat{\theta}_n = \bar{M}(q)$. If unity is not contained in the interval, we reject the null, and vice versa. Under the conditions of Lemma 3.4.2, the asymptotic validity of this subsampling approach follows easily from Theorem 3.2.1. The α-mixing condition is implied by the mixing condition (3.31) and Assumption 3.2.1 follows from Lemma 3.4.2 and the equivalence (3.35). In particular, note that an assumption like (A4) of Lo and MacKinley (1988) is not needed.

Example 3.4.7 (First marginal distribution). Let $F(\cdot)$ denote the common marginal distribution of the stationary sequence $\{X_t\}$ and let $\hat{F}(x) = n^{-1} \sum_{t=1}^{n} 1\{X_t \leq x\}$. Under regularity conditions (cf. Györfi et al., 1989), $\sqrt{n}(\hat{F}(x) - F(x))$ possesses a limiting normal distribution, and hence Assumption 3.2.1 is satisfied. Furthermore, $\sqrt{n}(\hat{F}(\cdot) - F(\cdot))$, viewed as a random function, converges weakly to a Gaussian process (cf. Deo, 1973).

Looking at the sup norm $\sup_x |\sqrt{n}(\hat{F}(x) - F(x))|$, uniform confidence bands for the unknown distribution $F(\cdot)$ can be set by the subsampling methodology similarly to the i.i.d. case of Subsection 2.5.2.

If it is known that $F(\cdot)$ is absolutely continuous with probability density $F'(\cdot)$, then $F'(\cdot)$ may be estimated by the derivative of a smoothed version of $\hat{F}(\cdot)$. The subsampling methodology will be useful here too, although the rate τ_n is no longer $\sqrt{n}$, and the problem of bias estimation and bias correction becomes important, in exact analogy to the example of the spectral density function.

Example 3.4.8 (Robust statistics from time series). Suppose that the first marginal of the stationary sequence $\{X_t\}$, that is, the distribution of the random variable X_1, is symmetric and unimodal, with unknown location $\theta(P)$. Much of the methodology of robustness can be applied to the case of dependent data as well (cf. Gastwirth and Rubin, 1975; Künsch, 1984; Martin and Yohai, 1986). Under regularity conditions, the median, the trimmed mean, the Hodges–Lehmann estimator, linear combinations of order statistics, etc., all possess asymptotic distributions, and hence Theorem 3.2.1 is directly applicable.

As an example, consider a Gaussian strong mixing sequence $\{X_t\}$, satisfying $\sum |R(k)| < \infty$, where $R(k) = Cov(X_1, X_{1+k})$. Then (cf. Gastwirth and Rubin, 1975), the Hodges–Lehmann estimator, that is, the median of all pairwise averages of the data, is asymptotically normal, with mean $\theta(P)$

and variance proportional to $2n^{-1} \sum_{k=-\infty}^{\infty} \arcsin(R(k)/2)$. It is apparent that to use this asymptotic normal distribution to set confidence intervals for $\theta(P)$, the constant $\sum \arcsin(R(k)/2)$ should be consistently estimated, which is a difficult task. To appreciate the difficulty recall than even estimating $\sum_{k=-\infty}^{\infty} R(k)$ is hard and amounts to estimation of the spectral density function at the origin. Using Theorem 3.2.1 to set approximate confidence intervals for $\theta(P)$ bypasses this difficult problem.

Example 3.4.9 (Mean of overdifferenced data). It is common practice in macroeconomics to estimate the expected growth rate of a series $\{Y_t\}$ using the sample mean of the differenced data after taking logs, that is, using the sample mean of $X_t = \ln Z_t - \ln Z_{t-1}$, $t = 1, \ldots, n$. A reasonable model is that

$$X_t = \mu + V_t - V_{t-1},$$

where $\{V_t\}$ is a stationary, strong mixing sequence. The goal is to construct a confidence interval for μ. It is easy to see that

$$\bar{X}_n = \mu + \frac{V_n - V_0}{n}.$$

Hence, it follows that $\bar{X}_n$ is a super-consistent estimator that converges at rate n. Indeed,

$$n(\bar{X}_n - \mu) \stackrel{\mathcal{L}}{\Longrightarrow} \mathcal{L}(V - V^*),$$

where V and V^* are two independent random variables having identical distribution equal to that of V_0. Unless one is willing to assume that this distribution belongs to some parametric family, asymptotic inference based on traditional methods seems futile.

Kocherlakota and Savin (1995) therefore propose to apply the subsampling method, since it only requires the existence of some limiting distribution at some known rate; though the latter can be relaxed (see Chapter 8). Cleary, both conditions are satisfied here. Some simulation studies in Kocherlakota and Savin (1995) show that subsampling performs reasonably well for samples of sizes $n = 100$ and $n = 200$.

3.5 Hypothesis Testing

In Section 2.6, it was discussed how to use subsampling for hypothesis testing when the null hypothesis does not translate into a null hypothesis on a parameter and thus the duality between hypothesis tests and confidence regions cannot be exploited. The discussion was limited to i.i.d. observations but the problem, of course, also exists for dependent observations. Goodness of fit tests are one of many examples. The approach presented here will be analogous to the one of Section 2.6. To provide a general framework, assume that $X_1, \ldots, X_n$ is a sample of stationary observations taking values in a sample space S. Denote the probability law governing the infinite, stationary sequence $\ldots, X_{-1}, X_0, X_1, \ldots$ by P. This unknown law P is assumed to belong to a certain class of laws $\mathbf{P}$. The null hypothesis H_0 asserts $P \in \mathbf{P_0}$, and the alternative hypothesis H_1 is $P \in \mathbf{P_1}$, where $\mathbf{P_i} \subset \mathbf{P}$ and $\mathbf{P_0} \bigcup \mathbf{P_1} = \mathbf{P}$. The goal is to construct an asymptotically valid test based on a given test statistic,

$$T_n = \tau_n t_n(X_1, \ldots, X_n),$$

where, as usual, τ_n is a fixed nonrandom normalizing sequence (again, this assumption could be relaxed; see Chapter 8). Let

$$G_n(x, P) = Prob_P\{\tau_n t_n(X_1, \ldots, X_n) \leq x\}.$$

As before, we will be assuming that $G_n(\cdot, P)$ converges in distribution, at least for $P \in \mathbf{P_0}$. The theorem we will present will assume t_n is constructed to satisfy the following: $t_n(X_1, \ldots, X_n) \to t(P)$ in probability, where $t(P)$ is a constant which satisfies $t(P) = 0$ if $P \in \mathbf{P_0}$ and $t(P) > 0$ if $P \in \mathbf{P_1}$.

To describe the test construction, let $t_{n,b,t}$ be equal to the statistic t_b evaluated at the block of data $\{X_t, \ldots, X_{t+b-1}\}$. The sampling distribution of T_n is then approximated by

$$\hat{G}_{n,b}(x) = \frac{1}{n-b+1} \sum_{t=1}^{n-b+1} 1\{\tau_b t_{n,b,t} \leq x\}. \quad (3.37)$$

Given the estimated sampling distribution, the critical value for the test is obtained as the $1 - \alpha$ quantile of $\hat{G}_{n,b}(\cdot)$; specifically, define

$$g_{n,b}(1-\alpha) = \inf\{x : \hat{G}_{n,b}(x) \geq 1 - \alpha\}. \quad (3.38)$$

Finally, the nominal level α test rejects H_0 if and only if $T_n > g_{n,b}(1-\alpha)$.

The following theorem gives results analogous to the ones of Theorem 2.6.1.

Theorem 3.5.1.

i. *Assume, for $P \in \mathbf{P_0}$, $G_n(P)$ converges weakly to a continuous limit law $G(P)$, whose corresponding cumulative distribution function is $G(\cdot, P)$ and whose $1 - \alpha$ quantile is $g(1 - \alpha, P)$. Assume $b/n \to 0$ and $b \to \infty$ as $n \to \infty$. Also, assume that $\alpha_X(m) \to 0$ as $m \to \infty$, where $\alpha_X(\cdot)$ is the mixing sequence corresponding to $\{X_t\}$. If $G(\cdot, P)$ is continuous at $g(1 - \alpha, P)$ and $P \in \mathbf{P_0}$, then*

$$g_{n,b}(1-\alpha) \to g(1-\alpha, P) \text{ in probability}$$

and

$$Prob_P\{T_n > g_{n,b}(1-\alpha)\} \to \alpha \text{ as } n \to \infty.$$

ii. *Assume the test statistic is constructed so that $t_n(X_1, \ldots, X_n) \to t(P)$ in probability, where $t(P)$ is a constant which satisfies $t(P) = 0$ if $P \in \mathbf{P_0}$ and $t(P) > 0$ if $P \in \mathbf{P_1}$. Assume $b/n \to 0$, $b \to \infty$, and $\tau_b/\tau_n \to 0$ as $n \to \infty$. Also, assume that $\alpha_X(m) \to 0$ as $m \to \infty$, where $\alpha_X(\cdot)$ is the mixing sequence corresponding to $\{X_t\}$. Then, if $P \in \mathbf{P_1}$, the rejection probability satisfies*

$$\text{Prob}_P\{T_n > g_{n,b}(1-\alpha)\} \to 1 \text{ as } n \to \infty.$$

iii. *Suppose P_n is a sequence of alternatives such that, for some $P_0 \in \mathbf{P_0}$, $\{P_n^{[n]}\}$ is contiguous to $\{P_0^{[n]}\}$. In this notation, $P_n^{[n]}$ denotes the law of the finite segment $X_1, \ldots, X_n$ when the law of the infinite sequence $\ldots, X_{-1}, X_0, X_1, \ldots$ is given by P_n. The meaning of $\{P_0^{[n]}\}$ is analogous. Assume $b/n \to 0$ and $b \to \infty$ as $n \to \infty$. Then,*

$$g_{n,b}(1-\alpha) \to g(1-\alpha, P_0) \text{ in } P_n^{[n]}\text{-probability}.$$

Hence, if T_n converges in distribution to T under P_n and $G(\cdot, P_0)$ is continuous at $g(1-\alpha, P_0)$, then

$$P_n^{[n]}\{T_n > g_{n,b}(1-\alpha)\} \to \text{Prob}\{T > g(1-\alpha, P_0)\}.$$

Proof. The proof mimics the proof of Theorem 2.6.1, with the differences being analogous to the differences of the proofs of Theorems 2.2.1 and 3.2.1. The details are left to the reader. ∎

Remark 3.5.1. Remarks 2.6.1 and 2.6.3 also apply here.

3.6 Data-Dependent Choice of Block Size

Theorem 2.7.1 can be generalized to the stationary time series case as well. Indeed, one can show that subsampling with a general data-driven choice of block size is consistent. In order to support this claim, one must show the convergence of $L_{n,b}(\cdot)$ to $J(\cdot, P)$ is uniform in a broad range of b values, say $j_n \leq b \leq k_n$ (as expressed in Theorem 2.7.1). As in the proof of Theorem 2.7.1, let $U_{n,b}(x)$ be the $U_n(x)$ considered in the proof of Theorem 3.2.1, but now the dependence on b is made clear. So,

$$U_{n,b}(x) = q_{n,b}^{-1} \sum_{t=1}^{q_{n,b}} 1\{\tau_b[\hat{\theta}_{n,b,t} - \theta(P)] \leq x\},$$

where $q_{n,b} = n - b + 1$. Then, the proof in the i.i.d. case goes through as long as we can bound

$$k_n \sup_{j_n \leq b \leq k_n} \text{Prob}_P\{|U_{n,b}(x) - J_b(x, P)| \geq t\}$$

by something tending to zero. The simplest strategy would be to apply Chebychev's inequality, which was used in the proof of Theorem 3.2.1 (but we did not have to worry about the k_n or the sup in front). The resulting bound is of order

$$O(\frac{k_n b}{q_{n,b}}) + O(\frac{k_n}{q_{n,b}}) q_{n,b}^{-1} \sum_{h=1}^{q_{n,b}-b} \alpha_X(h) \leq O(\frac{k_n^2}{q_{n,b}}) + O(\frac{k_n}{q_{n,b}}) o(1).$$

Hence, if k_n is assumed to satisfy $k_n = o(n^{1/2})$, the proof of Theorem 2.7.1 goes through. Unfortunately, this assumption on k_n is much stronger than the one used in the i.i.d. case (where it was essentially assumed that $k_n = o(n)$). Note, however, that the restriction to k_n satisfying $k_n = o(n^{1/2})$ means b cannot be too large, and this is substantiated by simulations and higher order considerations, as will be seen. On the other hand, one can essentially recover the i.i.d. result at the expense of a slightly stronger mixing condition. To do this, we appeal to an exponential type inequality for mixing sequences, as provided in Theorem 1.3 of Bosq (1996). Then, one can obtain uniform consistency over b in $\{b : j_n \leq b \leq k_n\}$ under the assumption $k_n = o(n)$ if one is willing to slightly strengthen the mixing assumption. All that is required is to assume, for some $\beta > 1$,

$$m^\beta \alpha_X(m) \to 0 \text{ as } m \to \infty.$$

In summary, the random block size result carries over in spirit from the i.i.d. case to the dependent case, though there is evidently a tradeoff between the amount of dependence assumed and how broad the range of b values can be. Even so, the added mixing assumption is extremely weak.

3.7 Bias Reduction

It is well known that statistics calculated from time series and random fields are often heavily biased. The subsampling methodology can be used for bias reduction, in the same vein as the original proposition of a "jackknife" by Quenouille (1949). Denote by $m_n^{(j)}$, $\mu_n^{(j)}$, and $\mu^{(j)}$ the j-th (noncentral) moments of the distributions $L_{n,b}(\cdot)$, $J_n(\cdot, P)$, and $J(\cdot, P)$. To outline the method, assume that Assumption 3.2.1 holds together with $\mu_n^{(1)} \to \mu^{(1)}$ and $m_n^{(1)} \to \mu^{(1)}$; usually, but not always, it will be the case that $\mu^{(1)} = 0$. Then, since $L_n(\cdot)$ and $J_n(\cdot, P)$ have the same limiting distribution $J(\cdot, P)$, (with first moments converging as well), one can approximate $Bias(\hat{\theta}_n) = E(\hat{\theta}_n - \theta(P))$ by a re-scaled version of the "empirical" bias, that is, by

$$\widehat{Bias}(\hat{\theta}_n) = \frac{1}{\tau_n} m_n^{(1)} = \frac{\tau_b}{\tau_n}(Ave(\hat{\theta}_{n,b,t}) - \hat{\theta}_n)$$

where $Ave(\hat{\theta}_{n,b,t}) = q^{-1} \sum_{t=1}^{q} \hat{\theta}_{n,b,t}$. Correspondingly, one can form the bias corrected estimator

$$\hat{\theta}_{n,BC} = \hat{\theta}_n - \widehat{Bias}(\hat{\theta}_n) = (1 + \frac{\tau_b}{\tau_n})\hat{\theta}_n - \frac{\tau_b}{\tau_n} Ave(\hat{\theta}_{n,b,t}). \quad (3.39)$$

It is obvious that this is an asymptotic bias correction. For example, in the simplest case where $\hat{\theta}_n$ is the sample mean (which is unbiased), due to edge effects, $\widehat{Bias}(\hat{\theta}_n)$; nevertheless $\widehat{Bias}(\hat{\theta}_n) \to 0$ as it should (cf. Politis and Romano, 1992a). In the following theorem, the conditions of Theorem 3.2.1 are strengthened to ensure that the bias correction suggested in equation (3.39) is indeed asymptotically valid. The argument is actually most relevant when $\mu^{(1)} \neq 0$, such as in the case of an optimally smoothed spectral density estimator (see Example 3.4.5).

Theorem 3.7.1. *Assume Assumption 3.2.1, strengthened to include $\mu_n^{(1)} \to \mu^{(1)}$; assume $b \to \infty$, $b/n \to 0$, and $\tau_b/\tau_n \to 0$, as $n \to \infty$. Also assume that $\alpha_X(k) \to 0$ and that $E|\hat{\theta}_{n,b,1}^*|^{2+\delta} < C$, where δ and C are two positive constants independent of n and $\hat{\theta}_{n,b,t}^* \equiv \hat{\theta}_{n,b,t}/\sqrt{Var(\hat{\theta}_{n,b,t})}$.*

Then, $|m_n^{(1)} - \mu_n^{(1)}| \to 0$ in probability.

Proof. Note that $E(m_n^{(1)}) = \tau_b E(\hat{\theta}_{n,b,1} - \hat{\theta}_n) = \mu_b^{(1)} - \frac{\tau_b}{\tau_n}\mu_n^{(1)} = \mu_b^{(1)} + o(1)$, and that $|\mu_b^{(1)} - \mu_n^{(1)}| \to 0$, by the (strengthened) Assumption 3.2.1. Now

$$Var(m_n^{(1)}) = Var(\tau_b(Ave(\hat{\theta}_{n,b,t}) - \hat{\theta}_n))$$
$$= Var(\tau_b(Ave(\hat{\theta}_{n,b,t}) - E(\hat{\theta}_{n,b,1})) - \tau_b(\hat{\theta}_n - E(\hat{\theta}_{n,b,1})))$$
$$= Var(\tau_b(Ave(\hat{\theta}_{n,b,t}) - E(\hat{\theta}_{n,b,1}))) + o(1),$$

because $Var(\tau_b(\hat{\theta}_n - E(\hat{\theta}_{n,b,1}))) \to 0$ as $\tau_b/\tau_n \to 0$. But,

$$Var(\tau_b(Ave(\hat{\theta}_{n,b,t}) - E(\hat{\theta}_{n,b,1}))) = \frac{\tau_b^2}{q}\sum_{t=0}^{q-1}(1-\frac{t}{q})Cov(\hat{\theta}_{n,b,1}, \hat{\theta}_{n,b,1+t})$$

and thus

$$|Var(\tau_b(Ave(\hat{\theta}_{n,b,t}) - E(\hat{\theta}_{n,b,1})))| \leq \frac{Const.}{q}\sum_{t=0}^{q-1}|Cov(\hat{\theta}_{n,b,1}^*, \hat{\theta}_{n,b,1+t}^*)|,$$

where it was taken into account that $\tau_b^2 = O(1/Var(\hat{\theta}_{n,b,1}))$. Finally, by a similar argument to the proof of Theorem 3.2.1, and using the mixing inequality

$$|Cov(\hat{\theta}_{n,b,1}^*, \hat{\theta}_{n,b,1+t}^*)| \leq 8C^2\{\alpha_X(\max(0, t-b+1))\}^{\delta/(2+\delta)}$$

(see Lemma A.0.1), it follows that $Var(m_n^{(1)}) \to 0$ and the theorem is proved. ∎

Remark 3.7.1. A bias correction identical to the one suggested in equation (3.39) can be employed in the i.i.d. setup of Chapter 2 as well; in that case of course, $Ave(\hat{\theta}_{n,b,t})$ should be redefined to be the average of all available $\hat{\theta}_{n,b,i}$.

Remark 3.7.2. Note that if a point estimate of the unknown parameter $\theta(P)$ is not of primary interest and the goal is only to construct a confidence interval or region for $\theta(P)$, no *explicit* bias correction is necessary. Even if the the estimator $\hat{\theta}_n$ is asymptotically biased, subsampling intervals or regions are still consistent. This follows immediately from the fact that the limiting distributions $J(P)$ in Assumptions 3.2.1 and 3.3.1 do not need to have mean zero. Therefore, the subsampling method can be thought of having a built-in, *implicit* bias correction, which is a great advantage in case the asymptotic bias is difficult or even impossible to estimate. For only one of many such examples, see Elliot (1998).

3.8 Variance Estimation

Consider the case where the parameter of interest $\theta(P)$ is univariate and we want to consistently estimate the variance of an estimator $\hat{\theta}_n$. This may come up when we have two unbiased estimators and want to estimate their relative efficiency. Another example is when we would like to base inference on $\theta(P)$ on the limiting distribution of $\hat{\theta}_n$ (properly normalized), which is known up to the exact value of the limiting variance; for example, in many cases we can approximate the distribution of $\hat{\theta}_n$ by $N(\theta(P), n^{-1/2}\sigma_\infty^2)$ with σ_∞^2 unknown.

We first discuss how the subsampling method can be used in a natural way for variance estimation for a general statistic; thereafter, we will look at the special case of the sample mean in more detail.

3.8.1 General Statistic Case

It turns out that, compared to the estimation of a distribution function, somewhat stronger conditions are required for application of the subsampling method. The obvious idea for an estimator of the variance of $\tau_n\hat{\theta}_n$ is to use the variance of a random variable with distribution given by $L_{n,b}(\cdot)$, as defined in (3.2). If we are interested in the variance of $\hat{\theta}_n$, we then simply divide the result by τ_n^2. This yields the estimator

$$\widehat{Var}_{n,b}(\hat{\theta}_n) = \frac{\tau_b^2}{\tau_n^2} \frac{1}{q} \sum_{t=1}^{q} (\hat{\theta}_{n,b,t} - \hat{\theta}_{n,b,\cdot})^2, \qquad (3.40)$$

where $q = n - b + 1$ and $\hat{\theta}_{n,b,\cdot} = \dfrac{1}{q}\sum_{t=1}^{q}\hat{\theta}_{n,b,t}$.

It should be pointed out that the notion of variance estimation via subsampling is due to Carlstein (1986) and predates the the notion of subsampling distribution estimation of Politis and Romano (1994b). Carlstein (1986) considered the problem of variance estimation for a general statistic. He proposed to divide the observed series $X_1, \ldots, X_n$ into *nonoverlapping* blocks of size b, corresponding to $Y_1, Y_{b+1}, \ldots, Y_{n-b+1}$ in our notation. Assuming $n = lb$, there are only l such blocks, rather than $n - b + 1$ for the overlapping approach of subsampling. The estimator for the variance of $\hat{\theta}_n$ is then simply the sample variance of $\hat{\theta}_b$ evaluated on the l nonoverlapping blocks, properly standardized:

$$\widehat{Var}_{n,b,CARL}(\hat{\theta}_n) = \frac{\tau_b^2}{\tau_n^2}\frac{1}{l}\sum_{k=0}^{l-1}(\hat{\theta}_{n,b,kb+1} - \hat{\theta}_{n,b,\cdot,CARL})^2, \qquad (3.41)$$

where

$$\hat{\theta}_{n,b,\cdot,CARL} = \frac{1}{l}\sum_{k=0}^{l-1}\hat{\theta}_{n,b,kb+1}.$$

The estimator $\widehat{Var}_{n,b,CARL}(\hat{\theta}_n)$ is easily seen to be an analogue to the subsampling variance estimator $\widehat{Var}_{n,b}(\hat{\theta}_n)$ as defined in (3.40), the only difference being nonoverlapping versus overlapping blocks. Künsch (1989) showed that the use of overlapping block results in higher efficiency; see our equation (3.46) in what follows.

The following result states the L_2 consistency of the Carlstein and subsampling variance estimators under a uniform integrability (U.I.) condition. For the estimation procedure to be meaningful, we require a positive, finite limiting variance of the standardized estimator, that is,

$$\lim_{n\to\infty} Var(\tau_n\hat{\theta}_n) = \sigma_\infty^2 \in (0, \infty).$$

Denote

$$\hat{\sigma}_{n,CARL}^2 = \tau_n^2\widehat{Var}_{n,b,CARL}(\hat{\theta}_n) \quad \text{and} \quad \hat{\sigma}_{n,SUB}^2 = \tau_n^2\widehat{Var}_{n,b}(\hat{\theta}_n). \qquad (3.42)$$

Also, denote $T_n = \tau_n(\hat{\theta}_n - E(\hat{\theta}_n))$.

Lemma 3.8.1. *Let $\{X_t\}$ be a stationary sequence of random variables defined on a common probability space. Denote the corresponding mixing coefficients by $\alpha_X(\cdot)$. Assume that $\tau_b/\tau_n \to 0$, $b/n \to 0$ and $b \to \infty$ as $n \to \infty$. Also, assume that*

- $\{(T_n)^4\}$ *are U.I.*, \hfill (3.43)
- $\alpha_X(m) \to 0$ *as $m \to \infty$.* \hfill (3.44)

Then, both $\hat{\sigma}_{n,CARL}^2$ and $\hat{\sigma}_{n,SUB}^2$ converge to σ_∞^2 in L_2 norm.

Proof. The consistency of $\hat{\sigma}^2_{n,CARL}$ is proven in Carlstein (1986), while the consistency of $\hat{\sigma}^2_{n,SUB}$ follows from Theorem 1 of Fukuchi (1997); see also Radulović (1998). Incidentally, Fukuchi (1997) also proved almost sure convergence of both variance estimators under stronger moment and mixing conditions. ∎

As seen in Section 2.8, in the i.i.d. case, the block size b can be taken of the same order as n under some extra conditions. This is due to the fact that the variance of $\hat{\sigma}^2_{n,SUB}$ is of order $O(1/n)$, independent of b. However, this is not possible in the dependent case.

To look at a specific example, consider the sample mean case where $\hat{\theta}_n = \bar{X}_n$. Then the variance estimator $\hat{\sigma}^2_{n,SUB}$ is actually asymptotically equivalent to the Bartlett kernel estimator of the spectral density at the origin (Künsch, 1989). It is well known (e.g., Priestley, 1981) that the variance of $\hat{\sigma}^2_{n,SUB}$ is of (exact) order $O(b/n)$, while its bias is of (exact) order $O(1/b)$ provided the $\{X_t\}$ is *not* uncorrelated. This of course implies that consistent variance estimation requires both $b \to \infty$ and $b/n \to 0$; in the next section, more details on this interesting case are given.

3.8.2 Case of the Sample Mean

In the interesting special case where $\hat{\theta}_n = \bar{X}_n$ is the usual sample mean of the stationary stretch $X_1, \ldots, X_n$, the estimators $\hat{\sigma}^2_{n,CARL}$ and $\hat{\sigma}^2_{n,SUB}$ have already been considered under different names in the literature of various disciplines besides statistics. For example, in operations research and industrial engineering, the estimators $\hat{\sigma}^2_{n,CARL}$ and $\hat{\sigma}^2_{n,SUB}$ go by the respective names "nonoverlapping" and "overlapping batch-means variance estimators," whereby "batch" indicates "block" or "subsample" in our notation (e.g., see Meketon and Schmeiser, 1984; Schmeiser, 1990; Welch, 1987). Similarly, in electrical engineering, $\hat{\sigma}^2_{n,CARL}$ and $\hat{\sigma}^2_{n,SUB}$ are related to Welch's (1967) spectral density estimation method by averaging "batch" periodograms, where $\hat{\sigma}^2_{n,CARL}$ and $\hat{\sigma}^2_{n,SUB}$ are estimators of the spectral density at the origin (see Subsection 3.4.8). Quite notably, Welch's (1967) main idea is essentially contained in very early papers by Bartlett (1946, 1950); in particular, $\hat{\sigma}^2_{n,SUB}$ is equivalent (up to some asymptotically negligible "end effects") to $2\pi \hat{f}_{BART,b}(0)$, where the Bartlett spectral density estimator $\hat{f}_{BART,b}(\omega)$ with block size b is defined by

$$\hat{f}_{BART,b}(\omega) = \frac{1}{2\pi} \sum_{k=-b}^{k=b} \left(1 - \frac{|k|}{b}\right) \hat{R}(k) e^{ik\omega}, \qquad (3.45)$$

for all $\omega \in [-\pi, \pi]$. Here,

$$\hat{R}(k) = \frac{1}{n} \sum_{t=1}^{n-|k|} (X_t - \bar{X}_n)(X_{t+|k|} - \bar{X}_n)$$

denotes the sample autocovariance at lag k.

In addition, $\hat{\sigma}^2_{n,SUB}$ is identical to the moving blocks bootstrap and/or jacknife variance estimator of the variance of the sample mean proposed by Künsch (1989) and Liu and Singh (1992) and discussed in the next section.

As a further important point, note that this body of previous work has demonstrated that $\hat{\sigma}^2_{n,SUB}$ is a more efficient estimator than $\hat{\sigma}^2_{n,CARL}$. In particular, while their biases are identical, we have that

$$\lim_{n\to\infty} \frac{Var(\hat{\sigma}^2_{n,SUB})}{Var(\hat{\sigma}^2_{n,CARL})} = \frac{2}{3} \qquad (3.46)$$

in the sample mean case (cf. Meketon and Schmeiser, 1984; Zhurbenko, 1986; Welch, 1987; Künsch, 1989; and our equation (9.3) in Chapter 10).

Finally, note that relation (3.46) remains valid even in the case where $\hat{\theta}_n$ is not the sample mean, but is a smooth, approximately linear statistic of the empirical distribution function of the observations (see Künsch, 1989; Liu and Singh, 1992, or Politis and Romano, 1993b) for more details.

3.9 Comparison with the Moving Blocks Bootstrap

Künsch (1989) and Liu and Singh (1992) independently introduced an extension of Efron's (1979) bootstrap to time series data. It was termed the "moving blocks" method by Liu and Singh. It should be mentioned that Künsch's (1989) paper was influenced by the previously mentioned work by Carlstein (1986). Politis and Romano (1992b, 1994a) proposed variants of the moving blocks bootstrap that possess interesting features such as random block sizes and a circular, periodic extension of the original data sequence.

To describe the moving blocks method, let Y_t the block of size b of data $\{X_t, \ldots, X_{t+b-1}\}$. For simplicity, assume for the moment that $n = lb$ for some integer l. Now, let P_n^* denote the empirical distribution of the blocks $Y_1, Y_2, \ldots, Y_{n-b+1}$. Denote $X_1^*, \ldots, X_n^*$ a pseudo time series that is generated by resampling l blocks $Y_1^*, \ldots, Y_l^*$ i.i.d. from P_n^* and concatenating them. Then let $\hat{\theta}_n^* = \hat{\theta}_n(X_1^*, \ldots, X_n^*)$, the estimator of $\theta(P)$ computed on the pseudo time series. The moving blocks method approximates the sampling distribution of $\tau_n(\hat{\theta}_n - \theta(P))$ by the distribution of $\tau_n(\hat{\theta}_n^* - \hat{\theta}_n)$ under P_n^*, that is, by

$$K_{n,b}(x) = Prob_{P_n^*}\{\tau_n(\hat{\theta}_n^* - \hat{\theta}_n) \leq x\}. \qquad (3.47)$$

In the general case, when the sample size n is not simply a multiple of the block size b, one can choose l as the smallest integer for which $lb \geq n$, generate a pseudo time series as above, and discard the last $lb - n$ pseudo-observations.

Since the normalizing constants are the same in the true sampling distribution and in the bootstrap approximation, namely, τ_n, it is not really necessary to include them. However, we do so to facilitate the comparison with the subsampling method. Also, as discussed below, the moving blocks bootstrap is generally only valid in regular situations where $\tau_n = \sqrt{n}$ and thus in situations where the nature of the normalizing constant is well known.

It can be seen that Efron's bootstrap is a special case of the moving blocks method, namely, if we take $b = 1$.

While the distribution of $\tau_n(\hat{\theta}_n^* - \hat{\theta}_n)$ under P_n^* is well defined and denoted by $K_{n,b}(\cdot)$ here, it usually cannot be obtained analytically and one has to resort to Monte Carlo approximations. To this end, a large number B of pseudo time series $X_1^*, \ldots, X_n^*$ are generated, by resampling from the observed blocks with replacement, and the statistic of interest $\hat{\theta}_n$ is computed from each series. This yields B pseudo statistics $\hat{\theta}_{n,1}^*, \ldots, \hat{\theta}_{n,B}^*$. The Monte Carlo approximation of the distribution $K_{n,b}(x)$ is then given by

$$\hat{K}_{n,b}(x) = \frac{1}{B} \sum_{h=1}^{B} 1\{\tau_n(\hat{\theta}_{n,h}^* - \hat{\theta}_n) \leq x\}. \qquad (3.48)$$

Given the approximation $\hat{K}_{n,b}$ of the sampling distribution of $\tau_n(\hat{\theta}_n - \theta(P))$, inference for $\theta(P)$ can be drawn analogously to the subsampling method. For example, confidence intervals for univariate parameters $\theta(P)$ are based on appropriate quantiles of $\hat{K}_{n,b}$. Also, the variance of $\tau_n\hat{\theta}_n$ can be approximated by the variance of a random variable with distribution $\hat{K}_{n,b}$.

Subsampling and moving blocks bootstrap are similar in that they utilize blocks of data Y_t of size b. The important difference is that subsampling looks upon these blocks as "subseries" or "mini time series," whereas moving blocks uses the blocks as "building stones" to construct new pseudo-time series. It turns out that the subsampling method works under weaker conditions. As seen previously, apart from moment and mixing conditions, the only requirement is the existence of a nondegenerate limiting distribution for the estimator $\hat{\theta}_n$. Moving blocks, in contrast, is generally only valid for asymptotically linear statistics, and parameters of an m-dimensional marginal distribution, with m fixed, such as the mean or an autocorrelation (see Künsch, 1989). Moreover, in order to show the validity of the moving blocks bootstrap, typically stronger conditions on the underlying sequence are needed than those that ensure a limiting distribution. A notable exception was provided by Radulović (1996), who showed that, for the case of the sample mean, the moving blocks bootstrap, studentized or not, works as long as the sequence is strong mixing and a central limit theorem holds. It should be pointed out, however, that in his theorems the necessary condition of the block size tending to infinity was inadvertently left out.

Note that an extension of the moving blocks bootstrap, the "blocks of blocks" bootstrap, has been given by Politis and Romano (1992a, 1993a,b) and followed up by Bühlmann and Künsch (1995). The "blocks of blocks" bootstrap handles the case of parameters of the whole, infinite dimensional law of the time series $\{X_t\}$, as opposed to just a finite dimensional marginal; a prime example of such a parameter is the spectral density function evaluated at a specific point. It also handles the ensuing τ_n rates that are less than $\sqrt{n}$, at the expense of assuming a special linear structure for the statistics employed.

Remark 3.9.1. The moving blocks bootstrap, being a generalization of Efron's (1979) bootstrap, also requires a "case by case" analysis to demonstrate its validity for particular applications. A general result like Theorem 3.2.1 for subsampling is not available. In some cases, however, subsampling can be used to prove consistency of the moving blocks bootstrap. As an example, in the next chapter we use results for subsampling to prove consistency of the moving block methods for making inference for the mean of a nonstationary time series (see Example 4.4.1).

3.10 Conclusions

In this chapter, it was demonstrated how the subsampling methodology can be extended to make asymptotically valid inferences in the context of dependent, stationary observations. The crux of the extension is to only allow blocks of size b of the observed sequence as proper subsamples, rather than any sets of b data points sampled without replacement. Thereby, the original dependence structure is preserved within each subsample.

It was described how to find confidence intervals for real-valued parameters and confidence regions for vector-valued parameters. Also, the issue of hypothesis testing in situations where the null hypothesis does not translate simply into a null hypothesis on a parameter was discussed. Finally, the issues of bias reduction and variance estimation were addressed.

Analogous to the i.i.d. case, the assumptions needed for subsampling to work are extremely weak. For example, in the case of making inference on a parameter, they essentially boil down to a mixing condition on the underlying sequence, the existence of a nondegenerate limiting distribution of a properly standardized statistic, and a weak condition on the block size b. Again, the "case-by-case analysis" of bootstrap methods, such as the moving blocks bootstrap, is avoided and consistency under less stringent assumptions ensues.

Last, but not least, the usefulness of subsampling for dependent data was demonstrated by illustrating it with a large number of examples.

4
Subsampling for Nonstationary Time Series

4.1 Introduction

Stationary time series are very convenient to work with from a mathematical point of view, but the assumption of stationarity is often violated when modeling real-life data. To mention only two examples, many economic time series exhibit seasonal fluctuations, while stock return data typically show time-dependent variability. The goal of this chapter is to demonstrate that the subsampling method is by no means restricted to stationary series. We will provide sufficient conditions under which asymptotically correct inference can be made even in the presence of nonstationarity. In outline and style, this chapter follows the previous one very closely. In particular, the subsampling methodology for nonstationary observations will be *identical* to the one for stationary observations. Many of the results derived under stationarity will be restated and reproven under weaker conditions. Much of what is presented is taken from Politis, Romano, and Wolf (1997).

At this point, we should mention that by nonstationary we mean "more or less stationary" or "asymptotically stationary." Two examples are an AR(1) process with parameter $|\rho| < 1$ and a heteroskedastic innovation sequence, and a Markov chain that does not start out in the equilibrium distribution. This is in contrast to other fields, such as econometrics, where a nonstationary time series behaves entirely different from a stationary one, the classic example being a random walk with or without drift.

Suppose $\{\ldots, X_{-1}, X_0, X_1, \ldots\}$ is a sequence of random variables taking values in an arbitrary sample space S, and defined on a common probability space. Denote the joint probability law governing the infinite sequence by P. The goal is to construct a confidence region for some parameter $\theta(P)$, on the basis of observing $\{X_1, \ldots, X_n\}$. The sequence $\{X_t\}$ will be assumed to satisfy a certain weak dependence condition, namely the condition of α-mixing (cf. Definition A.0.1). However, no assumption of stationarity, in the weak or the strict sense, is made.

In Section 4.2, the method is described for the general univariate parameter case and seen to be identical to the one for stationary data. A general theorem is stated, allowing for asymptotically valid inference under very weak conditions. An extension to the general multivariate parameter case is given in Section 4.3. The usefulness of the general theory is demonstrated by discussing a large number of examples in Section 4.4. Section 4.5 mentions briefly how subsampling can be used for testing purposes in the absence of a suitable parameter. Variance estimation is discussed in Section 4.6.

4.2 Univariate Parameter Case

Let $\hat{\theta}_n = \hat{\theta}_n(X_1, \ldots, X_n)$ be an estimator of $\theta(P) \in \mathbb{R}$, the parameter of interest. The method is identical to the stationary setting as outlined in Section 3.2, the crux being to compute the corresponding statistic on blocks of smaller size to get an approximation to the sampling distribution of $\hat{\theta}_n$. Thus, let $\hat{\theta}_{n,b,t} = \hat{\theta}_b(X_t, \ldots, X_{t+b-1})$, the estimator of $\theta(P)$ based on the subsample $\{X_t, \ldots, X_{t+b-1}\}$. Define $J_{b,t}(P)$ to be the sampling distribution of $\tau_b(\hat{\theta}_{n,b,t} - \theta(P))$, where τ_b is an appropriate normalizing constant. Also, define the corresponding cumulative distribution function:

$$J_{b,t}(x, P) = Prob_P\{\tau_b(\hat{\theta}_{n,b,t} - \theta(P)) \leq x\}. \tag{4.1}$$

Note that, unlike in the stationary setup of the previous chapter, $J_{b,t}(P)$ does depend on the index t. For notational convenience, let $J_n(P) = J_{n,1}(P)$ and $J_n(\cdot, P) = J_{n,1}(\cdot, P)$. Essentially, the only assumption that will be needed to consistently estimate $J_n(P)$ is the following.

Assumption 4.2.1. *There exists a limiting law $J(P)$ such that*

i. $J_n(P)$ *converges weakly to* $J(P)$ *as* $n \to \infty$.
ii. *For every continuity point x of $J(P)$ and for any sequences n, b with $n, b \to \infty$ and $b/n \to 0$, we have $\frac{1}{n-b+1} \sum_{t=1}^{n-b+1} J_{b,t}(x, P) \to J(x, P)$.*

Condition (i) states that the estimator, properly normalized, has a limiting distribution. It is hard to conceive of any asymptotic theory free of such a requirement.

Condition (ii) states that, for large n, the distribution functions of the normalized estimator based on the subsamples will be on average close to the distribution function of the normalized estimator based on the entire sample. This condition clearly allows for local effects, such as seasonality or other forms of heteroskedasticity.

The following assumption is somewhat stronger, but will be useful for later results.

Assumption 4.2.2. *There exists a limiting law $J(P)$ such that*

i. $J_n(P)$ *converges weakly to $J(P)$ as $n \to \infty$.*
ii. *For every continuity point x of $J(\cdot, P)$ and for any index sequence $\{t_b\}$, we have $J_{b,t_b}(x, P) \to J(x, P)$ as $b \to \infty$.*

Here, condition (ii) requires that the distribution function of the normalized statistic evaluated over a subsample converges to the same limiting law as the distribution function of the normalized estimator based on the entire sample, *uniformly* in the starting point of the subsample. Assuming (i), the condition will be clearly satisfied for stationary processes, but also for processes that exhibit approximate or asymptotic stationarity. For example, one can consider a Markov chain not starting out in the equilibrium distribution.

It is easy to see that Assumption 4.2.1 is implied by Assumption 4.2.2. If condition (ii) in Assumption 4.2.1 did not hold, there would have to exist a subsequence $\{b_k\}$ such that $J_{b_k, t_{b_k}}(x, P)$ is bounded away from $J(x, P)$ for some continuity point x of $J(\cdot, P)$. This contradicts condition (ii) in Assumption 4.2.2.

The approximation to $J_n(x, P)$ we study is identical to the one proposed in Subsection 3.2.2, namely,

$$L_{n,b}(x) = \frac{1}{n-b+1} \sum_{t=1}^{n-b+1} 1\{\tau_b(\hat{\theta}_{n,b,t} - \hat{\theta}_n) \leq x\}. \quad (4.2)$$

Theorem 4.2.1. *Assume Assumption 4.2.1 and that $\tau_b/\tau_n \to 0$, $b/n \to 0$, and $b \to \infty$ as $n \to \infty$. Also, assume that $\alpha_X(m) \to 0$ as $m \to \infty$.*

i. *If x is a continuity point of $J(\cdot, P)$, then $L_{n,b}(x) \to J(x, P)$ in probability.*
ii. *If $J(\cdot, P)$ is continuous, then $\sup_x |L_{n,b}(x) - J(x, P)|$ in probability.*
iii. *For $\alpha \in (0, 1)$, let*

$$c_{n,b}(1-\alpha) = \inf\{x : L_{n,b}(x) \geq 1-\alpha\}.$$

Correspondingly, define

$$c(1-\alpha, P) = \inf\{x : J(x, P) \geq 1-\alpha\}.$$

If $J(\cdot, P)$ is continuous at $c(1-\alpha, P)$, then

$$Prob_P\{\tau_n[\hat{\theta}_n - \theta(P)] \leq c_{n,b}(1-\alpha)\} \to 1-\alpha \text{ as } n \to \infty.$$

Thus, the asymptotic coverage probability under P of the interval $I_1 = [\hat{\theta}_n - \tau_n^{-1} c_{n,b}(1-\alpha), \infty)$ is the nominal level $1-\alpha$.

104   4. Subsampling for Nonstationary Time Series

Remark 4.2.1. In the stationary case, condition (ii) of Assumption 4.2.1 follows immediately from condition (i), and the theorem reduces to Theorem 3.2.1 of Chapter 3.

Remark 4.2.2. Corollary 2.2.1 carries over from the i.i.d. case to the nonstationary case in a similar manner. Thus, an asymptotically valid confidence interval can be constructed without the assumption $\tau_b/\tau_n \to 0$, as long as $b \to \infty$ and $b/n \to 0$.

Proof of Theorem 4.2.1. As before, we omit the b-subscripts in the proof and use $q \equiv q_n \equiv n - b + 1$. Let

$$U_n(x) = \frac{1}{q} \sum_{t=1}^{q} 1\{\tau_b[\hat{\theta}_{n,b,t} - \theta(P)] \leq x\}.$$

To prove (i), it suffices to show that $U_n(x)$ converges in probability to $J(x, P)$ for every continuity point x of $J(\cdot, P)$. This can be seen by noting that

$$L_n(x) = \frac{1}{q} \sum_{t=1}^{q} 1\{\tau_b[\hat{\theta}_{n,b,t} - \theta(P)] + \tau_b[\theta(P) - \hat{\theta}_n] \leq x\},$$

so that for every $\epsilon > 0$,

$$U_n(x - \epsilon)1\{E_n\} \leq L_n(x)1\{E_n\} \leq U_n(x + \epsilon),$$

where $1\{E_n\}$ is the indicator of the event $E_n = \{\tau_b|\theta(P) - \hat{\theta}_n| \leq \epsilon\}$. But the event E_n has probability tending to one. So, with probability tending to one,

$$U_n(x - \epsilon) \leq L_n(x) \leq U_n(x + \epsilon).$$

Thus, if $x + \epsilon$ and $x - \epsilon$ are continuity points of $J(\cdot, P)$, then $U_n(x \pm \epsilon)$ tending to $J(x \pm \epsilon, P)$ in probability implies

$$J(x - \epsilon, P) - \epsilon \leq L_n(x) \leq J(x + \epsilon, P) + \epsilon$$

with probability tending to one. Now, let $\epsilon \to 0$ such that $x \pm \epsilon$ are continuity points of $J(\cdot, P)$ to conclude that $L_n(x) \to J(x, P)$ in probability as well. Therefore, we may restrict our attention to $U_n(x)$.

Since $E(U_n(x)) = \frac{1}{q} \sum_{t=1}^{q} J_{b,t}(x)$, the proof of (i) reduces by Assumption 4.2.1 to showing that $Var(U_n(x))$ tends to zero (as n tends to infinity). Define

$$I_{b,t} = 1\{\tau_b[\hat{\theta}_{n,b,t} - \theta(P)] \leq x\}, \quad t = 1, \ldots, q,$$

$$s_{q,h} = \frac{1}{q} \sum_{t=1}^{q-h} Cov(I_{b,t}, I_{b,t+h}).$$

Then,

$$Var(U_n(x)) = \frac{1}{q}(s_{q,0} + 2\sum_{h=1}^{q-1} s_{q,h})$$

$$= \frac{1}{q}(s_{q,0} + 2\sum_{h=1}^{b-1} s_{q,h} + 2\sum_{h=b}^{q-1} s_{q,h})$$

$$\equiv A^* + A,$$

where $A^* = \frac{1}{q}(s_{q,0} + 2\sum_{h=1}^{b-1} s_{q,h})$ and $A = \frac{2}{q}\sum_{h=b}^{q-1} s_{q,h}$.

It is readily seen that $|A^*| = O(b/q)$. To handle A, we apply Lemma A.0.2. For $h \geq b$

$$|Cov(I_{b,t}, I_{b,t+h})| \leq 4\alpha_X(h-b+1)$$

and therefore,

$$|A| \leq \frac{8}{q}\sum_{h=1}^{q-b} \alpha_X(h).$$

By the monotonicity of mixing coefficients, $\alpha_X(m) \to 0$ (as $m \to \infty$) and therefore $\frac{1}{M}\sum_{m=1}^{M} \alpha_X(m) \to 0$ (as $M \to \infty$), which implies that A converges to zero. Thus, both A and A^* converge to zero, which completes the proof of (i).

The proofs of (ii) and (iii) are analogous to the proofs of (ii) and (iii) of Theorem 3.2.1 and are therefore omitted. ∎

At the end of Section 3.2, we discussed symmetric confidence intervals as an alternative to equal-tailed intervals. Of course, also for nonstationary observations, we have these two choices when two-sided confidence intervals are desired.

Recall that symmetric subsampling intervals are obtained by estimating the two-sided distribution function

$$J_{n,|\cdot|}(x, P) = Prob_P\{\tau_n \left|\hat{\theta}_n - \theta(P)\right| \leq x\}. \tag{4.3}$$

The subsampling approximation to $J_{n,|\cdot|}(x, P)$ is defined by

$$L_{n,b,|\cdot|}(x) = \frac{1}{n-b+1}\sum_{t=1}^{n-b+1} 1\{\tau_b \left|\hat{\theta}_{n,b,t} - \hat{\theta}_n\right| \leq x\}. \tag{4.4}$$

By Theorem 4.2.1 the asymptotic validity of two-sided symmetric subsampling intervals easily follows.

Corollary 4.2.1. *Make the same assumptions as in Theorem 3.2.1. Denote the limiting law of $J_{n,|\cdot|}(\cdot, P)$ by $J_{|\cdot|}(\cdot, P)$.*

i. *If x is a continuity point of $J_{|\cdot|}(\cdot, P)$, then $L_{n,b,|\cdot|}(x) \to J_{|\cdot|}(x, P)$ in probability.*

ii. If $J_{|\cdot|}(\cdot, P)$ is continuous, then $\sup_x |L_{n,|\cdot|}(x) - J_{|\cdot|}(x, P)| \to 0$ in probability.

iii. For $\alpha \in (0, 1)$, let

$$c_{n,b,|\cdot|}(1 - \alpha) = \inf\{x : L_{n,b,|\cdot|}(x) \geq 1 - \alpha\}.$$

Correspondingly, define

$$c_{|\cdot|}(1 - \alpha, P) = \inf\{x : J_{|\cdot|}(x, P) \geq 1 - \alpha\}.$$

If $J_{|\cdot|}(\cdot, P)$ is continuous at $c_{|\cdot|}(1 - \alpha, P)$, then

$$Prob_P\{\tau_n \left|\hat{\theta}_n - \theta(P)\right| \leq c_{n,b,|\cdot|}(1 - \alpha)\} \to 1 - \alpha \text{ as } n \to \infty.$$

Thus, the asymptotic coverage probability under P of the interval $I_{SYM} = [\hat{\theta}_n - \tau_n^{-1} c_{n,b,|\cdot|}(1 - \alpha), \hat{\theta}_n + \tau_n^{-1} c_{n,b,|\cdot|}(1 - \alpha)]$ is the nominal level $1 - \alpha$.

Proof. The proof follows immediately from Theorem 4.2.1 and the continuous mapping theorem. ∎

4.3 Multivariate Parameter Case

It is often necessary to construct confidence regions for multivariate parameters $\theta(P) \in \mathbb{R}^k$, with $k > 1$. For example, this problem arises when we want to draw simultaneous inference on the regression coefficients in a multivariate regression problem. The extension to higher dimensions is analogous to the stationary case.

Again, let $\hat{\theta}_{n,b,t} = \hat{\theta}_b(X_t, \ldots, X_{t+b-1})$ be an estimator of $\theta(P)$ based on the subsample $\{X_t, \ldots, X_{t+b-1}\}$. Define $J_{b,t}(P)$ to be the sampling distribution of $\tau_b(\hat{\theta}_{n,b,t} - \theta(P))$. Rather than working with multivariate distribution functions we will look at indicators of Borel sets. For any Borel set $A \in \mathbb{R}^k$, define

$$J_{b,t}(A, P) = Prob_P\{\tau_b(\hat{\theta}_{n,b,t} - \theta(P)) \in A\}. \tag{4.5}$$

Again, let $J_n(P) = J_{n,1}(P)$.

It is now obvious how Assumptions 4.2.1 and 4.2.2 need to be modified.

Assumption 4.3.1. *There exists a limiting law $J(P)$ such that*

i. $J_n(P)$ *converges weakly to $J(P)$ as $n \to \infty$. This means, for any Borel set A whose boundary has mass zero under $J(P)$, we have $J_n(A, P) \to J(A, P)$ as $n \to \infty$.*

ii. *for every Borel set A whose boundary has mass zero under $J(P)$, and for any sequences n, b with $n, b \to \infty$ and $b/n \to 0$, we have $\frac{1}{n-b+1} \sum_{t=1}^{n-b+1} J_{b,t}(A, P) \to J(A, P)$.*

Assumption 4.3.2. *There exists a limiting law $J(P)$ such that*

i. $J_n(P)$ converges weakly to $J(P)$ as $n \to \infty$. This means, for each Borel set A whose boundary has mass zero under $J(P)$, we have $J_n(A, P) \to J(A, P)$ as $n \to \infty$.
ii. for every Borel set A whose boundary has mass zero under $J(P)$ and for any index sequence $\{t_b\}$, we have $J_{b,t_b}(A, P) \to J(A)$ as $b \to \infty$.

Similarly to the univariate case, Assumption 4.3.2 implies Assumption 4.3.1.

Our method is analogous to the univariate case, and the approximation to $J_n(A, P)$ we study is defined by

$$L_{n,b}(A) = \frac{1}{n-b+1} \sum_{t=1}^{n-b+1} 1\{\tau_b(\hat{\theta}_{n,b,t} - \hat{\theta}_n) \in A\}. \quad (4.6)$$

Theorem 4.3.1. *Assume Assumption 4.3.1 and that $\tau_b/\tau_n \to 0$, $b/n \to 0$, and $b \to \infty$ as $n \to \infty$. Also, assume that $\alpha_X(m) \to 0$ as $m \to \infty$. Then,*

i. $L_{n,b}(A) \to J(A, P)$ in probability, for each Borel set A whose boundary has mass zero under $J(P)$.
ii. $\rho_k(L_{n,b}, J(P)) \to 0$ in probability for every metric ρ_k that metrizes weak convergence on $\mathbb{R}^k$.
iii. Let $\{Z_n\}$ and Z be random vectors with $\mathcal{L}(Z_n) = L_{n,b}$ and $\mathcal{L}(Z) = J(P)$. Then, for any almost everywhere $J(P)$ continuous real function f on $\mathbb{R}^k$ and any metric ρ_1 which metrizes weak convergence on $\mathbb{R}$,

$$\rho_1(\mathcal{L}(f(Z_n)), \mathcal{L}(f(Z))) \to 0 \text{ in probability.}$$

In particular, for any norm $\|\cdot\|$ on $\mathbb{R}^k$,

$$\rho_1(\mathcal{L}(\|Z_n\|), \mathcal{L}(\|Z\|)) \to 0 \text{ in probability.}$$

iv. Let Z be a random vector with $\mathcal{L}(Z) = J(P)$. For a norm $\|\cdot\|$ on $\mathbb{R}^k$, define univariate distributions $L_{n,\|\cdot\|}$ and $J_{\|\cdot\|}(P)$ in the following way:

$$L_{n,b,\|\cdot\|}(x) = \frac{1}{n-b+1} \sum_{t=1}^{n-b+1} 1\{\|\tau_b(\hat{\theta}_{n,b,t} - \hat{\theta}_n)\| \le x\},$$

$$J_{\|\cdot\|}(x, P) = \text{Prob}\{\|Z\| \le x\}.$$

For $\alpha \in (0, 1)$, let

$$c_{n,b,\|\cdot\|}(1-\alpha) = \inf\{x : L_{n,b,\|\cdot\|}(x) \ge 1-\alpha\}.$$

Correspondingly, define

$$c_{\|\cdot\|}(1-\alpha, P) = \inf\{x : J_{\|\cdot\|}(x, P) \ge 1-\alpha\}.$$

If $J_{\|\cdot\|}(\cdot, P)$ is continuous at $c_{\|\cdot\|}(1-\alpha, P)$, then

$$Prob_P\{\|T_n(\hat{\theta}_n - \theta(P))\| \leq c_{n,b,\|\cdot\|}(1-\alpha)\} \to 1 - \alpha \text{ as } n \to \infty.$$

Thus, the asymptotic coverage probability under P of the region $\{\theta : \|T_n(\theta - \hat{\theta}_n)\| \leq c_{n,b,\|\cdot\|}(1-\alpha)\}$ is the nominal level $1 - \alpha$.

Proof. The proof is very similar to the proof of Theorem 3.3.1, with the differences being analogous to the differences between the proofs of Theorem 3.2.1 and Theorem 4.2.1. ∎

4.4 Examples

This section extends many results of Section 3.4 to the nonstationary case. In addition, we present a result on the moving blocks bootstrap for the univariate mean.

Example 4.4.1 (Univariate mean). Suppose $\{X_t\}$ is a sequence of random variables with common univariate mean $\theta(P)$. The goal is to construct a confidence interval for $\theta(P)$, on the basis of observing $\{X_1, \ldots, X_n\}$.

Let $\hat{\theta}_{n,b,t} \equiv b^{-1}\sum_{a=t}^{t+b-1} X_a \equiv \bar{X}_{n,b,t}$ be the estimator of $\theta(P)$ based on the block of size b of the consecutive data $\{X_t, \ldots, X_{t+b-1}\}$. For convenience, denote $\bar{X}_n = \bar{X}_{n,n,1}$. Define $J_{b,t}(P)$ to be the sampling distribution of $b^{\frac{1}{2}}(\bar{X}_{n,b,t} - \theta(P))$. Also, define the corresponding cumulative distribution function:

$$J_{b,t}(x, P) = Prob_P\{b^{\frac{1}{2}}(\bar{X}_{n,b,t} - \theta(P)) \leq x\}. \tag{4.7}$$

The approximation to $J_n(x, P)$ we study is defined by

$$L_{n,b}(x) = \frac{1}{n-b+1} \sum_{t=1}^{n-b+1} 1\{b^{\frac{1}{2}}(\bar{X}_{n,b,t} - \bar{X}_n) \leq x\}. \tag{4.8}$$

The following theorem states sufficient moment and mixing conditions for which the subsampling technique allows us to draw a first-order correct inference for $\theta(P)$.

Theorem 4.4.1. Let $\{X_t\}$ be a sequence of random variables defined on a common probability space. Denote the corresponding mixing coefficients by $\alpha_X(\cdot)$. Define

$$T_{k,t} \equiv k^{-\frac{1}{2}} \sum_{a=t}^{t+k-1} X_a \text{ and } \sigma^2_{k,t} \equiv Var(T_{k,t}).$$

Assume the following conditions. For some $\delta > 0$:

- $E|X_t|^{2+2\delta} \leq \Delta$ for all t, $\tag{4.9}$

- $\sigma_{k,t}^2 \to \sigma^2 > 0$ *uniformly in t as $k \to \infty$,*   (4.10)

- $C(4,\delta) \equiv \sum_{k=1}^{\infty}(k+1)^2 \alpha_X^{\frac{\delta}{4+\delta}}(k) \leq K.$ (4.11)

Furthermore, assume that $b/n \to 0$ and $b \to \infty$ as $n \to \infty$, and let $J(P) = N(0,\sigma^2)$.

Then, conclusions (i–iii) of Theorem 4.2.1 are true.

Remark 4.4.1. Even for stationary data, we need the sample mean, properly normalized, to converge weakly to some nondegenerate limiting distribution (see Example 3.4.1). A reasonable condition for that to happen is $\sigma_{n,1}^2 \to \sigma^2$. Taking this into account, our condition (4.10) does not seem prohibitive. In fact, it should be difficult to imagine a reasonable situation where $\sigma_{n,1}^2 \to \sigma^2$, but condition (4.10) is violated.

Proof of Theorem 4.4.1. For the proof we will assume, without loss of generality, that true mean is equal to zero, that is, $\theta(P) = 0$. By Theorem 4.2.1 it is sufficient to verify our Assumption 4.2.2, as the mixing condition $\alpha_X(k) \to 0$ is implied by assumption (4.11).

Condition (i) of Assumption 4.2.2 follows directly from Theorem B.0.1. For condition (ii), let $\{t_b\}$ be an index sequence. Now apply Theorem B.0.1 to the triangular array, the n-th row of which is the block of size b of the consecutive data $\{X_{t_b}, \ldots, X_{t_b+b-1}\}$; recall that we consider b as a function of n. ∎

It is interesting to note that, using our result, we can fairly easily prove an analogous theorem for the moving blocks bootstrap. A general description of the moving blocks method can be found in Section 3.9. We will describe it here again for the special case of the univariate mean. As before, let Y_t be the block of size b of the consecutive data $\{X_t, \ldots, X_{t+b-1}\}$, let $l \equiv l_n \equiv \lfloor \frac{b}{n} \rfloor$, where $\lfloor \cdot \rfloor$ denotes the integer part, and let $q \equiv q_n \equiv n-b+1$. Conditional on the sample $\{X_1, \ldots, X_n\}$, denote the empirical distribution of $Y_1, \ldots, Y_q$ by P_n^*, that is, P_n^* puts mass $\frac{1}{q}$ on each of the Y_t. Assume for simplicity that $n = bl$. Generate a pseudo time series $\{X_1^*, \ldots, X_n^*\}$ in the following way: Let $Y_{b,1}^*, \ldots, Y_{b,l}^*$ i.i.d. $\sim P_n^*$ and join them together to one big block:

$$\{X_1^*, \ldots, X_n^*\} = \{Y_{1,1}^*, \ldots, Y_{1,b}^*, Y_{2,1}^*, \ldots, Y_{2,b}^*, \ldots, Y_{l,1}^*, \ldots, Y_{l,b}^*\}.$$

Here $Y_{j,t}^*$ denotes the t-th element of the block Y_j^*. In the general case when the sample size n is not simply a multiple of the block size b, one can choose l as the smallest integer for which $lb \geq n$, generate a pseudo time series as above, and discard the last $lb - n$ pseudo observations. The corresponding cumulative distribution function is given by

$$K_{n,b}(x) = Prob_{P_n^*}\{n^{\frac{1}{2}}(\bar{X}_n^* - \bar{X}_n) \leq x\}. \quad (4.12)$$

The following theorem states that the moving blocks method is also asymptotically valid, provided we strengthen the moment condition on the sequence $\{X_t\}$ somewhat compared to Theorem 4.4.1.

Theorem 4.4.2. *Let $\{X_t\}$ be a sequence of random variables defined on a common probability space. Denote the corresponding mixing coefficients by $\alpha_X(\cdot)$. Define*

$$T_{k,t} \equiv k^{-\frac{1}{2}} \sum_{a=t}^{t+k-1} X_a \quad \text{and} \quad \sigma_{k,t}^2 \equiv Var(T_{k,t}).$$

Assume the following conditions. For some $\delta > 0$:

- $E|X_t|^{4+2\delta} \leq \Delta$ *for all t,* (4.13)
- $\sigma_{k,t}^2 \to \sigma^2 > 0$ *uniformly in t as $k \to \infty$,* (4.14)
- $C(4,\delta) \equiv \sum_{k=1}^{\infty} (k+1)^2 \alpha_X^{\frac{\delta}{4+\delta}}(k) \leq K.$ (4.15)

Furthermore, assume that $b/n \to 0$ and $b \to \infty$ as $n \to \infty$, and let $J(P) = N(0, \sigma^2)$.

Then, conclusions (i–iii) of Theorem 4.2.1 are true if we replace $L_{n,b}(\cdot)$ by $K_{n,b}(\cdot)$ (as defined in (4.12)).

Remark 4.4.2. Relative to the result for the subsampling method, we need stronger moment conditions on the sequence $\{X_t\}$ here in order to show that the variance of the moving blocks distribution converges in probability to the proper limit. See the proof of the theorem below.

Remark 4.4.3. There have been previous results extending the moving blocks method for the sample mean to the nonstationary case.

A result similar to Theorem 4.4.2 was obtained by Fitzenberger (1997) in his Theorem 3.1 under a stronger condition on the block size b, namely $b = o(n^{1/2})$.

Under stronger assumptions, Lahiri (1992) not only showed a result similar to Theorem 4.4.2, but also obtained second order properties. However, he uses an even more stringent requirement on the block size, namely $b = o(n^{1/4})$. As will be discussed in Section 9.3, this seems to be a suboptimal condition.

Remark 4.4.4. As mentioned in Section 3.9, Radulović (1996) showed that, for stationary observations, the moving blocks bootstrap works for the sample mean whenever the underlying sequence is strong mixing and satisfies a Central Limit theorem (CLT), that is, results analogous to those for subsampling holds. This is a notable exception, since, compared to subsampling, stronger assumptions are typically needed to demonstrate the validity of the moving blocks bootstrap. It is conceivable that the arguments of Radulović can be extended to the nonstationary case, so that condition

(4.13) can be weakened to condition (4.9). The point of our proof, however, is to give an example of how results for subsampling can be utilized to prove consistency of the moving blocks bootstrap. This is a general idea that can simplify the justification of the moving blocks bootstrap in some cases (e.g., see Remark 7.4.4).

Proof of Theorem 4.4.2. Again, we will assume, without loss of generality, that $\theta(P) = 0$. First, we will state a result which we are going to need later on. It is a simple generalization of Proposition 1.3.1 (i), allowing for d_n variables in the n-th row rather than n. The proof is therefore omitted.

Proposition 4.4.1. *Let F and $\{F_n\}$ be distribution functions with corresponding means and variances μ, $\{\mu_n\}$, σ^2 and $\{\sigma_n^2\}$, respectively. Assume*
(1) $F_n \overset{\mathcal{L}}{\Longrightarrow} F$ and
(2) $\sigma_n^2 \to \sigma^2$.
Let $\{X_{n,t}, 1 \leq t \leq d_n\}$ be a triangular array of random variables, the n-th row of which satisfies: $X_n, \ldots, X_{n,d_n}$ i.i.d. $\sim F_n$ and define $\bar{X}_n = \frac{1}{d_n} \sum_{t=1}^{d_n} X_{n,t}$.
Then, $d_n^{\frac{1}{2}}(\bar{X}_n - \mu_n) \overset{\mathcal{L}}{\Longrightarrow} N(0, \sigma^2)$ as $d_n \to \infty$.

To prove the theorem, assume, without loss of generality, that $n = bl$. Note that

$$n^{\frac{1}{2}}(\bar{X}_n^* - \bar{X}_n) = l^{\frac{1}{2}}(\frac{1}{l} \sum_{t=1}^{l} b^{\frac{1}{2}} \{\bar{X}_{n,b,t}^* - \bar{X}_n\}) \equiv l^{\frac{1}{2}}(\frac{1}{l} \sum_{t=1}^{l} Z_{n,t}^*).$$

We see that the $Z_{n,t}^*$ are i.i.d. $\sim L_{n,b}$, where $L_{n,b}$ is defined as in (4.8). There, under weaker conditions, we have shown that $\rho_1(L_{n,b}, N(0, \sigma^2)) \to 0$ in probability, for any metric ρ_1 metrizing weak convergence on $\mathbb{R}$. The next step will be to check that $Var(L_{n,b}) \to \sigma^2$ in probability also. However, $Var(L_{n,b})$ is just the subsampling variance estimator $\hat{\sigma}_{n,SUB}^2$ of σ^2 as described in Section 4.6, and its consistency follows from Lemma 4.6.1.

Thus, both $Var(L_{n,b}) \to \sigma^2$ in probability and $\rho_1(L_{n,b}, N(0, \sigma^2)) \to 0$ in probability, where ρ_1 is any metric that metrizes weak convergence on $\mathbb{R}$. Hence, for any subsequence n_l, one can extract a further subsequence n_{l_j} such that both $\rho_1(L_{n_{l_j}, b_{l_j}}, N(0, \sigma^2)) \to 0$ and $Var(L_{n_{l_j}, b_{l_j}}) \to \sigma^2$ almost surely. By Proposition 4.4.1 it then follows that, on a set of probability one, $\rho_1(K_{n_{l_j}, b_{l_j}}, N(0, \sigma^2)) \to 0$. This proves that $\rho_1(K_{n,b}, N(0, \sigma^2)) \to 0$ in probability. ∎

Example 4.4.2 (Smooth function of the mean). In this example, we extend the result for the univariate mean to smooth functions of the multivariate mean.

Suppose $\{X_t\}$ is a sequence of multivariate random variables with common mean $\zeta \in \mathbb{R}^k$, and $k \geq 1$. Denote the joint probability law governing

4. Subsampling for Nonstationary Time Series

the sequence by P. Assume that on the basis of observing $\{X_1, \ldots, X_n\}$ we are interested in finding a confidence region for $\theta(P) \equiv f(\zeta)$, where $f(\cdot)$ is a smooth function from $\mathbb{R}^k$ to $\mathbb{R}^p$.

Let $\hat{\theta}_{n,b,t} = \hat{\theta}_b(X_t, \ldots, X_{t+b-1}) \equiv f(b^{-1}\sum_{a=t}^{t+b-1} X_a) \equiv f(\bar{X}_{n,b,t})$ be the estimator of $\theta(P)$ based on the block of size b of the consecutive data $\{X_t, \ldots, X_{t+b-1}\}$. Define $J_{b,t}(P)$ to be the sampling distribution of $b^{\frac{1}{2}}(f(\bar{X}_{n,b,t}) - \theta(P))$. Also define the corresponding analogue of a distribution function. For any Borel set $A \in \mathbb{R}^p$, let

$$J_{b,t}(x, P) = Prob_P\{b^{\frac{1}{2}}(f(\bar{X}_{n,b,t}) - \theta(P)) \in A\}. \qquad (4.16)$$

The approximation to $J_n(A, P)$ we study is defined by

$$L_{n,b}(A) = \frac{1}{n-b+1} \sum_{t=1}^{n-b+1} 1\{b^{\frac{1}{2}}(f(\bar{X}_{n,b,t}) - f(\bar{X}_n)) \in A\}. \qquad (4.17)$$

Theorem 4.4.3. *Let $\{X_t\}$ be a sequence of random vectors with common mean $\zeta \in \mathbb{R}^k$. Denote the corresponding mixing coefficients by $\alpha_X(\cdot)$. Define*

$$T_{l,t} \equiv l^{-\frac{1}{2}} \sum_{a=t}^{t+l-1} X_a \quad \text{and} \quad \Sigma_{l,t} \equiv Cov(T_{l,t}).$$

Assume the following conditions hold. For some $\delta > 0$,

- $E|X_{t,j}|^{2+\delta} \leq \Delta$ *for all t and all* $1 \leq j \leq k$, $\qquad (4.18)$
- $\Sigma_{l,t} \to \Sigma > 0$ *uniformly in t as* $l \to \infty$, $\qquad (4.19)$
- $C(4, \delta) \equiv \sum_{k=1}^{\infty}(k+1)^2 \alpha_X^{\frac{\delta}{4+\delta}}(k) \leq K.$ $\qquad (4.20)$
- $f : \mathbb{R}^k \to \mathbb{R}^p$ *is differentiable in a neighborhood of ζ.* $\qquad (4.21)$

Furthermore, assume that $b/n \to 0$ and $b \to \infty$ as $n \to \infty$, and let $J(P) = N(0, J'\Sigma J)$, where J is the $k \times p$ Jacobian matrix of f at ζ.

Then, conclusions (i–iv) of Theorem 4.3.1 are true.

Remark 4.4.5. In the case $p = 1$, $\theta(P)$ is a univariate parameter and the results of Section 4.2 apply alternatively. In particular, we construct both equal-tailed and symmetric two-sided confidence intervals for $\theta(P)$.

Proof of Theorem 4.4.3. For the proof we will assume, without loss of generality, that the true mean is equal to zero, i.e., $\zeta = 0$. The mixing condition in Theorem 4.3.1 is clearly implied by our assumption (4.20). Therefore it will be sufficient to show that Assumption 4.3.2 is satisfied.

To prove part (ii) of Assumption 4.3.2, let t_b be an index sequence that may depend on b. We will accomplish the proof in two steps.

Step 1: $T_{b,t_b} \overset{\mathcal{L}}{\Longrightarrow} N(0, \Sigma)$.

Proof: By the Cramér–Wold device, it will suffice to look at linear combinations of the type $\lambda'\Sigma^{-\frac{1}{2}}b^{-\frac{1}{2}}X_{t_b+a}$, where λ is a vector in $\mathbb{R}^p$. Without loss of generality, we may assume that λ has unit length. We observe that

i. $E(\lambda'\Sigma^{-\frac{1}{2}}X_{t_b+a}) = 0 \quad$ for $0 \le a \le b-1$,
ii. by assumption (4.18) and Minkowski, $E|\lambda'\Sigma^{-\frac{1}{2}}X_{t_b+a}|^{2+2\delta} \le \Delta^*$, where Δ^* is a constant that depends only on Δ and Σ (since $\|\lambda\| = 1$),
iii.
$$\sigma_{b,t_b}^2 \equiv Var(\lambda'\Sigma^{-\frac{1}{2}}b^{-\frac{1}{2}}\sum_{a=0}^{b-1}X_{t_b+a})$$
$$= \lambda'\Sigma^{-\frac{1}{2}}Cov(T_{b,t_b})\Sigma^{-\frac{1}{2}}\lambda$$
$$= \lambda'\Sigma^{-\frac{1}{2}}\Sigma_{b,t_b}\Sigma^{-\frac{1}{2}}\lambda \to 1 \quad \text{by assumption (4.19)}.$$

Apply Theorem B.0.1 to deduce that $\lambda'\Sigma^{-\frac{1}{2}}b^{-\frac{1}{2}}\sum_{a=0}^{b-1}X_{t_b+a} \overset{\mathcal{L}}{\Rightarrow} N(0,1)$ (to do so let $X_{n,a} = \lambda'\Sigma^{-\frac{1}{2}}X_{t_b+a-1}$ there). Since we did not specify λ, it follows from the Cramér–Wold device that $\Sigma^{-\frac{1}{2}}T_{b,t_b} \overset{\mathcal{L}}{\Rightarrow} N(0,I)$ or, equivalently, that $T_{b,t_b} \overset{\mathcal{L}}{\Rightarrow} N(0,\Sigma)$.

Step 2: $b^{\frac{1}{2}}(f(\bar{X}_{n,b,t_b}) - f(\zeta)) \overset{\mathcal{L}}{\Rightarrow} N(0, J'\Sigma J)$.

Proof: This follows immediately from Step 1, condition (4.21), and the multivariate δ-method.
Part (i) of Assumption 4.3.2 now follows trivially by letting $b = n$ and $t_b \equiv 1$ in (ii). ∎

Example 4.4.3 (Linear regression). Linear regression is a prime example where an extension of the subsampling method to the nonstationary case is of great interest. The field of econometrics is greatly concerned with inference methods for regression coefficients that are robust against both serial correlation and heteroskedasticity. The reason is that data are typically collected over time and/or exhibit nonconstant variance. Appropriate methods are often called HAC, which stands for heteroskedasticity and autocorrelation consistent. Linear regression is widely used in econometrics and HAC inference methods are deemed very important (see Hamilton, 1994, for a broad reference). We will demonstrate that the subsampling method provides an alternative HAC inference method and works under extremely weak conditions. Section 9.5.2 compares subsampling with some more standard HAC inference methods in terms of small sample behavior, using simulation studies.

Consider the linear model

$$y = X\beta + \epsilon, \qquad (4.22)$$

where y and ϵ are $n \times 1$ vectors, β is a $p \times 1$ vector and X an $n \times p$ matrix. Here X may be stochastic. The goal is to draw inference on the regression parameter β. In order to be able to apply the subsampling method we need to define subvectors and submatrices:

$$y_{b,t} \equiv (y_t, \ldots, y_{t+b-1})', \quad \epsilon_{b,t} \equiv (\epsilon_t, \ldots, \epsilon_{t+b-1})' \quad \text{and}$$

$$X_{b,t} \equiv \begin{pmatrix} x_t' \\ \vdots \\ x_{t+b-1}' \end{pmatrix}, \quad \text{where } X = \begin{pmatrix} x_1' \\ \vdots \\ x_n' \end{pmatrix}.$$

The estimator of β based on $X_{b,t}$ and $y_{b,t}$ is defined by

$$\hat{\beta}_{n,b,t} \equiv (X_{b,t}' X_{b,t})^{-1} X_{b,t}' y_{b,t} \tag{4.23}$$

Define $J_{b,t}(P)$ to be the sampling distribution of $b^{\frac{1}{2}}(\hat{\beta}_{n,b,t} - \beta)$. Also, for any Borel set $A \in \mathbb{R}^p$, let

$$J_{b,t}(A, P) = Prob_P\{b^{\frac{1}{2}}(\hat{\beta}_{n,b,t} - \beta) \in A\}. \tag{4.24}$$

The approximation to $J_n(A, P)$ we study is defined by

$$L_{n,b}(A) = \frac{1}{n-b+1} \sum_{t=1}^{n-b+1} 1\{b^{\frac{1}{2}}(\hat{\beta}_{n,b,t} - \hat{\beta}_n) \in A\}. \tag{4.25}$$

Theorem 4.4.4. *Let $\{(x_t, \epsilon_t)\}$ be a sequence of random vectors defined on a common probability space. Denote the mixing coefficients for the $\{(x_t, \epsilon_t)\}$ sequence by $\alpha(\cdot)$. Define*

$$T_{k,t} \equiv k^{-\frac{1}{2}} \sum_{a=t}^{t+k-1} x_a \epsilon_a, \quad V_{k,t} \equiv Cov(T_{k,t}), \quad \text{and} \quad M_{k,t} \equiv E(X_{k,t}' X_{k,t}/k).$$

Assume the following conditions hold. For some $\delta > 0$,

- $E(x_t \epsilon_t) = 0 \quad \text{for all } t,$ \hfill (4.26)
- $E|x_{t,j} \epsilon_t|^{2+2\delta} \leq \Delta_1 \quad \text{for all } t \text{ and all } 1 \leq j \leq p,$ \hfill (4.27)
- $E|x_{t,j}|^{4+2\delta} \leq \Delta_2 \quad \text{for all } t \text{ and all } 1 \leq j \leq p,$ \hfill (4.28)
- $V_{k,t} \to V > 0 \quad \text{uniformly in } t \text{ as } k \to \infty,$ \hfill (4.29)
- $M_{k,t} \to M > 0 \quad \text{uniformly in } t \text{ as } k \to \infty,$ \hfill (4.30)
- $C(4) \equiv \sum_{k=1}^{\infty} (k+1)^2 \alpha^{\frac{\delta}{4+\delta}}(k) \leq K.$ \hfill (4.31)

Furthermore assume that $b/n \to 0$ and $b \to \infty$ as $n \to \infty$, and let $J(P) = N(0, M^{-1}VM^{-1})$.

Then, conclusions (i–iv) of Theorem 4.3.1 are true, provided that we replace $\theta(P)$ by β, $\hat{\theta}_n$ by $\hat{\beta}_n$, and $\hat{\theta}_{n,b,t}$ by $\hat{\beta}_{n,b,t}$.

Remark 4.4.6. Regarding conditions (4.29) and (4.30), Remark 4.4.1 applies in spirit here as well.

Remark 4.4.7. Alternatively, the moving blocks bootstrap could be used for making inference on the regression parameter β. A related result concerning the validity of the moving blocks method for multivariate least squares linear regression in the context of nonstationary data was presented in Fitzenberger (1997).

Proof of Theorem 4.4.4. The mixing condition in Theorem 4.3.1 is clearly implied by our assumption (4.31). Therefore, it will be sufficient to show that Assumption 4.3.2 is satisfied.

To prove part (ii) of Assumption 4.3.2, let t_b be an index sequence that may depend on b. We will accomplish the proof in two steps.

Step 1: $b^{-\frac{1}{2}} X'_{b,t_b} \epsilon_{b,t_b} \overset{\mathcal{L}}{\Longrightarrow} N(0,V)$.

Proof: By the Cramér–Wold device, it will suffice to look at linear combinations of the type $\lambda' V^{-\frac{1}{2}} b^{-\frac{1}{2}} X'_{b,t_b} \epsilon_{b,t_b}$, where λ is a vector in $\mathbb{R}^p$. Without loss of generality, we may assume that λ has unit length. We observe that

i. $E(\lambda' V^{-\frac{1}{2}} x_{t_b+a} \epsilon_{t_b+a}) = 0$ for $0 \leq a \leq b-1$,

ii. by assumption (4.27) and Minkowski's inequality, we have that

$$E|\lambda' V^{-\frac{1}{2}} x_{t_b+a} \epsilon_{t_a+t}|^{2+2\delta} \leq \Delta_1^*,$$

where Δ_1^* is a constant that depends only on Δ_1 and V (since $\|\lambda\| = 1$),

iii.

$$\sigma_{b,t_b}^2 \equiv Var(\lambda' V^{-\frac{1}{2}} b^{-\frac{1}{2}} \sum_{a=0}^{b-1} x_{t_b+a} \epsilon_{t_b+a})$$

$$= \lambda' V^{-\frac{1}{2}} Cov(b^{-\frac{1}{2}} \sum_{a=0}^{b-1} x_{t_b+a} \epsilon_{t_b+a}) V^{-\frac{1}{2}} \lambda$$

$$= \lambda' V^{-\frac{1}{2}} V_{b,t_b} V^{-\frac{1}{2}} \lambda \to 1 \quad \text{by assumption (4.29)}.$$

Deduce from Theorem B.0.1 that $\lambda' V^{-\frac{1}{2}} b^{-\frac{1}{2}} \sum_{a=0}^{b-1} x_{t_b+a} \epsilon_{t_b+a} \overset{\mathcal{L}}{\Longrightarrow} N(0,1)$ (to do so let $X_{n,a} = \lambda' V^{-\frac{1}{2}} x_{t_b+a-1} \epsilon_{t_b+a-1}$ there). Since we did not specify λ, it follows that $V^{-\frac{1}{2}} b^{-\frac{1}{2}} X'_{b,t_b} \epsilon_{b,t_b} \overset{\mathcal{L}}{\Longrightarrow} N(0,I)$ by the Cramér–Wold device or, equivalently, that $b^{-\frac{1}{2}} X'_{b,t_b} \epsilon_{b,t_b} \overset{\mathcal{L}}{\Longrightarrow} N(0,V)$.

Step 2: $b^{\frac{1}{2}}(\hat{\beta}_{n,b,t_b} - \beta) \overset{\mathcal{L}}{\Longrightarrow} N(0, M^{-1}VM^{-1})$.

Proof: $b^{\frac{1}{2}}(\hat{\beta}_{n,b,t_b} - \beta) = (X'_{b,t_b} X_{b,t_b}/b)^{-1} b^{-\frac{1}{2}} X'_{b,t_b} \epsilon_{b,t_b}$. By the imposed mixing and moment conditions on the sequence $\{x_t\}$, $(X'_{b,t_b} X_{b,t_b}/b) - M_{b,t_b} \to 0$ in probability. In fact, denote the (i,j)-th entry of $[(X'_{b,t_b} X_{b,t_b}/b) - M_{b,t_b}]$ by $(D_b)_{i,j}$. Then,

$$(D_b)_{i,j} = \frac{1}{b} \sum_{l=0}^{b-1} x_{t_b+l,i} x_{t_b+l,j} - E(x_{t_b+l,i} x_{t_b+l,j}).$$

By definition, $E((D_b)_{i,j}) = 0$. Therefore, it will be sufficient to check that $Var((D_b)_{i,j}) \to 0$. Define

$$s_{b,d} = \frac{1}{b} \sum_{a=t_b}^{t_b+b-(d+1)} Cov(x_{a,i} x_{a,j}, x_{a+d,i} x_{a+d,j}).$$

Then,

$$Var((D)_{i,j}) = \frac{1}{b}(s_{b,0} + 2 \sum_{d=1}^{b-1} s_{b,d}).$$

By Lemma A.0.1,

$$Cov(x_{a,i} x_{a,j}, x_{a+d,i} x_{a+d,j}) \leq 8 \|x_{a,i} x_{a,j}\|_{2+\delta} \|x_{a+d,i} x_{a+d,j}\|_{2+\delta} \alpha(d)^{\frac{\delta}{2+\delta}}$$

$$\leq 8 \Delta_2^{\frac{2}{2+\delta}} \alpha(d)^{\frac{\delta}{2+\delta}} \quad \text{(by assumption (4.29))}.$$

Therefore, $s_{b,d} \leq 8 \Delta^{\frac{2}{2+\delta}} \alpha(d)^{\frac{\delta}{2+\delta}}$ and that implies

$$Var((D)_{i,j}) \leq \frac{16}{b} \Delta^{\frac{2}{2+\delta}} \sum_{d=0}^{b-1} \alpha(d)^{\frac{\delta}{2+\delta}} \to 0.$$

Recalling that $M_{b,t_b} \to M$ by assumption (4.30), we can can conclude that $(X'_{b,t_b} X_{b,t_b}/b) \to M$ in probability. By Step 1, $b^{-\frac{1}{2}} X'_{b,t_b} \epsilon_{b,t_b}$ is $O_P(1)$ and therefore

$$[(X'_{b,t_b} X_{b,t_b}/b)^{-1} - M^{-1}] b^{-\frac{1}{2}} X'_{b,t_b} \epsilon_{b,t_b} \to 0 \quad \text{in probability}.$$

But $M^{-1} b^{-\frac{1}{2}} X'_{b,t_b} \epsilon_{b,t_b} \xrightarrow{\mathcal{L}} N(0, M^{-1} V M^{-1})$ by Step 1 again and this completes the proof of part (ii) of Assumption 4.3.2.

Part (i) of Assumption 4.3.2 now follows trivially by letting $b = n$ and $t_b \equiv 1$ in (ii). ∎

Remark 4.4.8. Obviously, a theorem analogous to Theorem 4.4.3 allows us to make inference for a smooth function of β. The details are left to the reader.

Example 4.4.4 (Variance ratio and robust statistics). Variance ratios and robust statistics computed from time series were covered in Section 3.4. It is no surprise that these applications can also be extended to the nonstationary case.

Lo and MacKinley (1988) discuss variance ratio tests under the null hypothesis of uncorrelated and possibly heteroskedastic increments ϵ_t. More specifically,

$$H_0 : E(\epsilon_t \epsilon_{t-\tau}) = 0 \text{ for all } t, \tau \neq 0 \quad \text{and} \quad \lim_{n \to \infty} \frac{1}{n} \sum_{t=1}^{n} Var(\epsilon_t) = \sigma^2 < \infty.$$

(4.32)

The Lo and MacKinley test, which was introduced in Subsection 3.4.6, is robust to heteroskedasticity (see their 1988 paper). Of course, the problem of potential asymptotic correlations of the sample autocorrelations remains. On the other hand, the subsampling methodology also can handle heteroskedastic increments, but is robust against asymptotic correlations. Again, it is based on carrying out the test implicitly by computing a confidence interval for the q-period variance ratio $VR(q)$ and checking whether unity is contained in it. The extra conditions and notations necessary for the heteroskedastic case are comparable to the ones for the applications discussed previously in this section.

The analogy carries over to robust statistics computed from time series, such as the median, trimmed means, and the Hodges–Lehmann estimator. By imposing some sort of approximate stationarity conditions, Theorem 4.2.1 allows for the construction of approximate confidence intervals without having to estimate any limiting distributions.

4.5 Hypothesis Testing and Data-Dependent Choice of Block Size

Also in the nonstationary setup, there are many testing applications where the null hypothesis does not translate into a null hypothesis on a parameter. The method we propose is identical to the one described in Section 3.5 and a theorem generalizing Theorem 3.5.1 can be stated and proved, in the same way as Theorem 4.2.1 generalizes Theorem 3.2.1. The details are left to the reader.

Moreover, Theorem 2.7.1 can be generalized to the nonstationary time series case as well. The necessary modifications are analogous to those outlined in Section 3.6, and, again, the details are left to the reader.

4.6 Variance Estimation

Similarly to the estimation of a sampling distribution function, the problem of estimating the variance of a general statistic can be extended to nonstationary observations. As in Section 3.8, we consider the Carlstein and the subsampling variance estimators, using nonoverlapping and overlapping blocks, respectively.

For variance estimation to be meaningful, we require a limiting variance of the properly standardized estimator, that is,

$$\lim_n Var(\tau_n \hat{\theta}_n) = \sigma^2 \in (0, \infty).$$

The estimators $\hat{\sigma}^2_{n,CARL}$ and $\hat{\sigma}^2_{n,SUB}$ are defined in the same way as in Section 3.8.

Fukuchi (1997) proved L_2 consistency of both estimators. In the following lemma, denote $T_n = \tau_n(\hat{\theta}_n - E(\hat{\theta}_n))$ and $q = n - b + 1$. Recall that U.I. stands for uniformly integrable.

Lemma 4.6.1. *Let $\{X_t\}$ be a sequence of random variables defined on a common probability space. Denote the corresponding mixing coefficients by $\alpha_X(\cdot)$. Assume that $\tau_b/\tau_n \to 0$, $b/n \to 0$ and $b \to \infty$ as $n \to \infty$. Also, assume that*

- $\{(T_n)^4\}$ *are U.I.*, (4.33)

- $\dfrac{1}{q}\sum_{t=1}^{q} Var(\tau_b \hat{\theta}_{n,b,t}) \to \sigma^2$, (4.34)

- $\alpha_X(m) \to 0$ *as* $m \to \infty$. (4.35)

Then, both $\hat{\sigma}^2_{n,CARL}$ and $\hat{\sigma}^2_{n,SUB}$ converge to σ^2 in L_2 norm.

Proof. The result follows from Theorem 1 of Fukuchi (1997). Fukuchi also proved almost sure consistency of both variance estimators under stronger moment and mixing conditions.

Example 4.6.1 (MCMC simulation). Consider the setup where the observed sequence $\{X_1, X_2, \ldots, X_n\}$ is a sample path from a real-valued Markov chain that possesses a unique stationary distribution with mean $\theta(P)$. Note that the infinite sequence $\{X_1, X_2, \ldots\}$ will *not* be stationary unless X_1 is generated according to this unique stationary distribution; in general, the sequence $\{X_1, X_2, \ldots\}$ will only be asymptotically stationary.

Asymptotically stationary sequences of the above type are typically found in Markov chain Monte Carlo (MCMC) simulation experiments that are nowadays routinely used in the practice of Bayesian statistics (see, e.g., Berger, 1993). The objective here too is to use the sample mean $\bar{X}_n$ to consistently estimate $\theta(P)$, and the question that arises in such a context is "has the simulation (and/or our estimator $\bar{X}_n$) converged?"

Subsampling can be used to help answer this question. Notice that in the context of a simulation, the sequence $\{X_1, X_2, \ldots, X_n\}$ can be constructed for different values of n, say, $n_1 < n_2 < \cdots$. The estimator $\hat{\sigma}_{n_i,SUB}$ can be calculated for each n_i (using a specific recipe for choosing the block size b_i, e.g., $b_i = \lfloor n_i^{1/3} \rfloor$, $i = 1, 2, \ldots$). Therefore, a very simple way of addressing the convergence issue is to take n equal to the smallest n_i such that a 95% (say) confidence interval for $\theta(P)$ of the type $\bar{X}_{n_i} \pm 1.96 * \hat{\sigma}_{n_i,SUB}/\sqrt{n_i}$ has length that is less than or equal to some desired level of accuracy, say, 0.00001. To avoid a fluke, one can further require that the length of the 95% confidence interval *remain* under 0.00001 even for larger sample sizes; this is equivalent to taking n to be equal to the second (or third, etc.) smallest n_i corresponding to 95% confidence interval with length under

0.00001. More elaborate ways of addressing the same convergence question may be found in Giakoumatos et al. (1999).

4.7 Conclusions

In this chapter, the theory of the previous chapter was extended to nonstationary observations. These extensions are useful, since in many empirical applications, the hypothesis of stationarity can easily be rejected. The goal was to show that the subsampling method derived for dependent but stationary observations can still be employed, even if the assumption of stationarity is violated. The main theorems were reproved under weaker assumptions, allowing for considerable local heteroscedasticity. Also, some of the previous examples were reconsidered, relaxing the condition of stationarity.

5
Subsampling for Random Fields

5.1 Introduction and Definitions

Suppose $\{X(\mathbf{t}), \mathbf{t} \in \mathbb{G}^d\}$ is a random field in d dimensions, with $d \in \mathbb{Z}_+$; that is, $\{X(\mathbf{t}), \mathbf{t} \in \mathbb{G}^d\}$ is a collection of random variables $X(\mathbf{t})$ taking values in a state space S, defined on a probability space $(\Omega, \mathcal{A}, P)$, and indexed by the variable $\mathbf{t} \in \mathbb{G}^d$. Throughout this chapter, $\mathbb{G}$ will stand for either the set of real numbers $\mathbb{R}$, or the set of integers $\mathbb{Z}$; thus, the random field $\{X(\mathbf{t})\}$ is allowed to "run" in either continuous or discrete "time." Similarly, $\mathbb{G}_+$ will denote $\mathbb{R}_+$ or $\mathbb{Z}_+$ (the sets of positive real numbers and integers, respectively) according to whether $\mathbb{G} = \mathbb{R}$ or $\mathbb{G} = \mathbb{Z}$.

The random field $\{X(\mathbf{t})\}$ is assumed to be *homogeneous*, meaning that for any set $E \subset \mathbb{G}^d$, and for any point $\mathbf{i} \in \mathbb{G}^d$, the joint distribution of the random variables $\{X(\mathbf{t}), \mathbf{t} \in E\}$ is identical to the joint distribution of $\{X(\mathbf{t}), \mathbf{t} \in E+\mathbf{i}\}$. In the special case where $d = 1$, the random field $\{X(\mathbf{t})\}$ reduces to either a stationary continuous-time stochastic process, or a stationary time series, depending on whether $\mathbb{G} = \mathbb{R}$ or $\mathbb{G} = \mathbb{Z}$, respectively.

Here, and in the following chapter, we denote a point $\mathbf{t}$ in $\mathbb{G}^d$ by the row vector $\mathbf{t} = (t_1, \ldots, t_d)$. For two points $\mathbf{t} = (t_1, \ldots, t_d)$ and $\mathbf{u} = (u_1, \ldots, u_d)$ in $\mathbb{G}^d$, define the sup-distance in $\mathbb{G}^d$ by $\rho(\mathbf{t}, \mathbf{u}) = \sup_j |t_j - u_j|$, and for two sets E_1, E_2 in $\mathbb{G}^d$, define $\rho(E_1, E_2) = \inf\{\rho(\mathbf{t}, \mathbf{u}) : \mathbf{t} \in E_1, \mathbf{u} \in E_2\}$. Let $\mathbf{u}$ be in $\mathbb{G}_+^d$; here, and throughout this chapter, $E_\mathbf{u}$ will denote the rectangle consisting of the points $\mathbf{t} = (t_1, t_2, \ldots, t_d) \in \mathbb{G}^d$ such that $0 < t_k \leq u_k$, for

$k = 1, 2, \ldots, d$. In other words,

$$E_{\mathbf{u}} = \{\mathbf{t} \in \mathbb{G}^d : 0 < t_k \leq u_k, \text{ for } k = 1, 2, \ldots, d\}. \tag{5.1}$$

Note that $E_{\mathbf{u}}$ is either a solid (i.e., convex) "parallelopiped" (case $\mathbb{G} = \mathbb{R}$), or a rectangular lattice (case $\mathbb{G} = \mathbb{Z}$).

Our goal again is to construct a confidence region for a real or vector parameter $\theta(P)$, on the basis of observing the "stretch" of data $\{X(\mathbf{t}), \mathbf{t} \in E_{\mathbf{n}}\}$, where $\mathbf{n} = (n_1, n_2, \ldots, n_d) \in \mathbb{G}^d$. The sample size is again denoted by n, although now $n \equiv \prod_{i=1}^{d} n_i = |E_{\mathbf{n}}|$, where $|E|$ denotes either the cardinality of the set E (case $\mathbb{G} = \mathbb{Z}$), or Lebesgue measure (i.e., volume) of the set E (case $\mathbb{G} = \mathbb{R}$). Note that in the case $\mathbb{G} = \mathbb{R}$ we could have equally defined our "observation region" $E_{\mathbf{n}}$ as a closed set, e.g., by taking the closure of the set defined in (5.1); we chose the definition (5.1) because it allows us to treat the two cases ($\mathbb{G} = \mathbb{R}$ or $\mathbb{Z}$) simultaneously.

In Section 5.2, we introduce a useful variation of the well-known strong mixing coefficients of Rosenblatt (1956). Subsequently, the asymptotic validity of subsampling for random fields is shown in Section 5.3 under a mixing condition that is easy to satisfy in practice; both univariate and vector parameters will be considered, as well as asymptotics where the sample is assumed to "expand" to infinity in some (but not necessarily all) out of the d possible directions. The issues of variance estimation and bias reduction via subsampling are discussed in Section 5.4. Consistency of subsampling with maximum overlap in the case of continuous-time random fields is demonstrated in Section 5.5, while some illustrative examples are given in Section 5.6.

Finally, the case where the data is of the form $\{X(\mathbf{t}), \mathbf{t} \in E\}$, where the set E has a general shape (not necessarily rectangular like $E_{\mathbf{n}}$), but is still assumed to "expand" as the sample size increases, will be addressed in Chapter 6.

5.2 Some Useful Notions of Strong Mixing for Random Fields

The random field $\{X(\mathbf{t})\}$ will be assumed to satisfy a certain weak dependence condition. Following the recent literature on mixing for random fields (see, e.g., Doukhan, 1994), we define a collection of strong mixing coefficients by

$$\alpha_X(k; l_1, l_2) \equiv \sup_{E_1, E_2 \subset \mathbb{G}^d} \{|P(A_1 \cap A_2) - P(A_1)P(A_2)| :$$

$$A_i \in \mathcal{F}(E_i), \ |E_i| \leq l_i, i = 1, 2, \ \rho(E_1, E_2) \geq k\}, \tag{5.2}$$

where $\mathcal{F}(E)$ is the σ-algebra generated by $\{X(\mathbf{t}), \mathbf{t} \in E\}$. A weak dependence condition is formulated if $\alpha_X(k; l_1, l_2)$ is assumed to converge to zero at some rate, as k tends to infinity, and l_1, l_2 either remain fixed or tend to infinity as well. It is interesting to note that $\alpha_X(k; l_1, l_2)$ is increasing in each of its arguments l_1, l_2 separately. In particular, if we let $\alpha_X(k) = \alpha_X(k; \infty, \infty)$ denote the usual strong mixing coefficients of Rosenblatt (1956, 1985), it is apparent that $\alpha_X(k; l_1, l_2) \leq \alpha_X(k)$. If $\alpha_X(k) \to 0$ as $k \to \infty$, then the random field $\{X(\mathbf{t})\}$ is simply said to be "strong mixing."

Like the $d = 1$ case (of a stationary sequence or stationary stochastic process in continuous time), there are many examples of strong mixing random fields in the case $d > 1$ as well, e.g., Gaussian fields with continuous and strictly positive spectral density function (cf. Ibragimov and Rozanov, 1978, and Rosenblatt, 1985). However, in the case $d > 1$, some random fields of interest, (the so-called Gibbs states or Markov field models), are not necessarily strong mixing (cf. Dobrushin, 1968, for an example), but they do satisfy weak dependence conditions involving the general $\alpha_X(k; l_1, l_2)$ coefficients (see also Bolthausen, 1982; Zhurbenko, 1986; Bradley, 1991, 1992; and Doukhan, 1994).

We now introduce some slightly weaker notions of mixing coefficients for random fields that will be sufficient for most of our purposes. So let

$$\alpha_X(k; l_1) \equiv \sup_{\substack{E_2 = E_1 + \mathbf{t} \\ E_1 \subset \mathbb{G}^d, \mathbf{t} \in \mathbb{G}^d}} \{|P(A_1 \cap A_2) - P(A_1)P(A_2)| :$$

$$A_i \in \mathcal{F}(E_i), i = 1, 2, |E_1| \leq l_1, \rho(E_1, E_2) \geq k\}. \quad (5.3)$$

Note that, by definition, $\alpha_X(k; l_1)$ and $\alpha_X(k; l_1, l_1)$ are different entities; however, the only difference between them is that in the calculation of $\alpha_X(k; l_1)$, the sets E_1, E_2 are restricted to have the same shape, i.e., one is just a translate of the other. No such restriction exists for the calculation of $\alpha_X(k; l_1, l_1)$; consequently, $\alpha_X(k; l_1) \leq \alpha_X(k; l_1, l_1)$. Thus, a mixing condition based on the coefficients $\alpha_X(k; l_1)$ is easier to satisfy than a mixing condition based on the coefficients $\alpha_X(k; l_1, l_1)$.

An even weaker form of mixing coefficients is obtained if we further insist that the set E_1 in (5.3) be a parallelopiped of type (5.1); thus, let

$$\hat{\alpha}_X(k; l_1) \equiv \sup_{\substack{E_1 = E_\mathbf{u}, E_2 = E_1 + \mathbf{t} \\ \mathbf{t} \in \mathbb{G}^d, \mathbf{u} \in \mathbb{G}^d_+}} \{|P(A_1 \cap A_2) - P(A_1)P(A_2)| : i = 1, 2,$$

$$A_i \in \mathcal{F}(E_i), |E_1| \leq l_1, \rho(E_1, E_2) \geq k\}. \quad (5.4)$$

Note again that $\hat{\alpha}_X(k; l_1) \leq \alpha_X(k; l_1)$, so a mixing condition based on the coefficients $\hat{\alpha}_X(k; l_1)$ is easier to satisfy.

5.3 Consistency of Subsampling for Random Fields

5.3.1 Univariate Parameter Case

Let $\hat{\theta}_\mathbf{n} = \hat{\theta}_\mathbf{n}(X(\mathbf{t}), \mathbf{t} \in E_\mathbf{n})$ be a real-valued statistic employed to consistently estimate the unknown real-valued parameter $\theta(P)$. Let $J_\mathbf{n}(P)$ be the sampling probability law of $\tau_\mathbf{n}(\hat{\theta}_\mathbf{n} - \theta(P))$, and

$$J_\mathbf{n}(x, P) = Prob_P\{\tau_\mathbf{n}(\hat{\theta}_\mathbf{n} - \theta(P)) \leq x\}$$

be the corresponding sampling probability distribution function. Again, the only assumption that will be needed for consistency of subsampling distribution estimation is existence of an asymptotic weak limit for $J_\mathbf{n}(P)$, which is stated below.

Assumption 5.3.1. *$J_\mathbf{n}(P)$ converges weakly to a limit law $J(P)$ with corresponding distribution function $J(x, P)$, as $n_i \to \infty$ for $i = 1, \ldots, d$.*

Define the block $Y_\mathbf{j} = \{X(\mathbf{t}), \mathbf{t} \in E_{\mathbf{j},\mathbf{b},\mathbf{h}}\}$, where $\mathbf{j} = (j_1, j_2, \ldots, j_d)$ and $E_{\mathbf{j},\mathbf{b},\mathbf{h}}$ is the smaller (and displaced) rectangle consisting of the points $\mathbf{i} = (i_1, i_2, \ldots, i_d) \in \mathbb{G}^d$ such that $(j_k - 1)h_k < i_k \leq (j_k - 1)h_k + b_k$, for $k = 1, 2, \ldots, d$; here $\mathbf{b} = (b_1, \ldots, b_d)$ and $\mathbf{h} = (h_1, \ldots, h_d)$ are points in $\mathbb{G}_+^d$ that depend in general on $\mathbf{n}$. The point $\mathbf{b}$ indicates the shape and size of rectangle $E_{\mathbf{i},\mathbf{b},\mathbf{h}}$, and the point $\mathbf{h}$ indicates the amount of "overlap" between the rectangles $E_{\mathbf{i},\mathbf{b},\mathbf{h}}$ for neighboring $\mathbf{i}$'s, i.e., the size of their intersection. For example, if $\mathbf{h} = \mathbf{b}$, there is no overlap between $E_{\mathbf{i},\mathbf{b},\mathbf{h}}$ and $E_{\mathbf{j},\mathbf{b},\mathbf{h}}$ for $\mathbf{i} \neq \mathbf{j}$, while the case $\mathbf{h} = (1, 1, \ldots, 1)$ yields the maximum possible overlap if we are in the discrete case where $\mathbb{G} = \mathbb{Z}$; the maximum overlap case in continuous time ($\mathbb{G} = \mathbb{R}$ case) can only be achieved in the limit as $\mathbf{h} \to \mathbf{0}$, and will be addressed separately in Section 5.5. In what follows, it will generally be assumed that either $\mathbf{h}$ is a (nonzero) constant, or that as $b_i \to \infty$, $h_i/b_i \to a_i \in (0, 1]$, for $i = 1, 2, \ldots, d$.

For notational ease, denote $b = \prod_{i=1}^{d} b_i$, and $h = \prod_{i=1}^{d} h_i$, and observe that, with $E_\mathbf{n}$ and $\mathbf{n}$ fixed, $Y_\mathbf{j}$ is defined only for $\mathbf{j}$ such that $0 < j_k \leq q_k$, where $q_k = \lfloor \frac{n_k - b_k}{h_k} \rfloor + 1$; thus, the total number of the $Y_\mathbf{j}$ blocks available from the data is $q = \prod_{i=1}^{d} q_i$.

Let the subsample value $\hat{\theta}_{\mathbf{n},\mathbf{b},\mathbf{i}}$ be equal to the statistic $\hat{\theta}_\mathbf{b}$ evaluated at the data set $Y_\mathbf{i}$, i.e., $\hat{\theta}_{\mathbf{n},\mathbf{b},\mathbf{i}} = \hat{\theta}_\mathbf{b}(Y_\mathbf{i})$. The subsampling approximation to $J_\mathbf{n}(x, P)$ is now defined by

$$L_{\mathbf{n},\mathbf{b}}(x) = q^{-1} \sum_{i_1=1}^{q_1} \sum_{i_2=1}^{q_2} \cdots \sum_{i_d=1}^{q_d} 1\{\tau_\mathbf{b}(\hat{\theta}_{\mathbf{n},\mathbf{b},\mathbf{i}} - \hat{\theta}_\mathbf{n}) \leq x\}. \tag{5.5}$$

The following general theorem on consistency of subsampling distribution estimation was originally proved in Politis and Romano (1992c, 1994b) in the case $\mathbb{G} = \mathbb{Z}$ using the mixing coefficients $\alpha_X(k; l_1, l_2)$; we state it below using the weaker notion of the $\hat{\alpha}_X(k; l_1)$ coefficients, and for $\mathbb{G}$ being either $\mathbb{Z}$ or $\mathbb{R}$.

Theorem 5.3.1. *Assume Assumption 5.3.1, and that $\tau_\mathbf{b}/\tau_\mathbf{n} \to 0$, $b_i \to \infty$, and $n_i \to \infty$ for $i = 1, 2, \ldots, d$. Assume that $\mathbf{h}$ is either a (nonzero) constant, or that $h_i/b_i \to a_i \in (0, 1]$, for $i = 1, 2, \ldots, d$. Also assume that $\prod_{j=1}^d b_j/(n_j - b_j) \to 0$, and that $q^{-1} \sum_{k=1}^{q^*} k^{d-1} \hat{\alpha}_X(k; b) \to 0$, where $q^* = \max_i q_i$.*

i. *If x is a continuity point of $J(\cdot, P)$, then $L_{\mathbf{n},\mathbf{b}}(x) \to J(x, P)$ in probability.*
ii. *If $J(\cdot, P)$ is continuous, then $\sup_x |L_{\mathbf{n},\mathbf{b}}(x) - J_\mathbf{n}(x, P)| \to 0$ in probability, where $J_\mathbf{n}(x, P) = \text{Prob}_P\{\tau_\mathbf{n}(\hat{\theta}_\mathbf{n} - \theta(P)) \leq x\}$.*
iii. *Let*

$$c_{\mathbf{n},\mathbf{b}}(1 - \alpha) = \inf\{x : L_{\mathbf{n},\mathbf{b}}(x) \geq 1 - \alpha\}.$$

Correspondingly, define

$$c(1 - \alpha, P) = \inf\{x : J(x, P) \geq 1 - \alpha\}.$$

If $J(\cdot, P)$ is continuous at $c(1 - \alpha, P)$, then

$$\text{Prob}_P\{\tau_\mathbf{n}(\hat{\theta}_\mathbf{n} - \theta(P)) \leq c_{\mathbf{n},\mathbf{b}}(1 - \alpha)\} \to 1 - \alpha.$$

Thus, the asymptotic coverage probability under P of the interval $[\hat{\theta}_\mathbf{n} - \tau_\mathbf{n}^{-1} c_{\mathbf{n},\mathbf{b}}(1 - \alpha), \infty)$ is the nominal level $1 - \alpha$.

Proof. In what follows, $c_0, c_1, c_2, \ldots$ will denote some positive constants. As in the proof of Theorem 3.2.1, to prove (i) it suffices to show that $U_\mathbf{n}(x)$ converges in probability to $J(x, P)$, where

$$U_\mathbf{n}(x) = q^{-1} \sum_{i_1=1}^{q_1} \sum_{i_2=1}^{q_2} \cdots \sum_{i_d=1}^{q_d} 1\{\tau_\mathbf{b}(\hat{\theta}_{\mathbf{n},\mathbf{b},\mathbf{i}} - \theta(P)) \leq x\}.$$

Since $EU_\mathbf{n}(x) = J_\mathbf{b}(x, P)$, and $J_\mathbf{b}(x, P) \to J(x, P)$ as $b_i \to \infty$, for $i = 1, 2, \ldots, d$, (by Assumption 5.3.1), it suffices to look at $Var(U_\mathbf{n}(x))$. By the homogeneity of the random field $\{X(\mathbf{t})\}$, it follows that

$$Var(U_\mathbf{n}(x)) = q^{-1} \sum_{i_1=-q_1}^{q_1} \sum_{i_2=-q_2}^{q_2} \cdots \sum_{i_d=-q_d}^{q_d} (1 - \frac{|i_1|}{q_1})(1 - \frac{|i_2|}{q_2}) \cdots (1 - \frac{|i_d|}{q_d}) C(\mathbf{i}),$$

where $C(\mathbf{i})$ denotes the covariance between $1\{\tau_\mathbf{b}(\hat{\theta}_{\mathbf{n},\mathbf{b},\mathbf{1}} - \theta(P)) \leq x\}$ and $1\{\tau_\mathbf{b}(\hat{\theta}_{\mathbf{n},\mathbf{b},\mathbf{1}+\mathbf{i}} - \theta(P)) \leq x\}$; note that $C(\mathbf{i}) = C(-\mathbf{i})$. Let

$$E_q = \{\mathbf{i} \in \mathbb{G}^d : |i_j| \leq q_j, j = 1, 2, \ldots, d\}$$

and

$$E^* = \{\mathbf{i} \in \mathbb{G}^d : |i_j| \leq \lfloor b_j/h_j \rfloor, j = 1, 2, \ldots, d\},$$

5.3 Consistency of Subsampling for Random Fields

where $\lfloor \cdot \rfloor$ is the integer part. Then $Var(U_{\mathbf{n}}(x)) = A^* + A$, where

$$A^* = q^{-1} \sum_{\mathbf{i} \in E^*} (1 - \frac{|i_1|}{q_1})(1 - \frac{|i_2|}{q_2}) \cdots (1 - \frac{|i_d|}{q_d}) C(\mathbf{i}),$$

and

$$A = q^{-1} \sum_{\mathbf{i} \in E_q - E^*} (1 - \frac{|i_1|}{q_1})(1 - \frac{|i_2|}{q_2}) \cdots (1 - \frac{|i_d|}{q_d}) C(\mathbf{i}).$$

Looking at A^* it is seen that it is a sum of $\prod_{j=1}^{d} (2\lfloor b_j/h_j \rfloor + 1) \sim 2b/h$ terms of order $O(q^{-1})$; since $q = \prod_{j=1}^{d} (\lfloor \frac{n_j - b_j}{h_j} \rfloor + 1) \sim \prod_{j=1}^{d} (n_j - b_j)/h_j$, it follows that $|A^*| = O(\prod_{j=1}^{d} b_j/(n_j - b_j))$.

Now, by the well-known mixing inequality for the covariance between two bounded random variables (cf. Roussas and Ioannides, 1987), $|C(\mathbf{i})| \le c_0 \hat{\alpha}_X (i^* h^* - b^*; b)$, where $i^* = \max_k |i_k|$, $b^* = \max_i b_i$, and $h^* = \min_i h_i$. Therefore,

$$|A| \le c_0 q^{-1} \sum_{\mathbf{i} \in E_q - E^*} \hat{\alpha}_X (i^* h^* - b^*; b)$$

$$\le c_1 q^{-1} d \sum_{k=\lfloor b^*/h^* \rfloor + 1}^{q^*} W(k) \hat{\alpha}_X (k h^* - b^*; b),$$

where $W(k)$ represents either the cardinality or the volume (cases $\mathbb{G} = \mathbb{Z}$ or $\mathbb{G} = \mathbb{R}$, respectively) of the set

$$\{\mathbf{i} \in \mathbb{G}^d : i_1 = k, 0 < i_j \le i_1, j = 2, \ldots, d\}.$$

By an easy argument it now follows that $W(k) = O(k^{d-1})$, and therefore,

$$|A| \le c_2 \frac{1}{q} \sum_{k=\lfloor b^*/h^* \rfloor + 1}^{q^*} k^{d-1} \hat{\alpha}_X (k h^* - b^*; b).$$

It is obvious that by the imposed conditions both terms above converge to zero, and hence $Var(U_{\mathbf{n}}(x)) \to 0$, which completes the proof of (i).

The proofs of (ii) and (iii) are now exactly analogous to the proof of Theorem 3.2.1. ∎

It should be noted that, using stronger mixing assumptions, the fourth moments of $L_{\mathbf{n},\mathbf{b}}(x)$ could be appropriately bounded and convergence with probability one, that is, conclusion (iv) of Theorem 2.2.1 would hold here, too. Since, however, our emphasis is on obtaining asymptotically valid confidence regions under minimal assumptions, this approach will not be pursued.

In practice, since one gets to choose the design parameters $\mathbf{b}$ and $\mathbf{h}$ as functions of the given sample size, a realistic set of conditions would satisfy

$b_i \to \infty$, with $b_i/n_i \to 0$, as $n_i \to \infty$, and either **h** is a constant, or $h_i/b_i \to a_i \in (0,1]$, for $i = 1, 2, \ldots, d$. In the case where **h** is constant, the statement of the theorem simplifies and the following corollary is true.

Corollary 5.3.1. *Assume Assumption 5.3.1, and that $\tau_\mathbf{b}/\tau_\mathbf{n} \to 0$, $b_i \to \infty$, and $b_i/n_i \to 0$, as $n_i \to \infty$, for $i = 1, 2, \ldots, d$. Also assume that **h** is constant, and that $n^{-1} \sum_{k=1}^{n^*} k^{d-1} \hat{\alpha}_X(k; b) \to 0$, where $n^* = \max_i n_i$.*

Then, conclusions (i–iii) of Theorem 5.3.1 remain true.

Remark 5.3.1. It is easy to see that if the random field is actually strong mixing, then a sufficient weak dependence condition for Corollary 5.3.1 to hold is that $k^{d-1} \alpha_X(k) \to 0$ as $k \to \infty$. For the case $d > 1$, a sufficient condition is that $k^{d-1} \alpha_X(k)$ converges to some finite number as $k \to \infty$, and for the important special case of a time series or a stochastic process on the real line ($d = 1$), this sufficient condition is satisfied under the minimal assumption of strong mixing, i.e., that $\alpha_X(k) \to 0$ as $k \to \infty$.

Remark 5.3.2. Theorem 5.3.1 can also be extended to studentized roots as in Section 2.5.1; the details are obvious and are omitted. Nevertheless, the issue of constructing *accurate* variance estimators from dependent data is not trivial and will be revisited in Section 10.5.5.

Remark 5.3.3. It is also very important to note that subsampling for random fields is *not* limited to rectangular observation regions of the type $E_\mathbf{n}$. If $\mathbb{G} = \mathbb{R}$, the observation region can be any convex $K \subset \mathbb{R}^d$, and the asymptotics take place as K "expands" uniformly in all directions; if $\mathbb{G} = \mathbb{Z}$, then again we consider a convex $K \subset \mathbb{R}^d$ that expands in all directions, and our observation region is $K \cap \mathbb{Z}^d$. In case $\mathbb{G} = \mathbb{R}$, the subsample values of our statistic are recalculated over smaller subsets of K that are exact replicas of K at smaller size; in case $\mathbb{G} = \mathbb{Z}$, the intersection of the small replicas of K with $\mathbb{Z}^d$ yields the new "blocks" over which the subsample values of our statistic are recalculated (see, e.g., Sherman and Carlstein, 1996). We omit the details here as a complete description of the procedure outlined above is given in Chapter 6 in the context of random sampling.

Nevertheless, there is an interesting extension of Theorem 5.3.1 or Corollary 5.3.1 that can be easily addressed only in the setup of a rectangular observation region $E_\mathbf{n}$; this extension involves the case where the observation region can *not* be assumed to expand uniformly in all directions of d-dimensional space. So suppose that instead of having a limit theorem where $n_i \to \infty$, for $i = 1, \ldots, d$, we have a modified version of Assumption 5.3.1 that reads as follows.

Assumption 5.3.2. *$J_\mathbf{n}(P)$ converges weakly to a limit law $J(P)$, with corresponding distribution function $J(\cdot, P)$, as $n_i \to \infty$, for $i = 1, \ldots, d^*$, and $n_j \to Q_j$, for $j = d^* + 1, \ldots, d$, where $1 \leq d^* \leq d$, and the Q_j's are some fixed numbers in $\mathbb{G}_+$.*

This notation allows for the case of a limit theorem where not all dimensions n_i of the sample diverge to infinity; for an example of such a limit theorem in the sample mean case see Bradley (1992). To appreciate where such a limit theorem might be useful in practice, consider the case $d = 2$, and suppose the data are observed on a very long and thin strip on the plane; that is, suppose that n_2 is small for all practical purposes, whereas n_1 is large.

Since the index set cannot be thought to extend arbitrarily in all dimensions, it seems that $d^\star$ is the "effective" dimension, and the setup seems equivalent to a vector–valued random field in $d^\star$ dimensions. This point of view however obscures the fact that the probability structure is shift-invariant in *all* d dimensions, a fact that should be used in the analysis. The following corollary addresses this setup; its proof is analogous to the proof of Theorem 5.3.1.

Corollary 5.3.2. *Assume Assumption 5.3.2, and that $\tau_{\mathbf{b}}/\tau_{\mathbf{n}} \to 0$, $b_i \to \infty$, and $b_i/n_i \to 0$, as $n_i \to \infty$, for $i = 1, 2, \ldots, d^\star$, whereas $b_j \to Q_j$, and $n_j \to Q_j$, for $j = d^\star + 1, \ldots, d$. Also assume that $\mathbf{h}$ is constant, and that $n^{-1} \sum_{k=1}^{n^\star} k^{d^\star - 1} \hat{\alpha}_X(k; b) \to 0$, where $n^\star = \max_{i=1,\ldots,d^\star} n_i$.*

Then, conclusions (i–iii) of Theorem 5.3.1 remain true.

Remark 5.3.4. Corollary 2.2.1 can be easily generalized to the present setting as well.

5.3.2 Multivariate Parameter Case

Again, consider a statistic $\hat{\theta}_{\mathbf{n}} = \hat{\theta}_{\mathbf{n}}(X(\mathbf{t}), \mathbf{t} \in E_{\mathbf{n}})$ that is used in order to consistently estimate some parameter $\theta(P)$ of the underlying probability measure P. However, in this section we consider the more general case where $\hat{\theta}_{\mathbf{n}}$ and θ take values in some separable Banach space, endowed with a norm $\|\cdot\|$. Let $J_{\mathbf{n},\|\cdot\|}(P)$ be the sampling probability law of $\tau_{\mathbf{n}}\|\hat{\theta}_{\mathbf{n}} - \theta(P)\|$, and $J_{\mathbf{n},\|\cdot\|}(x, P) = Prob_P\{\tau_{\mathbf{n}}\|\hat{\theta}_{\mathbf{n}} - \theta(P)\| \leq x\}$ be the corresponding sampling probability distribution function. The required assumption for subsampling to 'work' asymptotically is existence of an asymptotic weak limit for $J_{\mathbf{n},\|\cdot\|}(P)$ as given below.

Assumption 5.3.3. $J_{\mathbf{n},\|\cdot\|}(P)$ *converges weakly to a limit law* $J_{\|\cdot\|}(P)$, *with corresponding distribution function* $J_{\|\cdot\|}(\cdot, P)$, *as* $n_i \to \infty$ *for* $i = 1, \ldots, d$.

The subsampling approximation to $J_{\mathbf{n},\|\cdot\|}(x, P)$ is now defined by

$$L_{\mathbf{n},\mathbf{b},\|\cdot\|}(x) = q^{-1} \sum_{i_1=1}^{q_1} \sum_{i_2=1}^{q_2} \cdots \sum_{i_d=1}^{q_d} 1\{\tau_{\mathbf{b}}\|\hat{\theta}_{\mathbf{n},\mathbf{b},\mathbf{i}} - \hat{\theta}_{\mathbf{n}}\| \leq x\}. \qquad (5.6)$$

The following general theorem on consistency of subsampling distribution estimation was also originally proved in Politis and Romano (1992c, 1994b) in the case $\mathbb{G} = \mathbb{Z}$ using the mixing coefficients $\alpha_X(k; l_1, l_2)$; we state it below using the weaker notion of the $\hat{\alpha}_X(k; l_1)$ coefficients, and for $\mathbb{G}$ being either $\mathbb{Z}$ or $\mathbb{R}$.

Theorem 5.3.2. *Assume Assumption 5.3.3, and that $\tau_b/\tau_n \to 0$, $b_i \to \infty$, and $n_i \to \infty$ for $i = 1, 2, \ldots, d$. Assume $\mathbf{h}$ is either a (nonzero) constant, or $h_i/b_i \to a_i \in (0, 1]$, for $i = 1, 2, \ldots, d$. Also assume $\prod_{j=1}^{d} b_j/(n_j - b_j) \to 0$, and that $q^{-1} \sum_{k=1}^{q^*} k^{d-1} \hat{\alpha}_X(k; b) \to 0$, where $q^* = \max_i q_i$.*

  i. *If x is a continuity point of $J_{\|\cdot\|}(\cdot, P)$, then $L_{\mathbf{n},\mathbf{b},\|\cdot\|}(x) \to J_{\|\cdot\|}(x, P)$ in probability.*
  ii. *If $J_{\|\cdot\|}(\cdot, P)$ is continuous, then $\sup_x |L_{\mathbf{n},\mathbf{b},\|\cdot\|}(x) - J_{\mathbf{n},\|\cdot\|}(x, P)| \to 0$ in probability.*
  iii. *Let*

$$c_{\mathbf{n},\mathbf{b},\|\cdot\|}(1-\alpha) = \inf\{x : L_{\mathbf{n},\mathbf{b},\|\cdot\|}(x) \geq 1-\alpha\}.$$

Correspondingly, define

$$c_{\|\cdot\|}(1-\alpha, P) = \inf\{x : J_{\|\cdot\|}(x, P) \geq 1-\alpha\}.$$

If $J_{\|\cdot\|}(\cdot, P)$ is continuous at $c_{\|\cdot\|}(1-\alpha, P)$, then

$$Prob_P\{\tau_\mathbf{n} \|\hat{\theta}_\mathbf{n} - \theta(P)\| \leq c_{\mathbf{n},\mathbf{b},\|\cdot\|}(1-\alpha)\} \to 1-\alpha.$$

Thus, the asymptotic coverage probability under P of the region $\{\theta : \tau_\mathbf{n} \|\hat{\theta}_\mathbf{n} - \theta\| \leq c_{\mathbf{n},\mathbf{b},\|\cdot\|}(1-\alpha)\}$ is the nominal level $1-\alpha$.

The proof of Theorem 5.3.2 is very similar to the proof of Theorem 5.3.1—with the addition of the triangle inequality argument as in Politis, Romano, and You (1993)—and thus is omitted. In the case where $\mathbf{h}$ is constant we have the following easy corollary.

Corollary 5.3.3. *Assume Assumption 5.3.3, and that $\tau_b/\tau_n \to 0$, $b_i \to \infty$, and $b_i/n_i \to 0$, as $n_i \to \infty$, for $i = 1, 2, \ldots, d$. Also assume that $\mathbf{h}$ is constant, and that $n^{-1} \sum_{k=1}^{n^*} k^{d-1} \hat{\alpha}_X(k; b) \to 0$, where $n^* = \max_i n_i$.*

Then, conclusions (i–iii) of Theorem 5.3.2 remain true.

As before, the case where not all (but at least one) of the n_i's diverge to infinity can be easily handled. Consider the following assumption and subsequent corollary to Theorem 5.3.2.

Assumption 5.3.4. *$J_{\mathbf{n},\|\cdot\|}(P)$ converges weakly to a limit law $J_{\|\cdot\|}(P)$, with corresponding distribution function $J_{\|\cdot\|}(x, P)$, as $n_i \to \infty$, for $i = 1, \ldots, d^*$, and $n_j \to Q_j$, for $j = d^* + 1, \ldots, d$, where $1 \leq d^* \leq d$, and the Q_j's are some fixed numbers in $\mathbb{G}_+$.*

Corollary 5.3.4. *Assume Assumption 5.3.4, and that $\tau_b/\tau_\mathbf{n} \to 0$, $b_i \to \infty$, and $b_i/n_i \to 0$, as $n_i \to \infty$, for $i = 1, 2, \ldots, d^*$, whereas $b_j \to Q_j$,* and $n_j \to Q_j$, for $j = d^\star + 1, \ldots, d$. Also assume that $\mathbf{h}$ is constant, and that $n^{-1} \sum_{k=1}^{n^\star} k^{d^\star - 1} \hat{\alpha}_X(k; b) \to 0$, where $n^\star = \max_{i=1,\ldots,d^\star} n_i$.

Then, conclusions (i–iii) of Theorem 5.3.2 remain true.

Remark 5.3.5. It is also interesting to observe that the hypothesis of homogeneity of the random field $\{X(\mathbf{t})\}$ may be relaxed just as the hypothesis of stationarity of the time series was relaxed in Chapter 4. However, our Assumptions 5.3.1 and 5.3.3 must be appropriately strengthened for subsampling to work in the case of nonhomogeneous random fields. For example, an easy way of strengthening Assumption 5.3.3 for the purpose of guaranteeing the asymptotic validity of subsampling is to add to Assumption 5.3.3 the requirement that

$$Prob_P\{\tau_\mathbf{b}\|\hat{\theta}_{\mathbf{n},\mathbf{b},\mathbf{i}} - \theta(P)\| \le x\} = J_{\mathbf{b},\|\cdot\|}(x, P) + o(1), \text{ uniformly in } \mathbf{i}.$$

Similarly, in the case of a real parameter θ, we can strengthen Assumption 5.3.1 (again for the purpose of guaranteeing the asymptotic validity of subsampling) by adding to it the requirement that

$$Prob_P\{\tau_\mathbf{b}(\hat{\theta}_{\mathbf{n},\mathbf{b},\mathbf{i}} - \theta(P)) \le x\} = J_\mathbf{b}(x, P) + o(1), \text{ uniformly in } \mathbf{i}.$$

5.4 Variance Estimation and Bias Reduction

Similarly to what was done in Sections 3.7 and 3.8, we can extend the notion of variance estimation and bias reduction via subsampling to the random fields case. Denote by $m_\mathbf{n}^{(j)}, \mu_\mathbf{n}^{(j)}$, and $\mu^{(j)}$ the jth (noncentral) moments of distributions $L_{\mathbf{n},\mathbf{b}}(\cdot), J_\mathbf{n}(\cdot, P)$, and $J(\cdot, P)$ respectively, assuming $\mu_\mathbf{n}^{(j)}$ and $\mu^{(j)}$ exist.

As mentioned earlier, statistics calculated from time series and random fields are often heavily biased; thus, the subsampling methodology could be used for bias reduction, in the same vein as the original proposition of Quenouille's (1949) precursor to the jackknife. Assume that Assumption 5.3.1 holds together with $\mu_\mathbf{n}^{(1)} \to \mu^{(1)}$ and $m_\mathbf{n}^{(1)} \to \mu^{(1)}$. Then, since $L_{\mathbf{n},\mathbf{b}}(\cdot)$ and $J_\mathbf{n}(\cdot, P)$ have the same limiting distribution $J(\cdot, P)$, with first moments converging as well, one can approximate $Bias(\hat{\theta}_\mathbf{n}) = E\hat{\theta}_\mathbf{n} - \theta(P)$ by a re-scaled version of the "empirical" bias, i.e., by

$$\widehat{Bias}(\hat{\theta}_\mathbf{n}) = \frac{1}{\tau_\mathbf{n}} m_\mathbf{n}^{(1)} = \frac{\tau_\mathbf{b}}{\tau_\mathbf{n}} (Ave(\hat{\theta}_{\mathbf{n},\mathbf{b},\mathbf{i}}) - \hat{\theta}_\mathbf{n})$$

where $Ave(\hat{\theta}_{\mathbf{n},\mathbf{b},\mathbf{i}}) = q^{-1} \sum_{i_1=1}^{q_1} \sum_{i_2=1}^{q_2} \cdots \sum_{i_d=1}^{q_d} \hat{\theta}_{\mathbf{n},\mathbf{b},\mathbf{i}}$; correspondingly, one can form the bias-corrected estimator

$$\hat{\theta}_{\mathbf{n},BC} = \hat{\theta}_\mathbf{n} - \widehat{Bias}(\hat{\theta}_\mathbf{n}) = (1 + \frac{\tau_\mathbf{b}}{\tau_\mathbf{n}})\hat{\theta}_\mathbf{n} - \frac{\tau_\mathbf{b}}{\tau_\mathbf{n}} Ave(\hat{\theta}_{\mathbf{n},\mathbf{b},\mathbf{i}}). \quad (5.7)$$

Note again that this is an asymptotic bias correction. For example, in the case where $\hat{\theta}_{\mathbf{n}}$ is the sample mean (which is unbiased), $\widehat{Bias}(\hat{\theta}_{\mathbf{n}}) \neq 0$ due to edge effects; nevertheless, $\widehat{Bias}(\hat{\theta}_{\mathbf{n}}) \to 0$ as it should (cf. Politis and Romano, 1993a). In the following theorem (which is due to Politis and Romano, 1994b), the conditions of Theorem 5.3.1 are strengthened to ensure that the bias correction suggested in equation (5.7) is indeed asymptotically valid. The argument is actually most relevant when $\mu^{(1)} \neq 0$, such as in the case of an optimally smoothed spectral density estimator of Example 3.4.5.

Theorem 5.4.1. *Assume Assumption 5.3.1 is strengthened to include $\mu_{\mathbf{n}}^{(1)} \to \mu^{(1)}$; assume $\tau_{\mathbf{b}}/\tau_{\mathbf{n}} \to 0$, $b_i \to \infty$, and $n_i \to \infty$, for $i = 1, 2, \ldots, d$. Also assume that $\prod_{j=1}^{d} b_j/(n_j - b_j) \to 0$, that $E|\hat{\theta}_{\mathbf{n},\mathbf{b},\mathbf{1}}|^{2+\delta} < C$, and that $q^{-1} \sum_{k=1}^{q^*} k^{d-1} \{\hat{\alpha}_X(k;b)\}^{\delta/(2+\delta)} \to 0$, where δ and C are two positive constants independent of $\mathbf{n}$, $\tilde{\theta}_{\mathbf{n},\mathbf{b},\mathbf{1}} \equiv \hat{\theta}_{\mathbf{n},\mathbf{b},\mathbf{1}}/\sqrt{Var(\hat{\theta}_{\mathbf{n},\mathbf{b},\mathbf{1}})}$, and $q^* = \max_i q_i$.*

Then, $|m_{\mathbf{n}}^{(1)} - \mu_{\mathbf{n}}^{(1)}| \to 0$ in probability.

Proof. First note that $Em_{\mathbf{n}}^{(1)} = \tau_{\mathbf{b}}(E\hat{\theta}_{\mathbf{n},\mathbf{b},\mathbf{1}} - E\hat{\theta}_{\mathbf{n}}) = \mu_{\mathbf{b}}^{(1)} - \frac{\tau_{\mathbf{b}}}{\tau_{\mathbf{n}}}\mu_{\mathbf{n}}^{(1)} = \mu_{\mathbf{b}}^{(1)} + o(1)$, and that $|\mu_{\mathbf{b}}^{(1)} - \mu_{\mathbf{n}}^{(1)}| \to 0$, by the (strengthened) Assumption 5.3.1. Now

$$Var(m_{\mathbf{n}}^{(1)}) = Var(\tau_{\mathbf{b}}(Ave(\hat{\theta}_{\mathbf{n},\mathbf{b},\mathbf{i}}) - \hat{\theta}_{\mathbf{n}}))$$

$$= Var(\tau_{\mathbf{b}}(Ave(\hat{\theta}_{\mathbf{n},\mathbf{b},\mathbf{i}}) - E\hat{\theta}_{\mathbf{n},\mathbf{b},\mathbf{1}}) - \tau_{\mathbf{b}}(\hat{\theta}_{\mathbf{n}} - E\hat{\theta}_{\mathbf{n},\mathbf{b},\mathbf{1}}))$$

$$= Var(\tau_{\mathbf{b}}(Ave(\hat{\theta}_{\mathbf{n},\mathbf{b},\mathbf{i}}) - E\hat{\theta}_{\mathbf{n},\mathbf{b},\mathbf{1}})) + o(1),$$

because $Var(\tau_{\mathbf{b}}(\hat{\theta}_{\mathbf{n}} - E\hat{\theta}_{\mathbf{n},\mathbf{b},\mathbf{1}})) \to 0$ as $\tau_{\mathbf{b}}/\tau_{\mathbf{n}} \to 0$. But

$$Var(\tau_{\mathbf{b}}(Ave(\hat{\theta}_{\mathbf{n},\mathbf{b},\mathbf{i}}) - E\hat{\theta}_{\mathbf{n},\mathbf{b},\mathbf{1}}))$$

$$= \tau_{\mathbf{b}}^2 q^{-1} \sum_{\mathbf{i} \in E_q} (1 - \frac{|i_1|}{q_1})(1 - \frac{|i_2|}{q_2}) \cdots (1 - \frac{|i_d|}{q_d}) Cov(\hat{\theta}_{\mathbf{n},\mathbf{b},\mathbf{1}}, \hat{\theta}_{\mathbf{n},\mathbf{b},\mathbf{1}+\mathbf{i}})$$

and thus

$$|Var(\tau_{\mathbf{b}}(Ave(\hat{\theta}_{\mathbf{n},\mathbf{b},\mathbf{i}}) - E\hat{\theta}_{\mathbf{n},\mathbf{b},\mathbf{1}}))| = O\left(q^{-1} \sum_{\mathbf{i} \in E_q} |Cov(\tilde{\theta}_{\mathbf{n},\mathbf{b},\mathbf{1}}, \tilde{\theta}_{\mathbf{n},\mathbf{b},\mathbf{1}+\mathbf{i}})|\right),$$

where it was taken into account that $Var(\hat{\theta}_{\mathbf{n},\mathbf{b},\mathbf{1}}) = O(1/\tau_{\mathbf{b}}^2)$, and E_q was defined in the proof of Theorem 5.3.1. Finally, by a similar argument to the proof of Theorem 5.3.1, and using the mixing inequality

$$|Cov(\tilde{\theta}_{\mathbf{n},\mathbf{b},\mathbf{1}}, \tilde{\theta}_{\mathbf{n},\mathbf{b},\mathbf{1}+\mathbf{i}})| \leq 10C^2 \{\hat{\alpha}_X(i^* h^* - b^*; b)\}^{\delta/(2+\delta)}$$

(cf. Roussas and Ioannides, 1987), it follows that $Var(m_\mathbf{n}^{(1)}) \to 0$ and the theorem is proved. ∎

Remark 5.4.1. In Section 10.5.5, we will use the above subsampling methodology for bias estimation in order to correct the subsampling variance estimator for bias; thus, turning subsampling back onto itself, a significant improvement in performance will be achieved.

In the same spirit, if we strengthen the assumptions of Theorem 5.3.1 to add that $m_\mathbf{n}^{(2)}$ converges to $\mu^{(2)}$, then the subsampling methodology can also be used for estimating the variance of the statistic $\hat{\theta}_\mathbf{n}$. For example, in the case where $\hat{\theta}_\mathbf{n}$ is the sample mean or a similarly "well-behaving" statistic, convergence of $m_\mathbf{n}^{(2)}$ to $\mu^{(2)}$ can actually be proven under stronger moment and mixing conditions and does not have to be explicitly included in the assumptions (see e.g., Possolo, 1991; Politis and Romano, 1993a,b; Sherman, 1992, 1996; and Sherman and Carlstein, 1994, 1996, for variance estimation results similar to those in Section 3.8).

Nevertheless, looking at the problem of variance estimation can yield useful insights. For example, a most interesting question for practical applications is how to choose **b** and **h** as functions of **n**. In the case of sample mean type statistics from discrete-time data (i.e., $\mathbb{G} = \mathbb{Z}$), it turns out that to have a most accurate (from the point of view of asymptotic mean squared error) variance estimator, one should let $\mathbf{h} = (1, 1, \ldots, 1)$ (i.e., maximum overlap of blocks), and $b \sim An^{\frac{d}{d+2}}$, (cf. Politis and Romano, 1993a,b); more details on the important issue of block size choice will be given in Chapter 9. Note here that the constant $A > 0$ can be calculated given the specifics of the problem, e.g., the form of the estimator and a quantification of the strength of dependence. Typically, however, the strength of dependence is unknown in practice, and the constant A has to be estimated from the data at hand; this is doable in principle, albeit requiring huge sample sizes to give accurate results.

It is interesting to note that, in the sample mean and related examples, the variance of the variance estimator $m_\mathbf{n}^{(2)}$ can be shown to be of order $O(b/n)$ *regardless* of the choice of **h** (cf. Künsch, 1989; Raïs and Moore, 1990; Raïs, 1992; or Politis and Romano, 1992a, 1993a). However, taking $\mathbf{h} = (1, \ldots, 1)$ in the case $\mathbb{G} = \mathbb{Z}$ is preferred because it decreases the variance of $m_\mathbf{n}^{(2)}$ by a constant factor. Intuitively this makes sense, since letting $\mathbf{h} = (1, 1, \ldots, 1)$ corresponds to *maximum overlap* between the rectangles $E_{\mathbf{i},\mathbf{b},\mathbf{h}}$, and has the effect of maximizing q, the number of subsamples available from the data, making it equal to $\prod_{i=1}^{d}(n_i - b_i + 1)$. On the other hand, taking $h_i/b_i \to a_i \in (0,1]$ would imply that a proportion of the $\prod_{i=1}^{d}(n_i - b_i + 1)$ available $\hat{\theta}_{\mathbf{n},\mathbf{b},\mathbf{i}}$'s are thrown away when computing the "empirical" estimate $L_{\mathbf{n},\mathbf{b}}$ and its variance; for example, this is what happens in the nonoverlapping case of Carlstein (1986) where $h_i = b_i$. The "maximum overlap" notion of blocking in the continuous-parameter case is addressed in the next section.

5.5 Maximum Overlap Subsampling in Continuous Time

In Section 5.4, it was mentioned that, in the discrete case ($\mathbb{G} = \mathbb{Z}$), taking the subsamples with maximum overlap (i.e., letting $\mathbf{h} = (1, 1, \ldots, 1)$) leads to improved efficiency of our subsampling variance estimators; see also equation (9.3) and the corresponding discussion in our Chapter 10. A similar argument shows that, for any given block size b, the variance of the subsampling distribution estimator $L_{\mathbf{n},\mathbf{b}}(x)$ is minimal in the case of maximum overlap of the blocks used in subsampling (still in the discrete case where $\mathbb{G} = \mathbb{Z}$).

Therefore, it seems worthwhile to examine the "maximum block overlap" case in continuous time, that is, $\mathbb{G} = \mathbb{R}$. Nevertheless, maximum overlap in continuous time is achieved only in the limit as $\mathbf{h} \to \mathbf{0}$, in which case the sums defining $L_{\mathbf{n},\mathbf{b}}(x)$ in (5.6) have to be substituted with integrals.

So we now explicitly address the case where $\{X(\mathbf{t}), \mathbf{t} \in \mathbb{R}^d\}$ is a continuous parameter random field with $d \in \mathbb{Z}_+$, and the observed data constitute the stretch $\{X(\mathbf{t}), \mathbf{t} \in E_\mathbf{n}\}$ where $E_\mathbf{n} \subset \mathbb{R}^d$, where

$$E_\mathbf{n} = \{\mathbf{t} \in \mathbb{R}^d : 0 < t_k \leq n_k, \text{ for } k = 1, 2, \ldots, d\};$$

note that the subsampling methodology could equally handle "observation regions" that are closed sets. As before, we are interested in using a statistic $\hat{\theta}_\mathbf{n} = \hat{\theta}_\mathbf{n}(X(\mathbf{t}), \mathbf{t} \in E_\mathbf{n})$ in order to estimate some parameter $\theta(P)$ of the underlying probability measure P. Here, we will assume the general case where $\hat{\theta}_\mathbf{n}$ and θ take values in some separable Banach space, endowed with a norm $\|\cdot\|$. Let $J_{\mathbf{n},\|\cdot\|}(P)$ be the sampling probability law of $\tau_\mathbf{n}\|\hat{\theta}_\mathbf{n} - \theta(P)\|$, and $J_{\mathbf{n},\|\cdot\|}(x, P) = Prob_P\{\tau_\mathbf{n}\|\hat{\theta}_\mathbf{n} - \theta(P)\| \leq x\}$ be the corresponding sampling probability distribution function. The required assumption is again simply the existence of an asymptotic weak limit for $J_{\mathbf{n},\|\cdot\|}(P)$ as given in Assumption 5.3.3.

Let the block $Y_\mathbf{u} = \{X_t, t \in E_{\mathbf{u},\mathbf{b}}\}$, where

$$E_{\mathbf{u},\mathbf{b}} = \{\mathbf{t} = (t_1, \ldots, t_d) \, ; \, u_j < t_j \leq u_j + b_j \, , j = 1, \ldots, d\}.$$

Let $E_{\mathbf{n}-\mathbf{b}}$ be the rectangle consisting of the points $\mathbf{t} = (t_1, t_2, \ldots, t_d) \in \mathbb{R}^d$ such that $0 < t_k \leq n_k - b_k$, for $k = 1, 2, \ldots, d$. Note that, as $\mathbf{u}$ varies, $E_{\mathbf{u},\mathbf{b}}$ corresponds to a shift (by $\mathbf{u}$) of the first block defined by the rectangle $E_{\mathbf{0},\mathbf{b}} = E_\mathbf{b}$.

As before, the subsample value of our statistic on each block is given by $\hat{\theta}_{\mathbf{n},\mathbf{b},\mathbf{u}} = \hat{\theta}_\mathbf{b}(Y_\mathbf{u})$, but the "maximum overlap" subsampling distribution is now defined by the integral

$$L_{\mathbf{n},\mathbf{b},||\cdot||}^{\max}(x) = |E_{\mathbf{n}-\mathbf{b}}|^{-1} \int_{E_{\mathbf{n}-\mathbf{b}}} 1\{\tau_{\mathbf{b}}\|\hat{\theta}_{\mathbf{n},\mathbf{b},\mathbf{u}} - \hat{\theta}_{\mathbf{n}}\| \leq x\}d\mathbf{u}.$$

As the following theorem will make precise, $L_{\mathbf{n},\mathbf{b},||\cdot||}^{\max}(\cdot)$ is consistent as an estimator of $J_{||\cdot||}(\cdot, P)$.

Theorem 5.5.1. *Assume Assumption 5.3.3, and that $\tau_\mathbf{b}/\tau_\mathbf{n} \to 0$, $b_i \to \infty$, and $n_i \to \infty$ for $i = 1, 2, \ldots, d$. Also assume that there exists a vector $\delta = (\delta_1, \ldots \delta_d)$, depending on $\mathbf{n}$, and such that $2 \leq \delta_i \leq (n_i - b_i)/b_i$, for all $i = 1, \ldots, d$, as well as $|\delta| = \prod_i \delta_i \to \infty$, and*

$$|\delta|\,\alpha_X\left(\min_i\{((n_i - b_i)/\delta_i) - b_i\}; (2^{-d}|\delta| - 1)C(\mathbf{n},\mathbf{b},\delta), 2^d C(\mathbf{n},\mathbf{b},\delta)\right) \to 0, \tag{5.8}$$

where $C(\mathbf{n},\mathbf{b},\delta) = \prod_i(\frac{n_i-b_i}{\delta_i} + b_i)$.

i. *If x is a continuity point of $J_{||\cdot||}(\cdot, P)$, then $L_{\mathbf{n},\mathbf{b},||\cdot||}^{\max}(x) \to J_{||\cdot||}(x, P)$ in probability.*
ii. *If $J_{||\cdot||}(\cdot, P)$ is continuous, then $\sup_x |L_{\mathbf{n},\mathbf{b},||\cdot||}^{\max}(x) - J_{\mathbf{n},||\cdot||}(x, P)| \to 0$ in probability.*
iii. *Let*

$$c_{\mathbf{n},\mathbf{b},||\cdot||}(1-\alpha) = \inf\{x : L_{\mathbf{n},\mathbf{b},||\cdot||}^{\max}(x) \geq 1 - \alpha\}.$$

Correspondingly, define

$$c_{||\cdot||}(1-\alpha, P) = \inf\{x : J_{||\cdot||}(x, P) \geq 1 - \alpha\}.$$

If $J_{||\cdot||}(\cdot, P)$ is continuous at $c_{||\cdot||}(1-\alpha, P)$, then

$$Prob_P\{\tau_\mathbf{n}\|\hat{\theta}_\mathbf{n} - \theta(P)\| \leq c_{\mathbf{n},\mathbf{b},||\cdot||}(1-\alpha)\} \to 1 - \alpha.$$

Thus, the asymptotic coverage probability under P of the region $\{\theta : \tau_\mathbf{n}\|\hat{\theta}_\mathbf{n} - \theta\| \leq c_{\mathbf{n},\mathbf{b},||\cdot||}(1-\alpha)\}$ is the nominal level $1 - \alpha$.

The proof of Theorem 5.5.1 is based on a new Bernstein inequality and is given in Bertail, Politis, and Rhomari (1996). Note that the existence of δ such that $2 \leq \delta_i \leq (n_i - b_i)/b_i$, and $|\delta| \to \infty$ immediately implies that $b/n \to 0$ so the latter does not have to be explicitly assumed. Also, note that although condition (5.8) seems to be complicated, it is actually very weak. For example, if the process is strong mixing, i.e., $\alpha_X(k) \to 0$ as $k \to \infty$, condition (5.8) is immediately satisfied.

If θ is real valued, then we have the option to look at the subsampling distribution

$$L_{\mathbf{n},\mathbf{b}}^{\max}(x) = |E_{\mathbf{n}-\mathbf{b}}|^{-1} \int_{E_{\mathbf{n}-\mathbf{b}}} 1\{\tau_\mathbf{b}(\hat{\theta}_{\mathbf{n},\mathbf{b},\mathbf{u}} - \hat{\theta}_\mathbf{n}) \leq x\}d\mathbf{u},$$

and the following theorem ensues.

Theorem 5.5.2. *Assume Assumption 5.3.1, and that $\tau_\mathbf{b}/\tau_\mathbf{n} \to 0$, $b_i \to \infty$, and $n_i \to \infty$ for $i = 1, 2, \ldots, d$. Also assume that there exists a vector $\delta = (\delta_1, \ldots \delta_d)$, depending on $\mathbf{n}$, and such that $2 \leq \delta_i \leq (n_i - b_i)/b_i$, for all $i = 1, \ldots, d$, as well as $|\delta| = \prod_i \delta_i \to \infty$, and*

$$|\delta| \, \alpha_X \left(\min_i \{((n_i - b_i)/\delta_i) - b_i\} ; (2^{-d}|\delta| - 1) C(\mathbf{n},\mathbf{b},\delta), 2^d C(\mathbf{n},\mathbf{b},\delta) \right) \to 0, \tag{5.9}$$

where $C(\mathbf{n},\mathbf{b},\delta) = \prod_i (\frac{n_i - b_i}{\delta_i} + b_i)$.

i. *If x is a continuity point of $J(\cdot, P)$, then $L_{\mathbf{n},\mathbf{b}}^{\max}(x) \to J(x, P)$ in probability.*
ii. *If $J(\cdot, P)$ is continuous, then $\sup_x |L_{\mathbf{n},\mathbf{b}}^{\max}(x) - J_\mathbf{n}(x, P)| \to 0$ in probability.*
iii. *Let*

$$c_{\mathbf{n},\mathbf{b}}(1-\alpha) = \inf\{x : L_{\mathbf{n},\mathbf{b}}^{\max}(x) \geq 1 - \alpha\}.$$

Correspondingly, define

$$c(1-\alpha, P) = \inf\{x : J(x, P) \geq 1 - \alpha\}.$$

If $J^\star(\cdot, P)$ is continuous at $c(1-\alpha, P)$, then

$$Prob_P\{\tau_\mathbf{n}(\hat{\theta}_\mathbf{n} - \theta(P)) \leq c_{\mathbf{n},\mathbf{b}}(1-\alpha)\} \to 1 - \alpha.$$

Thus, the asymptotic coverage probability under P of the interval $[\hat{\theta}_\mathbf{n} - \tau_\mathbf{n}^{-1} c_{\mathbf{n},\mathbf{b}}(1-\alpha), \infty)$ is the nominal level $1 - \alpha$.

5.6 Some Illustrative Examples

Most of the examples in Chapter 3 concerning stationary time series have their analogues in the case of random fields in discrete or continuous time. Thus, subsampling can be seen to be applicable for large-sample statistical inference concerning the mean, median, or other robust estimators of location, the autocovariance function, the spectral distribution and density functions, probability density estimators, etc., all in the context of random field data. Rather than revisiting all those examples, we go into detail for just one of them (the sample mean), and offer two new examples connected in particular with continuous-time data.

Example 5.6.1 (Univariate mean). Assume $\{X(\mathbf{t}), \mathbf{t} \in \mathbb{G}^d\}$ is a real-valued random field, and let $\hat{\mu} = n^{-1} \sum_{\mathbf{t} \in E_\mathbf{n}} X(\mathbf{t})$ or $\hat{\mu} = n^{-1} \int_{\mathbf{t} \in E_\mathbf{n}} X(\mathbf{t}) d\mathbf{t}$, respectively, denote the sample mean in the two cases $\mathbb{G} = \mathbb{Z}$ or $\mathbb{G} = \mathbb{R}$. In the discrete case $\mathbb{G} = \mathbb{Z}$, Bolthausen (1982) has shown that, under some moment and mixing assumptions, $\hat{\mu}$ is consistent for $\mu = EX(\mathbf{t})$, and asymptotically normal at rate $\sqrt{n}$, that is, $\sqrt{n}(\hat{\mu} - \mu) \xrightarrow{\mathcal{L}} N(0, \sigma^2)$, as $n_i \to \infty$ for $i = 1, \ldots, d$. Therefore, Assumption 5.3.1 is satisfied, and confidence intervals for μ may be set using subsampling, and thus bypassing the issue of explicit estimation of the asymptotic variance σ^2.

Still in the discrete case, Bradley (1992) has shown, under some regularity assumptions, $\sqrt{n}(\hat{\mu} - \mu)$ is asymptotically normal with mean zero even if not all of the n_is (but at least one) diverge to infinity. In this case, Assumption 5.3.2 is seen to be satisfied, and subsampling may again be used to set confidence intervals for μ. In the continuous case ($\mathbb{G} = \mathbb{R}$), mixing and moment conditions under which the central limit theorem for $\hat{\mu}$ holds can be found in Yadrenko (1983) and Ivanov and Leonenko (1986). Therefore, our Assumption 5.3.1 is again satisfied, and subsampling-based statistical inference for μ is possible in the continuous-time case.

Example 5.6.2 (Kernel regression). Suppose the process $\{X_t, t \in \mathbb{R}\}$ is bivariate, i.e., $X_t = [X_t^{(1)}, X_t^{(2)}]$, where the two processes $\{X_t^{(1)}\}$ and $\{X_t^{(2)}\}$ are assumed real valued. For some point x of interest, our objective is to estimate the regression function $\rho(x) \equiv E(m(X_t^{(1)})|X_t^{(2)} = x)$ given the data $\{X_t, 0 \le t \le n\}$; here $m(\cdot)$ is a (well-behaved) function of interest. In Bosq (1996, p. 130), the following kernel estimator of $\rho(x)$ is suggested:

$$\rho_n(x) = \int_0^n p_{t,n}(x) m(X_t^{(1)}) dt,$$

where

$$p_{t,n}(x) = K\left(\frac{x - X_t^{(2)}}{h_n}\right) / \int_0^n K\left(\frac{x - X_t^{(2)}}{h_n}\right) dt,$$

the kernel $K(\cdot)$ is a strictly positive probability density function, and h_n is a bandwidth parameter such that $h_n \to 0$ as $n \to \infty$.

Under some regularity assumptions (including geometric strong mixing for the process $\{X_t, t \in \mathbb{R}\}$), Bosq (1996, p. 137) showed that $\rho_n(x)$ is a $\sqrt{nh_n}$-consistent and asymptotically normal estimator of $\rho(x)$; therefore, Assumption 5.3.1 is verified with $\tau_n = \sqrt{nh_n}$, and continuous-time subsampling as presented in this chapter can be used to yield confidence intervals for the unknown function $\rho(x)$. Furthermore, there are two reasons for actually resorting to subsampling (as opposed to the asymptotic normality) for obtaining confidence intervals for $\rho(x)$; namely, (a) the asymptotic variance of $\rho_n(x)$ is messy to compute and difficult to estimate, and (b) if h_n is chosen in a way to minimize the mean squared error of $\rho_n(x)$, then the bias of $\rho_n(x)$ is nonnegligible and has to be accounted for in the confidence interval for $\rho(x)$. Although this bias term is very cumbersome to estimate, subsampling succeeds in automatically accounting for the bias in the construction of confidence intervals. This is because the limiting law of Assumption 5.3.1 is not assumed to have a nonzero mean.

Besides bypassing the difficult issues of bias and variance estimation in connection with the setting of confidence intervals, subsampling has the additional advantage of handling multivariate statistics with no extra effort. For example, consider simultaneous estimation of $(\rho(x_1),\ldots,\rho(x_K))$ by $(\rho_n(x_1),\ldots,\rho_n(x_K))$, where $x_1,\ldots,x_K$ are some points of interest. The results of Bosq (1996) together with the Cramér–Wold device immediately imply the asymptotic multivariate normality of vector $(\rho_n(x_1),\ldots,\rho_n(x_K))$; therefore, our Assumption 5.3.3 is now verified with $\tau_n = \sqrt{nh_n}$, and the norm $||\cdot||$ being the sup norm (say). Consequently, we can set a confidence region for the vector $(\rho(x_1),\ldots,\rho(x_K))$, by estimating the distribution of $\sqrt{nh_n}\sup_{i=1,\ldots,K}|\rho_n(x_i) - \rho(x_i)|$ via subsampling.

Example 5.6.3 (Diffusion process). We now give an application to parameter estimation in a class of diffusion processes in real time. Let the observed real random process $\{X_t, 0 \leq t \leq n\}$ obey the differential equation

$$dX_t = m(\theta, X_t) + \sigma(X_t)dw_t, \quad X_0 = x_0,\ 0 \leq t \leq n,$$

where $\theta = \theta(P) \in \Theta$ is an unknown (and to-be-estimated) parameter, $\{w_t\}$ is a real Wiener process, and $m(\cdot,\cdot)$ and $\sigma(\cdot)$ are real (and known) Borel functions such that $m(\cdot,\cdot)$ is locally bounded,

$$0 < \sigma_0 \leq \sigma^2(x) \leq \sigma_1 < \infty \quad \text{and} \quad \limsup_{x \to \infty} m(\theta, x)/|x| < \infty.$$

The above assumptions actually imply that the process $\{X_t\}$ is geometrically strong mixing, i.e., the strong mixing coefficients decrease exponentially to zero (see Leblanc, 1994, and Veretennikov, 1994, Propositions 1 and 2).

Consider the maximum-likelihood estimator of θ, say, $\hat{\theta}_n$. Under some additional conditions, this estimator is asymptotically normal,

$$n^{1/2}(\hat{\theta}_n - \theta) \overset{\mathcal{L}}{\Longrightarrow} N(0, I(\theta)^{-1})$$

where $I(\theta) = E_\theta[\sigma^{-2}(\xi)(\frac{\partial}{\partial \theta}m(\theta,\xi))^2]$ and ξ is a random variable with the invariant probability distribution, i.e., the stationary marginal distribution of the process. For more details, see Kutoyants (1984, Theorem 3.3.4. and Lemma 3.4.7, p. 95). So Assumption 5.3.1 is verified with $\tau_n = n^{1/2}$ and b may be chosen such that $b/n = o(1)$ and $b \to \infty$. Hence, continuous-time subsampling may be used to obtain confidence intervals (unilateral or bilateral) for θ.

Note that if we alternatively wanted to use the asymptotic Gaussian distribution to set confidence intervals for θ, we would require a consistent estimator of the asymptotic variance $I(\theta)^{-1}$, which is not trivial to obtain. The reason is that the expectation in computing $I(\theta)$ is taken with respect to the invariant (stationary) distribution of the process that in general may be difficult to compute. The same arguments hold for k-dimensional diffusion-type processes and the case of multivariate parameters and yield confidence regions with asymptotically correct coverage without explicit estimation of the variance–covariance matrix.

5.7 Conclusions

In this chapter, it was demonstrated how the subsampling methodology can be extended to make asymptotically valid inference in the context of homogeneous (that is, stationary) random fields. Both discrete-parameter as well as continuous-parameter random fields were considered, and a useful variation of the well-known strong mixing coefficients was defined, yielding a mixing condition that is sufficient for the asymptotic validity of subsampling while being easy to satisfy in practice.

The subsampling methodology for random fields has a direct analogy to subsampling time series in that the subsample replicates are given by our statistic computed from "blocks" of adjacent data. For convenience of presentation, it was assumed that the sample has a rectangular shape, and the "blocks" were taken to be of rectangular shape (but smaller); nevertheless, this is not necessary (see Remark 5.3.3). Notably, assuming that the sample (and the subsampling blocks) are rectangular affords the opportunity to consider asymptotics where the sample "expands" to infinity in some (but not necessarily all) out of the d possible directions. General convex shapes for the sample and the subsampling "blocks" will be considered in Chapter 6 in the context of marked point process data. Similarly, the assumption of homogeneity for the random field in question can be relaxed without destroying the asymptotic validity of subsampling (as long as some other conditions are appropriately strengthened); see Remark 5.3.5.

It was described how to find confidence intervals for real-valued parameters and confidence regions for vector-valued parameters. As in the time series case, it was argued that maximum overlap between the blocks yields efficient subsampling estimators. In the case of continuous-time random fields, subsampling with maximum overlap has to be defined through an integral as opposed to a sum; this was shown not to create any difficulty. Finally, the issues of bias reduction and variance estimation were addressed, and some more examples of the applicability of subsampling with dependent data were given.

6
Subsampling Marked Point Processes

6.1 Introduction

In this chapter, we assume that $\{X(\mathbf{t}), \mathbf{t} \in \mathbb{R}^d\}$ is a homogeneous random field in d dimensions, with $d \in \mathbb{Z}_+$; that is, $\{X(\mathbf{t}), \mathbf{t} \in \mathbb{R}^d\}$ is a collection of random variables $X(\mathbf{t})$ taking values in an arbitrary state space S, and indexed by the continuous variable $\mathbf{t} \in \mathbb{R}^d$. However, for reasons to be apparent shortly, the probability law of the random field $\{X(\mathbf{t}), \mathbf{t} \in \mathbb{R}^d\}$ will be denoted by P_X (and not P) throughout this chapter.

Our objective is statistical inference pertaining to the unknown probability law P_X on the basis of data. In case the data are of the form $\{X(\mathbf{t}), \mathbf{t} \in E\}$, with E being a finite subset either of the rectangular lattice $\mathbb{Z}^d$ or of $\mathbb{R}^d$, general methodologies for large-sample nonparametric inference were developed in our Chapter 5 (see also Hall, 1985; Politis and Romano, 1994b; Sherman, 1992, 1996; Sherman and Carlstein, 1994, 1996; and Shao and Tu, 1995) for related subsampling ideas in the context of dependent and independent data. However, in many important cases, e.g., queueing theory, spatial statistics, mining and geostatistics, meteorology, etc., the data correspond to observations of $X(\mathbf{t})$ at nonlattice, irregularly spaced points. For instance, if $d = 1$, $X(\mathbf{t})$ might represent the required service time for a customer arriving at a service station at time t. If $d = 2$, $X(\mathbf{t})$ might represent a measurement of the quality or quantity of the ore found in location $\mathbf{t}$, or a measurement of precipitation at location t during a fixed time interval, etc. As a matter of fact, in case $d > 1$, irregularly spaced data seem to be the rule rather than the exception (see, for example, Ripley, 1981; Karr, 1991; and Cressie, 1993).

A useful and parsimonious way to model the irregularly scattered **t**-points is to assume they are generated by a Poisson point process observable on a compact subset $K \in \mathbb{R}^d$, and assumed independent of the random field $\{X(\mathbf{t})\}$. So let N denote such a Poisson process on $\mathbb{R}^d$ with mean measure Λ; i.e., $EN(A) = \Lambda(A)$ for any set $A \subset \mathbb{R}^d$, and independent of $\{X(\mathbf{t})\}$. The point process N can be expressed as $N = \sum_i \epsilon_{\mathbf{t}_i}$, where $\epsilon_\mathbf{t}$ is a point mass at $\mathbf{t}$; i.e., $\epsilon_\mathbf{t}(A)$ is 1 or 0 according to whether $\mathbf{t} \in A$ or not. In other words, N is a random (counting) measure on $\mathbb{R}^d$. The expected number of **t**-points to be found in A is $\Lambda(A)$, whereas the actual number of **t**-points found in set A is given by $N(A)$. The joint (product) probability law of the random field $\{X(\mathbf{t})\}$ and the point process N will be denoted by P. The observations then are described via the "marked point process" $\widetilde{N} = \sum_i \epsilon_{\{\mathbf{t}_i, X(\mathbf{t}_i)\}}$, that is, the point process N where each **t**-point is "marked" by the value of X at that point.

The Poisson assumption can of course be tested with a particular dataset at hand; for example, in the precipitation example previously mentioned, the locations of a rain gauge for a region would usually be found in the nearest small town, and locations of towns can plausibly be thought as conforming to Poisson models (cf. Karr, 1986). In general, there are many reasons to justify the popularity of the Poisson assumption. To name just a few:

i. Parsimony: the law of the point process is completely specified by prescribing the mean measure Λ.
ii. Conditional uniformity or "spatial randomness" of the **t**-points: given the value of $N(K)$, i.e., given the number of the **t**-points (which plays the role of our sample size here) the locations of the **t**-points are i.i.d. from the probability measure ξ, where $\xi(A) = \Lambda(A \cap K)/\Lambda(K)$ (see Cressie, 1993, or Karr, 1991). In particular, if the Poisson process is homogeneous, i.e., if Λ is proportional to Lebesgue measure, then the **t**-points are uniformly distributed over K.
iii. Alias-free measurement or identifiability of P_X (rather: of all finite-dimensional distributions corresponding to P_X) on the basis of Poisson samples (see Theorem 10.18 in Karr, 1991).

Under the assumption of a homogeneous Poisson process generating the **t**-points, Karr (1986) has developed an elegant nonparametric estimation methodology for standard problems in the analysis of time series and random fields such as estimation of the mean μ and the autocovariance $R(\cdot)$, and he proved consistency and asymptotic normality of his estimators (see also Karr, 1991, ch. 10; Lii and Masry, 1994; Lii and Tsou, 1995; Masry, 1978, 1983, 1988; Krickeberg, 1982; and Jolivet, 1981) for related developments, including estimation of the spectrum and bispectrum of the random field $\{X(\mathbf{t})\}$ based on random sampling.

In this chapter, we will develop a nonparametric methodology for the construction of subsampling-based confidence regions for general parameters of the probability law P_X under the assumption of a homogeneous Poisson process generating the **t**-points; our exposition will largely follow Politis, Paparoditis, and Romano (1998). We should mention here that our subsampling methodology for marked point processes would still "work" if the Poisson assumption were relaxed to an appropriate weak dependence (mixing) assumption on the homogeneous point process N; however, the details are omitted for the sake of simplicity.

In Section 6.2, the different notions of strong mixing coefficients are revisited, and a new variation is proposed. Subsequently, the asymptotic validity of subsampling for marked point processes are shown in Section 6.3; as usual, both univariate and vector parameters are considered. Different ways to consistently approximate the integral associated with the subsampling distribution are discussed in Section 6.4. The issue of variance estimation via subsampling is discussed in Section 6.5, while some illustrative examples are given in Section 6.6.

6.2 Definitions and Some Different Notions on Mixing

Recall that the random field $\{X(\mathbf{t})\}$ is called *stationary* or *homogeneous* if for any set $E \subset \mathbb{R}^d$, and for any point $\mathbf{h} \in \mathbb{R}^d$, the joint distribution of the random variables $\{X(\mathbf{t}), \mathbf{t} \in E\}$ is identical to the joint distribution of $\{X(\mathbf{t}), \mathbf{t} \in E+\mathbf{h}\}$, where $E+\mathbf{h} = \{\mathbf{y} : \mathbf{y} = \mathbf{t}+\mathbf{h}, \text{ with } \mathbf{t} \in E\}$. If $\{X(\mathbf{t})\}$ is real valued (case $S = \mathbb{R}$), then we can define a weaker notion of stationarity based on second order moments only: the real-valued random field $\{X(\mathbf{t})\}$ is called *weakly stationary* or L_2-*stationary* if, for any $\mathbf{t}, \mathbf{h} \in \mathbb{R}^d$, $EX(\mathbf{t})^2 < \infty$, $EX(\mathbf{t}) = \mu$, and $Cov(X(\mathbf{t}), X(\mathbf{t}+\mathbf{h})) = R(\mathbf{h})$, in other words if $EX(\mathbf{t})$ and $Cov(X(\mathbf{t}), X(\mathbf{t}+\mathbf{h}))$ do not depend on $\mathbf{t}$ at all. In the presence of finite second moments, stationarity implies weak stationarity.

Denote by $|A|$ the volume (i.e., Lebesgue measure) of the set $A \subset \mathbb{R}^d$, and by $d\mathbf{t}$ the corresponding volume element situated at $\mathbf{t} \in \mathbb{R}^d$. Let $\mathbb{R}_+$ denote the positive part of the real axis, and let K be a compact, convex subset of $\mathbb{R}_+^d$ that will be our observation region, i.e., the region where the point process N is observed. For two row vectors $\mathbf{t} = (t_1, \ldots, t_d)$ and $\mathbf{u} = (u_1, \ldots, u_d)$ in $\mathbb{R}^d$, denote the l_∞ or sup distance in $\mathbb{R}^d$ by $\rho(\mathbf{t}, \mathbf{u}) = \sup_j |t_j - u_j|$, and for two sets E_1, E_2 in $\mathbb{R}^d$, define $\rho(E_1, E_2) = \inf\{\rho(\mathbf{t}, \mathbf{u}) : \mathbf{t} \in E_1, \mathbf{u} \in E_2\}$. Let $\delta(K)$ denote the supremum of the diameters of all l_∞ balls (i.e., hypercubes) contained in K, and let $\Delta(K)$ denote the infimum of the diameters of all l_∞ balls that contain K.

Observe that, because $N(K)$ is Poisson distributed with mean $\Lambda(K)$, if

$$\Lambda(K) \to \infty \text{ as } |K| \to \infty, \tag{6.1}$$

6.2 Definitions and Some Different Notions on Mixing

then this is sufficient to ensure that

$$N(K)/\Lambda(K) \xrightarrow{a.s.} 1; \tag{6.2}$$

in other words, our sample size $N(K)$ will tend to infinity with probability one as $|K| \to \infty$ provided $\Lambda(K) \to \infty$. The Poisson process N is homogeneous if its mean measure Λ is proportional to Lebesgue measure, that is, if $\Lambda(d\mathbf{t}) = \lambda d\mathbf{t}$, for some constant $\lambda > 0$. For a homogeneous Poisson process it follows immediately that equation (6.1), and thus equation (6.2) as well, hold true.

It is easy to construct examples where, without some form of weak dependence holding, statistical inference may be inaccurate even for large samples. Therefore, the (continuous time) random field $\{X(\mathbf{t})\}$ will be assumed to satisfy a certain weak dependence condition. In order to quantify the strength of dependence, we could make use of the strong mixing coefficients defined in equation (5.2) of Chapter 5.

Nevertheless, we now introduce a yet weaker form of a strong mixing coefficient for random fields that will be sufficient for our purposes. Recall the definition

$$\alpha_X(k; l_1) \equiv \sup_{\substack{E_1 \subset \mathbb{R}^d \\ E_2 = E_1 + \mathbf{t}, \mathbf{t} \in \mathbb{R}^d}} \{|P(A_1 \cap A_2) - P(A_1)P(A_2)| :$$

$$A_i \in \mathcal{F}(E_i), i = 1, 2, \ |E_1| \leq l_1, \ \rho(E_1, E_2) \geq k\};$$

we now define the following:

$$\bar{\alpha}_X(k; l_1) \equiv \sup_{\substack{E_1 \text{convex} \subset \mathbb{R}^d \\ E_2 = E_1 + \mathbf{t}, \mathbf{t} \in \mathbb{R}^d}} \{|P(A_1 \cap A_2) - P(A_1)P(A_2)| :$$

$$A_i \in \mathcal{F}(E_i), i = 1, 2, \ |E_1| \leq l_1, \ \rho(E_1, E_2) \geq k\}. \tag{6.3}$$

Again note that, by definition, $\alpha_X(k; l_1)$ and $\alpha_X(k; l_1, l_1)$ are different entities, and that $\alpha_X(k; l_1) \leq \alpha_X(k; l_1, l_1)$. Thus, a mixing condition based on the coefficients $\alpha_X(k; l_1)$ is easier to satisfy than a mixing condition based on the coefficients $\alpha_X(k; l_1, l_1)$. Similarly, $\bar{\alpha}_X(k; l_1) \leq \alpha_X(k; l_1)$ because of the convexity restriction in calculation of $\bar{\alpha}_X(k; l_1)$.

The essence of the following lemma is that the strong mixing properties of the random field $\{X(\mathbf{t})\}$ are passed on to the marked point process $\tilde{N}$, at least in the sense that covariances of arbitrary functions can be bounded by the mixing coefficients; the lemma is patterned after Lemma 4.1 of Masry (1988).

Lemma 6.2.1. *Let N be a Poisson process on $\mathbb{R}^d$ independent of the random field $\{X(\mathbf{t}), \mathbf{t} \in \mathbb{R}^d\}$; neither N nor $\{X(\mathbf{t})\}$ are assumed homogeneous here. Let E_1, E_2 be two subsets of $\mathbb{R}^d$ such that $|E_i| = l_i, i = 1, 2$, and $\rho(E_1, E_2) = k > 0$. Also let V_i be a random variable measurable with respect to the σ-algebra generated by the part of the marked point process*

$\widetilde{N}$ whose **t**-points happen to fall inside the set E_i; also assume that $|V_i| \leq 1$ almost surely for $i = 1, 2$.

Then,
$$|Cov(V_1, V_2)| \leq 4\alpha_X(k; l_1, l_2).$$

Proof. Consider first the identity
$$E^{N(E_1),N(E_2)}V_1V_2 - E^{N(E_1),N(E_2)}V_1 E^{N(E_1),N(E_2)}V_2 =$$
$$= E^{N(E_1),N(E_2)}\{E(V_1V_2|N) - E(V_1|N)E(V_2|N)\}$$
$$+ E^{N(E_1),N(E_2)}\{E(V_1|N)E(V_2|N)\} - E^{N(E_1),N(E_2)}V_1 E^{N(E_1),N(E_2)}V_2,$$

where $E^{N(E_1),N(E_2)}V$ is a short-hand for the conditional expectation $E(V|N(E_1), N(E_2))$.

Note that $E(V_i|N)$ is a function of only the $\mathbf{t}_1^{(i)}, \mathbf{t}_2^{(i)}, \ldots, \mathbf{t}_{N(E_i)}^{(i)}$ points. Also note that $\mathbf{t}_1^{(1)}, \mathbf{t}_2^{(1)}, \ldots, \mathbf{t}_{N(E_1)}^{(1)}$ and $\mathbf{t}_1^{(2)}, \mathbf{t}_2^{(2)}, \ldots, \mathbf{t}_{N(E_2)}^{(2)}$ are two *different* collections of **t**-points; i.e., there are no common **t**-points to both collections. Finally, by the Poisson assumption, the generated **t**-points are conditionally (given $N(E_1), N(E_2)$) i.i.d. Thus, $E(V_1|N)$ and $E(V_2|N)$ are conditionally (given $N(E_1), N(E_2)$) independent as functions of the distinct, independent collections of **t**-points $\mathbf{t}_1^{(1)}, \ldots, \mathbf{t}_{N(E_1)}^{(1)}$ and $\mathbf{t}_1^{(2)}, \ldots,$ $\mathbf{t}_{N(E_2)}^{(2)}$ respectively. Hence,
$$E^{N(E_1),N(E_2)}\{E(V_1|N)E(V_2|N)\} =$$
$$= E^{N(E_1),N(E_2)}\{E(V_1|N)\}E^{N(E_1),N(E_2))}\{E(V_2|N)\}$$
$$= E^{N(E_1),N(E_2)}V_1 E^{N(E_1),N(E_2)}V_2,$$

where $E^{N(E_1),N(E_2)}\{E(V|N)\} = E^{N(E_1),N(E_2)}V$ was used, since the σ-algebra generated by the random variables $N(E_1), N(E_2)$ is coarser than that generated by the whole process N.

Thus we have shown that
$$E^{N(E_1),N(E_2)}V_1V_2 - E^{N(E_1),N(E_2)}V_1 E^{N(E_1),N(E_2))}V_2$$
$$= E^{N(E_1),N(E_2)}\{E(V_1V_2|N) - E(V_1|N)E(V_2|N)\}. \qquad (6.4)$$

Now note that $E^{N(E_1),N(E_2)}V_i = E^{N(E_i)}V_i$, and that $E^{N(E_i)}V_i$ is a function of the random variable $N(E_i)$ only. Also note that because E_1, E_2 are assumed disjoint (since $\rho(E_1, E_2) > 0$), the Poisson process properties imply that the random variables $N(E_1), N(E_2)$ are independent. Hence,
$$E\{E^{N(E_1),N(E_2)}V_1 E^{N(E_1),N(E_2))}V_2\} = E\{E^{N(E_1)}V_1 E^{N(E_2))}V_2\}$$
$$= E\{E^{N(E_1)}V_1\}E\{E^{N(E_2)}V_2\} = EV_1 EV_2.$$

Therefore, taking expectations on both sides of (6.4) we finally arrive at the relation

$$EV_1V_2 - EV_1EV_2 = E\{E(V_1V_2|N) - E(V_1|N)E(V_2|N)\}. \quad (6.5)$$

But, by a well-known mixing inequality (e.g., see Roussas and Ioannides, 1987) we have that $|E(V_1V_2|N) - E(V_1|N)E(V_2|N)| \leq 4\alpha_X(k;l_1,l_2)$, since, conditionally on the realization of the point process N, events generated by the random variable V_i are included in the σ-algebra $\mathcal{F}(E_i)$; thus, the assertion of the lemma follows immediately. ∎

As a clarification, we note that V_i is some function of the **t**-points generated by N that happen to fall inside the set E_i, along with their corresponding $X(\mathbf{t})$ "marks." The ordering of the **t**-points is immaterial, and therefore arbitrary, as long as each **t**-point is properly paired with its corresponding $X(\mathbf{t})$ mark. For example, if $g_{i,k}$ is a real-valued, bounded function of k two-dimensional arguments, then we may say that $V_i = g_{i,N(E_i)}([\mathbf{t}_1^{(i)}, X(\mathbf{t}_1^{(i)})], [\mathbf{t}_2^{(i)}, X(\mathbf{t}_2^{(i)})], \ldots, [\mathbf{t}_{N(E_i)}^{(i)}, X(\mathbf{t}_{N(E_i)}^{(i)})])$. Ignoring momentarily the assumed boundedness, V_i can be thought of as being a general statistic calculated from the data $\{[\mathbf{t}_1^{(i)}, X(\mathbf{t}_1^{(i)})], [\mathbf{t}_2^{(i)}, X(\mathbf{t}_2^{(i)})], \ldots [\mathbf{t}_{N(E_i)}^{(i)}, X(\mathbf{t}_{N(E_i)}^{(i)})]\}$.

If the sets E_1, E_2 happen to be simple translates of one another, then the following lemma is also true; its proof is analogous to the proof of Lemma 6.2.1.

Lemma 6.2.2. *Let N be a Poisson process on $\mathbb{R}^d$ independent of the random field $\{X(\mathbf{t}), \mathbf{t} \in \mathbb{R}^d\}$; neither N nor $\{X(\mathbf{t})\}$ are assumed homogeneous here. Let E_1 be a subset of $\mathbb{R}^d$, and let $E_2 = E_1 + \mathbf{t}$ for some $\mathbf{t} \in \mathbb{R}^d$. Let $|E_1| = l_1$, and $\rho(E_1, E_2) = k > 0$. Also let V_i be a random variable measurable with respect to the σ-algebra generated by the part of the marked point process $\widetilde{N}$ whose $\mathbf{t}$-points happen to fall inside the set E_i; also assume that $|V_i| \leq 1$ almost surely for $i = 1, 2$.*

Then,

$$|Cov(V_1, V_2)| \leq 4\alpha_X(k;l_1).$$

If the set E_1 is convex, then also

$$|Cov(V_1, V_2)| \leq 4\bar{\alpha}_X(k;l_1).$$

6.3 Subsampling Stationary Marked Point Processes

6.3.1 Sampling Setup and Assumptions

Now, and for the remainder of this chapter, we will assume that the random field $\{X(\mathbf{t})\}$ is stationary and independent of the homogeneous Poisson process N of rate $\lambda > 0$. Note that λ is in general unknown but estimable by $N(K)/|K|$; see (6.2). Our goal will be to construct a confidence region for a parameter $\theta(P_X)$, on the basis of observing $\{X(\mathbf{t})\}$ for the $\mathbf{t}$-points generated by the Poisson process N over the compact, convex set K, i.e., observing the marked point process $\widetilde{N}$ over the corresponding region. The parameter θ will be assumed to generally take values in a normed linear space (e.g., a Banach space) Θ endowed with norm $\|\cdot\|$; the case where $\Theta = \mathbb{R}$ is an important special case.

Let $\hat{\theta}_K = \hat{\theta}_K(\widetilde{N})$ be a random variable measurable with respect to the σ-algebra generated by the part of the marked point process $\widetilde{N}$ whose $\mathbf{t}$-points happen to fall inside the set K. Similarly to the V_i appearing in Lemma 6.2.1, $\hat{\theta}_K$ is a function of the arguments $[\mathbf{t}_1, X(\mathbf{t}_1)], [\mathbf{t}_2, X(\mathbf{t}_2)], \ldots, [\mathbf{t}_{N(K)}, X(\mathbf{t}_{N(K)})]$, where $\mathbf{t}_1, \mathbf{t}_2, \ldots, \mathbf{t}_{N(K)}$ are the $\mathbf{t}$ points generated by N (and ordered by some fashion) that happen to fall inside the set K; the functional form of this function only depends on K and on $N(K)$. Note that, because of homogeneity of the random field $\{X(\mathbf{t})\}$, and of the Poisson process N, the functional form of $\hat{\theta}_K(\cdot)$ does not depend on the absolute location of the set K within $\mathbb{R}^d$; thus, for example, the probability law of $\hat{\theta}_K(\widetilde{N})$ is identical to that of $\hat{\theta}_{K+\mathbf{y}}(\widetilde{N})$, for any $\mathbf{y} \in \mathbb{R}^d$, although of course only $\hat{\theta}_K(\widetilde{N})$ is actually observed in our experiment.

The statistic $\hat{\theta}_K$ is employed to consistently estimate $\theta(P_X)$ at some known rate $\tau_{\Lambda(K)}$. Note that the requirement that the shape of function τ. be known can be side-stepped (see Remark 6.3.2). For simplicity, although this can also be appropriately side-stepped, we also assume in this chapter that τ. is a regularly varying function of a given index ζ, i.e., for $u \in \mathbb{R}_+$

$$\tau_u = u^\zeta s(u), \tag{6.6}$$

with $s(\cdot)$ being a known, normalized slowly varying function, such that $\lim_{x\to\infty} \frac{s(lx)}{s(x)} = 1$ for any $l > 0$, (see Bingham, Goldie, and Teugels, 1987), and $\zeta > 0$ a known constant. Assumption (6.6) in connection with (6.2) will be used in our substituting $\tau_{N(K)}$ in place of $\tau_{\Lambda(K)}$ since then $\tau_{N(K)}/\tau_{\Lambda(K)} \xrightarrow{a.s.} 1$ as $|K| \to \infty$. Note that because of the homogeneity of the point process $\Lambda(K) = \lambda|K|$, and—in view of equation (6.6)—we could drop the constant λ and say that $\hat{\theta}_K$ is consistent for $\theta(P_X)$ at rate $\tau_{|K|}$; nevertheless, it seems more intuitive to have the rate of convergence depend on the (actual or expected) sample size directly.

6.3 Subsampling Stationary Marked Point Processes

Although this is not a crucial point, let us here add that considerations of invariance also lead us to prefer having the rate as a function of $\Lambda(K) = \lambda|K|$ or $N(K)$ rather than just $|K|$; to see this, observe that $\Lambda(K)$ and $N(K)$ are dimensionless quantities, that is, pure numbers, whereas $|K|$ has the dimensions (and units) of volume. For example, suppose we have that asymptotically $Var(\tau_{\Lambda(K)}\hat{\theta}_K) \approx Var(\tau_{N(K)}\hat{\theta}_K) \approx \sigma^2$, and also that $Var(\tilde{\tau}_{|K|}\hat{\theta}_K) \approx \tilde{\sigma}^2$, for two different functions $\tau.$ and $\tilde{\tau}.$; if the units of measurement of volume are changed (in other words, a scaling is performed), then $\tau_{\Lambda(K)}, \tau_{N(K)}$ and σ^2 all remain unchanged (invariant), whereas both $\tilde{\tau}_{|K|}$ and $\tilde{\sigma}^2$ have to change (in opposing directions) to accomodate the fact that $Var(\hat{\theta}_K)$ has to remain unchanged.

Denote by $J_{K,\|\cdot\|}(P)$ the sampling law of the real-valued random variable $\tau_{\Lambda(K)}\|\hat{\theta}_K - \theta(P_X)\|$, with corresponding sampling distribution function $J_{K,\|\cdot\|}(\cdot, P)$. The only assumption that will essentially be needed is existence of a large-sample distribution for this random variable stated as assumption 6.3.1 below.

Assumption 6.3.1. $J_{K,\|\cdot\|}(P)$ *converges weakly to a limit law* $J_{\|\cdot\|}(P)$, *with corresponding distribution function* $J_{\|\cdot\|}(\cdot, P)$, *as* $\delta(K) \to \infty$.

Note that one of the implications of Assumption 6.3.1 is that, as $\delta(K)$ increases, $J_{K,\|\cdot\|}(P)$ progressively loses its dependence on the shape of K.

If $\Theta = \mathbb{R}$, we can also define $J_K(P)$ to be the sampling law and $J_K(\cdot, P)$ the sampling distribution function of $\tau_{\Lambda(K)}(\hat{\theta}_K - \theta(P_X))$. Then, we can also formulate Assumption 6.3.2 stated below:

Assumption 6.3.2. $J_K(P)$ *converges weakly to a limit law* $J(P)$, *with corresponding distribution function* $J(\cdot, P)$, *as* $\delta(K) \to \infty$.

Many practically important statistics have been shown to possess asymptotic distributions under some conditions, usually in connection with homogeneous Poisson sampling. For example, Karr (1986) in the case of homogeneous Poisson sampling of a real-valued random field $\{X(\mathbf{t})\}$ showed that the sample mean $\hat{\mu} = N(K)^{-1} \int_K X(\mathbf{t})N(d\mathbf{t})$ is consistent for μ, and asymptotically normal at rate $\sqrt{|K|}$ and asymptotic variance equal to $\int R(\mathbf{t})d\mathbf{t} + \lambda^{-1}R(\mathbf{0})$; here, and in what follows, if the region of integration is left unspecified, it is implied that integration takes place over the whole of $\mathbb{R}^d$. Incidentally, note the obvious typographical mistake in formula (3.10) of Karr (1986) for the asymptotic variance. Similarly, Karr (1986) showed asymptotic normality at some bandwidth-dependent rate for the nonparametric estimate of the autocovariance $R(\mathbf{t})$. Other examples of statistics of interest possessing well-defined large-sample distributions are given by Karr (1991), Krickeberg (1982), Lii and Tsou (1995), Masry (1978, 1983, 1988), and Jolivet (1981).

In any case, Assumptions 6.3.1 and 6.3.2 are minimal in the sense that in order to conduct inference with no parametric assumptions on P_X one is inevitably led to large-sample theory, and existence of an asymptotic distribution is an almost *sine qua non* condition. One should also note that we pose no restrictions on the statistic $\hat{\theta}_K$ other, in essence, that it be a proper statistic, i.e., a (measurable) function of our data at hand. Nonetheless, as will be seen shortly, we will be able to derive an estimator of the large-sample distribution of $\hat{\theta}_K$ with no further assumptions other than mixing; this unique generality is afforded to us only via the subsampling methodology.

6.3.2 Main Consistency Result

Recall that our observation region K is a compact, convex subset of $\mathbb{R}^d_+$. Let $c = c(K)$ be a number in $(0,1)$ depending on K, and define a scaled-down replica of K by $B = \{c\mathbf{t} : \mathbf{t} \in K\}$, where $\mathbf{t} = (t_1, \ldots, t_d)$ and $c\mathbf{t} = (ct_1, \ldots, ct_d)$; B has the same shape as K but smaller dimensions. Also define the displaced sets $B + \mathbf{y}$ and the set of "allowed" displacements $K_{1-c} = \{\mathbf{y} \in K : B + \mathbf{y} \subset K\}$. By temporarily "ignoring" all data with locations outside of $B + \mathbf{y}$ we can compute the statistic $\hat{\theta}_{B+\mathbf{y}} = \hat{\theta}_{B+\mathbf{y}}(\widetilde{N})$, for $\mathbf{y} \in K_{1-c}$, as if the point process N with its attached "marks" were observed only over $B+\mathbf{y}$. So let $\hat{\theta}_{K,B,\mathbf{y}} = \hat{\theta}_{B+\mathbf{y}}(\widetilde{N})$ denote this "subsample" value.

In the spirit of Section 5.5, define subsampling estimators of $J_K(x, P)$ and $J_{K, \|\cdot\|}(x, P)$ by $L_{K,B}(x)$ and $L_{K,B,\|\cdot\|}(x)$ respectively, where

$$L_{K,B,\|\cdot\|}(x) = |K_{1-c}|^{-1} \int_{K_{1-c}} 1\{\tau_{N(B+\mathbf{y})} \|\hat{\theta}_{K,B,\mathbf{y}} - \hat{\theta}_K\| \leq x\} d\mathbf{y} \quad (6.7)$$

and whereas, if θ and $\hat{\theta}_K$ are real valued,

$$L_{K,B}(x) = |K_{1-c}|^{-1} \int_{K_{1-c}} 1\{\tau_{N(B+\mathbf{y})}(\hat{\theta}_{K,B,\mathbf{y}} - \hat{\theta}_K) \leq x\} d\mathbf{y}. \quad (6.8)$$

The following theorem was originally proved in Politis, Paparoditis, and Romano (1998); it establishes asymptotic consistency of $L_{K,B}$, and asymptotic validity of confidence intervals for $\theta(P)$ that are constructed based on quantiles of $L_{K,B}$.

Theorem 6.3.1. *Let θ be real valued and assume Assumption 6.3.2 and (6.6). Let $\delta(K) \to \infty$, and let $c = c(K) \in (0,1)$ be such that $c \to 0$ but $c\delta(K) \to \infty$. Finally, assume that*

$$|K_{1-c}|^{-1} \int_0^{(1-c)\Delta(K)} u^{d-1} \bar{\alpha}_X(u; c^d|K|) du \to 0. \quad (6.9)$$

i. $L_{K,B}(x) \xrightarrow{P} J(x, P)$ *for every continuity point x of $J(x, P)$.*
ii. *If $J(\cdot, P)$ is continuous, then $\sup_x |L_{K,B}(x) - J(x, P)| \xrightarrow{P} 0$.*

6.3 Subsampling Stationary Marked Point Processes

iii. Let
$$c_{K,B}(1-\alpha) = \inf\{x : L_{K,B}(x) \geq 1-\alpha\}.$$
If $J(x,P)$ is continuous at $x = \inf\{x : J(x,P) \geq 1-\alpha\}$, then
$$Prob_P\{\tau_{\Lambda(K)}(\hat{\theta}_K - \theta(P_X)) \leq c_{K,B}(1-\alpha)\} \to 1-\alpha.$$

Thus the asymptotic coverage probability under P of the interval $[\hat{\theta}_K - \tau_{N(K)}^{-1} c_{K,B}(1-\alpha), \infty)$ is the nominal level $1-\alpha$.

Proof. First note that since $c\delta(K) \to \infty$, it follows that $|B+\mathbf{y}| \to \infty$ for any $\mathbf{y}$. Because of $c \to 0$ however, it follows that $|B+\mathbf{y}|/|K| \to 0$. Now by Assumption (6.6) for $\zeta > 0$ we have that $\tau_{\Lambda(B)}/\tau_{\Lambda(K)} \to 0$, and that $\tau_{\Lambda(B)}/\tau_{N(B+\mathbf{y})} \xrightarrow{a.s.} 1$.

Let x be a continuity point of $J(x, P)$. Now by an argument similar to the one used in the proof of Theorem 3.2.1, and using the fact that by Assumption (6.6), $\tau_{N(A)}/\tau_{\Lambda(A)} \xrightarrow{a.s.} 1$ for any set A such that $|A| \to \infty$, it follows that to prove (i) it is sufficient to show that $U_K(x) \xrightarrow{P} J(x, P)$, where
$$U_K(x) = |K_{1-c}|^{-1} \int_{K_{1-c}} 1\{\tau_{\Lambda(B+\mathbf{y})}(\hat{\theta}_{K,B,\mathbf{y}} - \theta) \leq x\} d\mathbf{y}.$$

Note that
$$EU_K(x) = |K_{1-c}|^{-1} \int_{K_{1-c}} Prob_P\{\tau_{\Lambda(B+\mathbf{y})}(\hat{\theta}_{K,B,\mathbf{y}} - \theta) \leq x\} d\mathbf{y}.$$

But,
$$Prob_P\{\tau_{\Lambda(B+\mathbf{y})}(\hat{\theta}_{K,B,\mathbf{y}} - \theta) \leq x\} = J_{B+\mathbf{y}}(x, P) = J_B(x, P),$$
the latter equality holding being due to homogeneity (translation invariance of the probability laws). So the integrand does not depend on $\mathbf{y}$, and it follows immediately that
$$EU_K(x) = J_B(x, P) \to J(x, P),$$
the latter convergence being due to Assumption 6.3.2 in connection with $\delta(B) = c\delta(K) \to \infty$.

Now look at
$$Var(U_K(x)) = |K_{1-c}|^{-2} \int_{K_{1-c}} \int_{K_{1-c}} Cov(I(\mathbf{x}), I(\mathbf{y})) d\mathbf{x} d\mathbf{y},$$
where
$$I(\mathbf{x}) \equiv 1\{\tau_{\Lambda(B+\mathbf{x})}(\hat{\theta}_{K,B,\mathbf{x}} - \theta) \leq x\}$$
and
$$I(\mathbf{y}) \equiv 1\{\tau_{\Lambda(B+\mathbf{y})}(\hat{\theta}_{K,B,\mathbf{y}} - \theta) \leq x\}.$$

Again by the homogeneity concerning translations of the domain of $\mathbf{x}$, the random field $\{I(\mathbf{x}), \mathbf{x} \in \mathbf{R}^d\}$ is stationary as well, and we have

$$Cov(I(\mathbf{x}), I(\mathbf{y})) = C(\mathbf{x} - \mathbf{y}) \text{ for some function } C(\cdot).$$

A straightforward calculation now gives

$$Var(U_K(x)) = |K_{1-c}|^{-1} \int_{K^{\pm}_{1-c}} C(\mathbf{y}) \frac{|K_{1-c} \cap K_{1-c} - \mathbf{y}|}{|K_{1-c}|} d\mathbf{y}$$

$$\leq |K_{1-c}|^{-1} \int_{K^{\pm}_{1-c}} |C(\mathbf{y})| d\mathbf{y}$$

$$= |K_{1-c}|^{-1} \int_{B^{\pm}} |C(\mathbf{y})| d\mathbf{y} + |K_{1-c}|^{-1} \int_{K^{\pm}_{1-c} - B^{\pm}} |C(\mathbf{y})| d\mathbf{y}$$

where

$$K^{\pm}_{1-c} = \{\mathbf{t} : \mathbf{t} = \mathbf{x} - \mathbf{y}, \text{ with } \mathbf{x}, \mathbf{y} \in K_{1-c}\}$$

and

$$B^{\pm} = \{\mathbf{t} : \mathbf{t} = \mathbf{x} - \mathbf{y}, \text{ with } \mathbf{x}, \mathbf{y} \in B\}.$$

Note that $|C(\mathbf{y})| \leq 1$, and thus:

$$|K_{1-c}|^{-1} \int_{B^{\pm}} |C(\mathbf{y})| d\mathbf{y} = O(|B^{\pm}|/|K_{1-c}|) = O(c^d) = o(1).$$

Finally, for $\mathbf{y} \in K^{\pm}_{1-c} - B^{\pm}$, we have by Lemma 6.2.2 that

$$|C(\mathbf{y})| \leq 4\bar{\alpha}_X(\max_i |y_i - b_i|; |B|).$$

Therefore,

$$|K_{1-c}|^{-1} \int_{K^{\pm}_{1-c} - B^{\pm}} |C(\mathbf{y})| d\mathbf{y} =$$

$$= O\left(|K_{1-c}|^{-1} \int_{K^{\pm}_{1-c} - B^{\pm}} \bar{\alpha}_X(\max_i |y_i - b_i|; |B|) d\mathbf{y}\right)$$

$$= O\left(|K_{1-c}|^{-1} \int_0^{(1-c)\Delta(K)} u^{d-1} \bar{\alpha}_X(u; |B|) du\right),$$

which is $o(1)$ by assumption. Thus $Var(U_K(x)) \to 0$ and (i) is proved.

The proofs of (ii) and (iii) now proceed exactly as in the proof of Theorem 2.2.1. Note that substituting $\tau_{N(K)}$ in place of $\tau_{\Lambda(K)}$ in the confidence limit is permissible asymptotically because Equation (6.6) implies $\tau_{N(K)}/\tau_{\Lambda(K)} \to 1$ almost surely as $|K| \to \infty$. ∎

Remark 6.3.1. If the limit law $J(x, P)$ happens to be known, then a mixture of the subsampling distribution estimator $L_{K,B}(x)$ with $J(x, P)$ may well possess higher-order accuracy properties as in Booth and Hall (1993) and Bertail (1997); see Chapter 10 for more details on this interpolation idea.

The following theorem establishes asymptotic consistency of $L_{K,B,||\cdot||}$, and asymptotic validity of confidence regions for $\theta(P_X)$ that are constructed based on quantiles of $L_{K,B,||\cdot||}$.

Theorem 6.3.2. *Assume Assumption 6.3.1 as well as (6.6) and (6.9). Let $\delta(K) \to \infty$, and let $c = c(K) \in (0,1)$ be such that $c \to 0$ but $c\delta(K) \to \infty$. Then,*

i. $L_{K,B,||\cdot||}(x) \xrightarrow{P} J_{||\cdot||}(x, P)$ *for every continuity point x of $J_{||\cdot||}(x, P)$.*
ii. *If $J_{||\cdot||}(\cdot, P)$ is continuous, then $\sup_x |L_{K,B,||\cdot||}(x) - J_{||\cdot||}(x, P)| \xrightarrow{P} 0$.*
iii. *Let*

$$c_{K,B,||\cdot||}(1-\alpha) = \inf\{x : L_{K,B,||\cdot||}(x) \geq 1 - \alpha\}$$

and let

$$c_{||\cdot||}(1-\alpha, P) = \inf\{x : J_{||\cdot||}(x, P) \geq 1 - \alpha\}.$$

If $J_{||\cdot||}(x, P)$ is continuous at $x = c_{||\cdot||}(1-\alpha, P)$, then

$$Prob_P\{\tau_{\Lambda(K)}||\hat{\theta}_K - \theta(P_X)|| \leq c_{K,B,||\cdot||}(1-\alpha)\} \to 1 - \alpha.$$

Thus the asymptotic coverage probability under P of the region $\{\theta \in \Theta : \tau_{N(K)}||\hat{\theta}_K - \theta|| \leq c_{K,B,||\cdot||}(1-\alpha)\}$ is the nominal level $1 - \alpha$.

Proof. The proof of Theorem 6.3.2 is similar to the proof of Theorem 6.3.1, taking into account the arguments presented in Politis, Romano, and You (1993). ■

Remark 6.3.2. If ζ and $s(\cdot)$ appearing in (6.6) happen to be unknown, they can be estimated consistently via a preliminary round of subsampling in the same fashion as discussed in Chapter 8. Then, the estimated rate ($\hat{\tau}_{\Lambda(K)}$, say) can be used in place of the true rate $\tau_{\Lambda(K)}$ in the construction of subsampling confidence intervals and regions without destroying their asymptotic validity.

6.3.3 Nonstandard Asymptotics

It is of interest to consider what happens if K does not expand towards $\mathbb{R}^d$ in all directions as implied by the assumption $\delta(K) \to \infty$. If $d = 2$, for example, one can envision a situation where the practitioner is faced with an observation region K that is long and thin; notwithstanding the thinness in one direction, if $|K|$ is large enough, the sample size $N(K)$ will be large enough so accurate estimation may still be possible.

150   6. Subsampling Marked Point Processes

To avoid cumbersome notations, and because this actually may be a most useful setup in practice, let us now assume that K and B are the rectangles given by
$$K = [0,k_1] \times [0,k_2] \times \cdots \times [0,k_d], \text{ and } B = [0,b_1] \times [0,b_2] \times \cdots \times [0,b_d],$$
where $b_i = ck_i$, for $i = 1, 2, \ldots, d$, and c is a number in $(0,1)$. Therefore,
$$K_{1-c} = [0, (1-c)k_1] \times [0, (1-c)k_2] \times \cdots \times [0, (1-c)k_d]$$
is a rectangle with the property that for all $\mathbf{y} \in K_{1-c}$, $B + \mathbf{y}$ is a subset of K.

Now assume that
$$0 < k_i \uparrow \infty \text{ for } i = 1, \ldots, d^* \text{ and } 0 < k_i \uparrow K_i < \infty \text{ for } i = d^* + 1, \ldots, d \tag{6.10}$$

where d^* is some positive integer less than or equal to d. Then, the following two corollaries to Theorems 6.3.1 and 6.3.2, respectively, can be derived.

Corollary 6.3.1. *Let θ be real valued and assume (6.6) and (6.10). Assume that $J_K(P)$ converges weakly to a limit law $J(P)$, with corresponding distribution function $J(x,P)$, as $|K| \to \infty$ under (6.10). Let $c = c(K) \in (0,1)$ be such that $c \to 0$ but $c^d|K| \to \infty$ as $|K| \to \infty$ under (6.10). Finally, assume that $|K_{1-c}|^{-1} \int_0^{(1-c)\max_i k_i} u^{d^*-1} \bar{\alpha}_X(u; c^d|K|) du \to 0$.*

Then, conclusions (i–iii) of Theorem 6.3.1 remain true.

Corollary 6.3.2. *Assume (6.6), (6.10), and that $J_{K,\|\cdot\|}(P)$ converges weakly to a limit law $J_{\|\cdot\|}(P)$, with corresponding distribution function $J_{\|\cdot\|}(x,P)$, as $|K| \to \infty$ under (6.10). Let $c = c(K) \in (0,1)$ be such that $c \to 0$ but $c^d|K| \to \infty$ as $|K| \to \infty$ under (6.10). Finally, assume that $|K_{1-c}|^{-1} \int_0^{(1-c)\max_i k_i} u^{d^*-1} \bar{\alpha}_X(u; c^d|K|) du \to 0$.*

Then, conclusions (i–iii) of Theorem 6.3.2 remain true.

To appreciate that the mixing condition
$$|K_{1-c}|^{-1} \int_0^{(1-c)\max_i k_i} u^{d^*-1} \bar{\alpha}_X(u; c^d|K|) du \to 0,$$

as well as the analogous condition (6.9) appearing in the assumptions of Theorems 6.3.1 and 6.3.2, are actually quite weak, let us consider the one-dimensional case; it is quite obvious that, if $d = 1$, the assumed mixing conditions are indeed satisfied provided the process $\{X(\mathbf{t})\}$ is simply strong mixing, i.e., if $\alpha_X(u) \to 0$ as $u \to \infty$. Therefore, the following corollary is immediate.

Corollary 6.3.3. *Let $d = 1$, and assume that the stochastic process $\{X(\mathbf{t})\}$ is strong mixing; also assume (6.6). Let $K = [0,k_1]$, and let $c = c(k_1) \in (0,1)$ be such that $c \to 0$ but $ck_1 \to \infty$ as $k_1 \to \infty$.*

  a. *If Assumption 6.3.1 holds, then conclusions (i), (ii), and (iii) of Theorem 6.3.2 are true.*

b. *If θ happens to be real valued and Assumption 6.3.2 holds, then conclusions (i), (ii), and (iii) of Theorem 6.3.1 are true.*

Remark 6.3.3. Some discussion on proper choice of c is in order here. Note that the (rather vague) assumption that $c = c(K) \to 0$ but $c\delta(K) \to \infty$ as $\delta(K) \to \infty$ in Theorems 6.3.1 and 6.3.2 (or its analogue $c \to 0$ but $c^d|K| \to \infty$ as $|K| \to \infty$ in our corollaries) suffices for consistency of our subsampling estimators and asymptotic validity of the corresponding confidence regions. Nevertheless, to implement the subsampling method in a practical finite-sample situation, a particular choice of c is required. The situation here is quite analogous to the problem of choosing the block size in subsampling or resampling stationary dependent observations as discussed in Chapter 9, and in general requires higher-order considerations. Note also that the "quest" for the "optimal" block size ultimately depends on the optimality criterion employed; for example, it might be a criterion of accuracy of variance, distribution function, or quantile estimation. Even if the vague assumption that $c = c(K) \to 0$ but $c\delta(K) \to \infty$ is "narrowed" to something like $c = A\delta(K)^{-a}$ for some well-defined constants $a \in (0,1)$ and $A > 0$, the situation typically is that a is known, but A is unknown because it depends on unknown parameters of the underlying probability law. Thus, to use effectively a "recipe" such as $c = A\delta(K)^{-a}$ requires estimation of the underlying unknown parameters that influence A, and this typically is a problem harder by an order of magnitude than the original estimation problem at hand; see Chapter 9 for more discussion on this difficult issue.

6.4 Stochastic Approximation

In the previous section, it was shown that the subsampling distributions $L_{K,B,\|\cdot\|}$ and $L_{K,B}$ are consistent estimators of their corresponding large-sample distributions, and can be used to derive asymptotically valid confidence regions for the unknown $\theta(P_X)$. Nevertheless, equations (6.7) and (6.8) involve integrals which, to be of practical use, have to be approximated by finite sums. There are two general ways of performing this approximation; namely,

i. Deterministic approximation: For example, the region K_{1-c} can be "tiled" by a grid consisting of "small" cells, and then the integral can be approximated by the corresponding Riemann sum.
ii. Monte Carlo or stochastic approximation: Points can be dropped "at random" on K_{1-c}, and the averages of $1\{\tau_{N(B+\mathbf{y})}\|\hat{\theta}_{K,B,\mathbf{y}} - \hat{\theta}_K\| \leq x\}$ and $1\{\tau_{N(B+\mathbf{y})}(\hat{\theta}_{K,B,\mathbf{y}} - \hat{\theta}_K) \leq x\}$ over the dropped $\mathbf{y}$-points will be used to approximate $L_{K,B,\|\cdot\|}$ and $L_{K,B}$, respectively.

Both ways are valid, provided the number of points over which the approximations are computed is big enough (see also Politis and Romano, 1994b,

152   6. Subsampling Marked Point Processes

for a related discussion). Since the method of deterministic approximation is quite obvious, we will only elaborate on the second method of stochastic approximation

Let n be a positive integer, and generate n i.i.d. random variables $\mathbf{z}_1, \mathbf{z}_2, \ldots, \mathbf{z}_n$ with distribution that is uniform on K_{1-c}; this construction (which is performed independently of the random field $\{X(\mathbf{t})\}$ and of the Poisson process N) essentially amounts to generating a realization of the point process $H_n = \sum_{i=1}^n \epsilon_{\mathbf{z}_i}$. Stochastic approximations $\widetilde{L}_{K,B,\|\cdot\|}$ and $\widetilde{L}_{K,B}$ to $L_{K,B,\|\cdot\|}$ and $L_{K,B}$ can now be constructed by

$$\widetilde{L}_{K,B,\|\cdot\|}(x) = n^{-1}\int_{K_{1-c}} 1\{\tau_{N(B+\mathbf{y})}\|\hat{\theta}_{K,B,\mathbf{y}} - \hat{\theta}_K\| \le x\}H_n(d\mathbf{y}) \quad (6.11)$$

and,

$$\widetilde{L}_{K,B}(x) = n^{-1}\int_{K_{1-c}} 1\{\tau_{N(B+\mathbf{y})}(\hat{\theta}_{K,B,\mathbf{y}} - \hat{\theta}_K) \le x\}H_n(d\mathbf{y}), \quad (6.12)$$

the latter being defined only if θ and $\hat{\theta}_K$ are real valued. Define also the corresponding quantiles $\widetilde{c}_{K,B,\|\cdot\|}(1-\alpha) = \inf\{x : \widetilde{L}_{K,B,\|\cdot\|}(x) \ge 1-\alpha\}$, and $\widetilde{c}_{K,B}(1-\alpha) = \inf\{x : \widetilde{L}_{K,B}(x) \ge 1-\alpha\}$.

The following theorem shows that the proposed stochastic approximations $\widetilde{L}_{K,B}(x)$ and $\widetilde{L}_{K,B,\|\cdot\|}(x)$ may indeed be used in place of $L_{K,B}(x)$ and $L_{K,B,\|\cdot\|}(x)$ for constructing asymptotically valid confidence regions for θ.

Theorem 6.4.1. *Let $n \to \infty$ as $\delta(K) \to \infty$.*

a. *Under the assumptions of Theorem 6.3.1, if $J(x, P)$ is continuous at $x = \inf\{x : J(x,P) \ge 1-\alpha\}$, then the asymptotic coverage probability of the interval $[\hat{\theta}_K - \tau_{N(K)}^{-1}\widetilde{c}_{K,B}(1-\alpha), \infty)$ is the nominal level $1-\alpha$.*

b. *Under the assumptions of Theorem 6.3.2, if $J_{\|\cdot\|}(x, P)$ is continuous at $x = c_{\|\cdot\|}(1-\alpha, P)$, then the asymptotic coverage probability of the confidence region $\{\theta \in \Theta : \tau_{N(K)}\|\hat{\theta}_K - \theta\| \le \widetilde{c}_{K,B,\|\cdot\|}(1-\alpha)\}$ is the nominal level $1-\alpha$.*

Proof. We will show (a) only, and in particular we will just show that $\widetilde{L}_{K,B}(x) \to J(x,P)$ in probability; showing (b) is similar.

Let x be a continuity point of $J(x, P)$. Let $\widetilde{P}$ denote the joint probability law of $H_n, \{X(\mathbf{t})\}$ and N, and let $\widetilde{E}, \widetilde{Var}$ denote the corresponding expectation and variance; also let P^c, E^c, Var^c denote conditional probability, expectation, and variance, respectively, where we condition on the realization of the random field $\{X(\mathbf{t})\}$ and that of the Poisson process N. As before, P denotes the joint probability law of $\{X(\mathbf{t})\}$ and N, and E, Var denote the corresponding expectation and variance.

Now let $I(\mathbf{y}) \equiv 1\{\tau_{\Lambda(B+\mathbf{y})}(\hat{\theta}_{K,B,\mathbf{y}} - \theta) \le x\}$, which is a new homogeneous random field for $\mathbf{y} \in \mathbb{R}^d$. Again, by an argument similar to the one used

6.4 Stochastic Approximation

in the proof of Theorem 3.2.1, it will be sufficient for our purposes to just show $\tilde{U}_K(x) = J(x,P) + o_{\tilde{P}}(1)$, where

$$\tilde{U}_K(x) = n^{-1} \int_{K_{1-c}} I(\mathbf{y}) H_n(d\mathbf{y}).$$

But it is obvious that $E^c \tilde{U}_K(x) = U_K(x)$, where $U_K(x)$ was defined in the proof of Theorem 6.3.1; thus, $\tilde{E}\tilde{U}_K(x) = E\{E^c\tilde{U}_K(x)\} = EU_K(x) = J_B(x,P) \to J(x,P)$. So we look at $\tilde{Var}(\tilde{U}_K(x))$.

Observe that

$$\tilde{E}\{(\tilde{U}_K(x))^2|H_n\} = \tilde{E}\{n^{-2}\int_{K_{1-c}}\int_{K_{1-c}} I(\mathbf{y})I(\mathbf{x})H_n(d\mathbf{y})H_n(d\mathbf{x})|H_n\}$$

$$= n^{-2}\int_{K_{1-c}}\int_{K_{1-c}} \tilde{E}\{I(\mathbf{y})I(\mathbf{x})|H_n\}H_n(d\mathbf{y})H_n(d\mathbf{x}).$$

But $\tilde{E}\{I(\mathbf{y})I(\mathbf{x})|H_n\} = EI(\mathbf{y})I(\mathbf{x}) = C(\mathbf{y}-\mathbf{x}) + (J_B(x,P))^2$, where $C(\mathbf{y}-\mathbf{x}) = Cov(I(\mathbf{y}), I(\mathbf{x}))$ as defined in the proof of Theorem 6.3.1. So

$$\tilde{E}\{(\tilde{U}_K(x))^2|H_n\} = n^{-2}\int_{K_{1-c}}\int_{K_{1-c}} C(\mathbf{y}-\mathbf{x})H_n(d\mathbf{y})H_n(d\mathbf{x}) + (J_B(x,P))^2.$$

It is easy to see that for any set $A \subset K_{1-c}$ we have $Prob_{\tilde{P}}\{H_n(A) = l|H_n(K_{1-c}) = n\} = \frac{n!}{l!(n-l)!}p^l(1-p)^{n-l}$, where $p = |A|/|K_{1-c}|$, from which it follows that $\tilde{E}H_n(A) = np$, $\tilde{Var}(H_n(A)) = np(1-p)$, and that $\tilde{Cov}(H_n(A), H_n(A')) = 0$ if A and A' are disjoint. The facts regarding $\tilde{E}H_n(d\mathbf{x})$ and $\tilde{E}H_n(d\mathbf{y})H_n(d\mathbf{x})$ that are stated below are a consequence of the Poisson approximation to this binomial; thus, we have

$$\tilde{E}H_n(d\mathbf{x}) = \frac{n}{|K_{1-c}|}d\mathbf{x},$$

and

$$\tilde{E}H_n(d\mathbf{y})H_n(d\mathbf{x}) = \frac{n^2}{|K_{1-c}|^2}d\mathbf{y}d\mathbf{x} + \frac{n}{|K_{1-c}|}d\mathbf{x}\epsilon_{\mathbf{x}}(d\mathbf{y}),$$

from which it now follows that

$$\tilde{E}(\tilde{U}_K(x))^2 = n^{-2}\int_{K_{1-c}}\int_{K_{1-c}} C(\mathbf{y}-\mathbf{x})\frac{n^2}{|K_{1-c}|^2}d\mathbf{y}d\mathbf{x}$$

$$+ n^{-1}C(\mathbf{0}) + (J_B(x,P))^2.$$

Observe now that in the course of the proof of Theorem 6.3.1 we had shown that

$$\frac{1}{|K_{1-c}|^2}\int_{K_{1-c}}\int_{K_{1-c}} C(\mathbf{y}-\mathbf{x})d\mathbf{y}d\mathbf{x} = \widetilde{Var}(U_K) = o(1).$$

Hence $\widetilde{Var}(\tilde{U}_K(x)) = \tilde{E}(\tilde{U}_K(x))^2 - (J_B(x,P))^2 = o(1)$, and (i) of part (a) is proved. ∎

It is perhaps intuitive to have n depend on the original sample size $N(K)$, say, for instance, to take $n = N(K)$. The construction of the new point process H_n is then modified to read: conditionally on the value of n, but independently of the random field $\{X(\mathbf{t})\}$ and of the locations of the points generated by the original Poisson process N, generate n i.i.d. random variables $\mathbf{z}_1, \mathbf{z}_2, \ldots, \mathbf{z}_n$ with uniform distribution on K_{1-c}. The following corollary to Theorem 6.4.1 is now easily available.

Corollary 6.4.1. *Let $n = h(N(K))$, where $h(\cdot)$ is a monotone, strictly increasing function.*

  a. *Under the assumptions of Theorem 6.3.1, and if $J(x,P)$ is continuous at $x = \inf\{x : J(x,P) \geq 1 - \alpha\}$, then the asymptotic coverage probability of the interval $[\hat{\theta}_K - \tau_{N(K)}^{-1} \tilde{c}_{K,B}(1 - \alpha), \infty)$ is the nominal level $1 - \alpha$.*
  b. *Under the assumptions of Theorem 6.3.2, and if $J_{\|\cdot\|}(x,P)$ is continuous at $x = c_{\|\cdot\|}(1-\alpha, P)$, then the asymptotic coverage probability of the confidence region $\{\theta \in \Theta : \tau_{N(K)}\|\hat{\theta}_K - \theta\| \leq \tilde{c}_{K,B,\|\cdot\|}(1-\alpha)\}$ is the nominal level $1 - \alpha$.*

Remark 6.4.1. Note that H_n is in some sense "mimicking" the original point process N; this is especially apparent if we let $n = N(K)$ as in Corollary 6.4.1 because in that case H_n becomes itself a Poisson process, identically distributed to N.

6.5 Variance Estimation via Subsampling

We now address the issue of subsampling variance estimation from marked point process data. As before, we let $\hat{\theta}_K(\tilde{N})$ be a real-valued statistic that estimates a univariate parameter of interest $\theta(P_X)$. For concreteness, assume now that the rate of convergence of $\hat{\theta}_K(\tilde{N})$ to $\theta(P_X)$ is the regular $\tau_u = u^{1/2}$. To be more specific, we assume that

$$Var(|K|^{1/2}\hat{\theta}_K(\tilde{N})) \to \sigma_{\hat{\theta}}^2 \geq 0 \text{ as } \delta(K) \to \infty. \tag{6.13}$$

The issue now is to estimate consistently the asymptotic variance $\sigma_{\hat{\theta}}^2$. Since the subsampling distribution $L_{K,B}(\cdot)$ is a consistent estimator of the distribution of the (centered and normalized) estimator $\hat{\theta}_K(\tilde{N})$, a natural estimator of $\sigma_{\hat{\theta}}^2$ is given by the variance of a random variable having distribution $L_{K,B}(\cdot)$. Again to be more specific, define

$$\hat{\sigma}^2_{K,SUB} \equiv \frac{c^d}{(1-c)^d} \int_{K_{1-c}} \left(\hat{\theta}_{K,B,\mathbf{y}} - \hat{\theta}_{K,B,\cdot}\right)^2 d\mathbf{y}, \qquad (6.14)$$

where $\hat{\theta}_{K,B,\cdot} \equiv \int_{K_{1-c}} \hat{\theta}_{K,B,\mathbf{y}} d\mathbf{y}/|K_{1-c}|$ is the average of the $\hat{\theta}_{K,B,\mathbf{y}}$ replicates.

The following theorem establishes the asymptotic L_2 consistency properties of the subsampling variance estimator $\hat{\sigma}^2_{K,SUB}$; its proof, together with a generalization to non-Poisson sampling schemes, may be found in Politis and Sherman (1998).

Theorem 6.5.1. *Assume (6.13), and that for all compact and convex sets K we have*

$$E\left(|K|^{1/2}|\hat{\theta}_K(\tilde{N}) - E\hat{\theta}_K(\tilde{N})|\right)^{4+\epsilon} \leq C'_\epsilon,$$

for some $\epsilon > 0$, and some finite constant C'_ϵ. Let $\delta(K) \to \infty$, and let $c = c(K) \in (0,1)$ be such that $c \to 0$ but $c\delta(K) \to \infty$.

If $\bar{\alpha}_X(l; l^d) \to 0$, as $l \to \infty$, then $\hat{\sigma}^2_{K,SUB} \xrightarrow{L_2} \sigma^2_{\hat{\theta}}$.

6.6 Examples

We conclude this chapter by presenting some illustrative examples where the subsampling methodology may be of practical interest; for the examples, we assume that $X(\mathbf{t})$ is real valued.

Example 6.6.1 (Univariate mean). Let $X(\mathbf{t})$ be real valued, and let $\hat{\mu} = N(K)^{-1} \int_K X(\mathbf{t}) N(d\mathbf{t})$ be the sample mean. As mentioned in Section 6.3, Karr (1986) has shown that, under some regularity assumptions, $\hat{\mu}$ is consistent for μ, and asymptotically normal at rate $\sqrt{|K|}$, with asymptotic variance equal to $\int R(\mathbf{t}) d\mathbf{t} + \lambda^{-1} R(\mathbf{0})$. Nevertheless, to actually use this asymptotic normality to construct confidence intervals for the mean, the asymptotic variance must be estimated. While it is relatively easy to estimate $R(\mathbf{0})$ (see our Example 6.6.2 that follows), and λ is a.s. consistently estimable by $N(K)/|K|$, consistent estimation of $\int R(\mathbf{t}) d\mathbf{t}$ is not a trivial matter. To see this, note that $\int R(\mathbf{t}) d\mathbf{t} = f(\mathbf{0})$, where $f(\mathbf{w}) \equiv \int R(\mathbf{t}) e^{-i(\mathbf{t}\cdot\mathbf{w})} d\mathbf{t}$ is the spectral density function defined for $\mathbf{w} \in \mathbf{R}^d$, and $(\mathbf{t}\cdot\mathbf{w})$ denotes the inner product.

Since an asymptotic distribution exists, Assumption 6.3.2 is satisfied and the subsampling methodology is immediately seen to be applicable under a mixing condition. Obtaining confidence intervals for μ using the subsampling distribution $L_{K,B}(x)$, rather than the asymptotic normal distribution, side-steps the thorny issue of having to perform nonparametric estimation of the spectral density function at the origin. Nevertheless, as discussed

156   6. Subsampling Marked Point Processes

in Section 5.4, the variance corresponding to the subsampling distribution $L_{K,B}(x)$ may be, under additional moment and mixing conditions, a consistent estimator of the asymptotic variance of $\hat{\theta}_K$, although we will not pursue this here.

Example 6.6.2 (Autocovariance $R(\mathbf{t})$). The usual estimator of $R(\mathbf{t})$ at some point $\mathbf{t}$ is given as

$$\hat{R}(\mathbf{t}) = \frac{1}{\lambda^2 |K|} \int_K \int_K W_K(\mathbf{t} - \mathbf{t}_1 + \mathbf{t}_2) X(\mathbf{t}_1) X(\mathbf{t}_2) N(d\mathbf{t}_1)(N - \epsilon_{\mathbf{t}_1})(d\mathbf{t}_2),$$

where $W_K(\mathbf{t}) = a_K^{-d} W(\mathbf{t}/a_K)$, and the kernel W is a positive, bounded, isotropic probability function on $\mathbb{R}^d$; if λ is unknown, its estimator $N(K)/|K|$ may be used instead. Under regularity conditions, and if the bandwidth $a_K \to 0$ but with $a_K^d |K| \to \infty$ as $\delta(K) \to \infty$, Karr (1986) showed asymptotic normality of $\hat{R}(\mathbf{t})$ at rate $\sqrt{a_K^d |K|}$. Karr (1986) also calculated the asymptotic variance, which is found to depend on the fourth-order cumulant function of the random field $X(\mathbf{t})$. Thus, unless the fourth-order cumulant function is known to vanish, e.g., if the random field $X(\mathbf{t})$ is Gaussian, to obtain confidence intervals for $R(\mathbf{t})$ using the asymptotic distribution, it is necessary to estimate the 4th order cumulant function; this is a task quite more formidable than estimation of $R(\mathbf{t})$. Subsampling again side-steps this difficulty, and immediately yields valid confidence intervals for $R(\mathbf{t})$ since again Assumption 6.3.2 holds.

Example 6.6.3 (Integral $\int R(\mathbf{t}) g(\mathbf{t}) d\mathbf{t}$). Let g be a continuous function with compact support satisfying $g(\mathbf{0}) = 0$, and suppose it is required to estimate the integral $\int R(\mathbf{t}) g(\mathbf{t}) d\mathbf{t}$. Karr (1986, 1991) proposed the estimator

$$\frac{1}{\lambda^2 |K|} \int_K M(d\mathbf{t}_1) \int g(\mathbf{t}_2 - \mathbf{t}_1) M(d\mathbf{t}_2),$$

where $M(d\mathbf{t}) = Y(\mathbf{t}) N(d\mathbf{t})$, and showed its asymptotic normality at rate $\sqrt{|K|}$. In this case, however, the asymptotic variance is too difficult to calculate, let alone estimate. Again, subsampling comes to the rescue, yielding valid confidence intervals for $\int_A R(\mathbf{t}) d\mathbf{t}$, since Assumption 6.3.2 holds.

Example 6.6.4 (Probability and spectral density). Let $X(\mathbf{t})$ have a distribution that admits a probability density function $p(x)$, and let $f(\mathbf{w})$ denote the spectral density function as defined in Example 6.6.1. Different smoothing methods have been developed for consistent estimation of $p(x)$ and $f(\mathbf{w})$ (at some given points $x \in \mathbb{R}$ and $\mathbf{w} \in \mathbb{R}^d$) and asymptotic normality of the proposed estimates at well-defined rates has been established (see, e.g., Lii and Masry, 1994, and Masry, 1978, 1983, 1988). Nevertheless, besides the issue of complicated asymptotic variances, there is a more serious issue here if our objective is to use the asymptotic normal distribution to obtain confidence intervals for $p(x)$ and $f(\mathbf{w})$ on the basis of their estimators denoted by $\hat{p}(x)$ and $\hat{f}(\mathbf{w})$; this is the issue of bias. It is well known that the asymptotic distribution of an optimally smoothed probability or spectral density estimator has a remaining bias term; the term "optimally" here implies a bandwidth choice that "balances" the bias and the standard deviation of the nonparametric estimator at hand, therefore minimizing the mean squared error of estimation. Consequently, the use of the asymptotic normal distribution of an optimally smoothed $\hat{p}(x)$ or $\hat{f}(\mathbf{w})$ will only yield confidence intervals for $E\hat{p}(x)$ and $E\hat{f}(\mathbf{w})$, respectively, rather than for $p(x)$ and $f(\mathbf{w})$, unless the bias term is explicitly accounted for, which is usually difficult (see also Politis and Romano, 1994b for a more detailed discussion on the bias issue). Nevertheless, the asymptotic distribution $J(P)$ of Assumption 6.3.2 is not required to have zero mean for our subsampling methodology to apply. Valid confidence intervals for $p(x)$ and $f(\mathbf{w})$ can immediately be obtained from conclusion (iii) of Theorem 6.3.1 here; an intuitive explanation of this phenomenon is that subsampling provides an automatic built-in estimation of the bias, with subsequently bias-corrected confidence intervals.

Example 6.6.5 (Bispectrum and test for normality). As the spectral density function $f(\mathbf{w})$ is simply the Fourier transform of the second-order cumulant function of the random field $\{X(\mathbf{t})\}$ (i.e., the autocovariance), the bispectrum $f(\mathbf{w},\mathbf{l})$ is the Fourier transform of the third-order cumulant function; the latter notably is a function of only two, rather than three, arguments) by the assumed stationarity of $\{X(\mathbf{t})\}$. For the time series case $d = 1$, Lii and Tsou (1995) developed nonparametric, kernel-smoothed, consistent estimators $\hat{f}(\mathbf{w},\mathbf{l})$ of the bispectrum $f(\mathbf{w},\mathbf{l})$. Under some regularity conditions, Lii and Tsou (1995) also showed asymptotic multivariate normality at some well-defined, bandwidth-dependent rate we will denote by $\tau_{|K|}$. To elaborate, if

$$A = \{(\mathbf{w}_j,\mathbf{l}_j) \text{ for } j = 1,\ldots,J\}$$

is a discrete set of points, we can define the (multivariate) statistic $\hat{\theta}_K$ to have as its jth coordinate the value $\hat{f}(\mathbf{w}_j,\mathbf{l}_j)$, and the (multivariate) parameter $\theta(P_X)$ to have as its jth coordinate the value $\theta_j \equiv f(\mathbf{w}_j,\mathbf{l}_j)$. Note now that asymptotic (multivariate) normality of $\hat{\theta}_K$ implies (by the continuous mapping theorem) existence of an asymptotic distribution for $||\hat{\theta}_K - \theta(P_X)||_\infty \equiv \sup_j |\hat{f}(\mathbf{w}_j,\mathbf{l}_j) - f(\mathbf{w}_j,\mathbf{l}_j)|$; in other words, letting $\Theta = \mathbb{R}^J$ equipped with the l_∞ or sup norm $||\cdot||_\infty$, the results of Lii and Tsou (1995) imply that our Assumption 6.3.1 is satisfied, and thus subsampling may be used to construct an approximate $(1-\alpha)100\%$ confidence region for $\theta(P_X)$ of the following type: $\tau_{|K|} \sup_j |\hat{f}(\mathbf{w}_j,\mathbf{l}_j) - \theta_j| \leq c_{K,B,||\cdot||_\infty}(1-\alpha)$.

Now since the parameter θ is really a function evaluated at different points, and because we are using the sup distance (uniform distance) coordinate-wise on θ, the confidence region for $\theta(P_X)$ will have the interpretation of a $(1-\alpha)100\%$ uniform confidence band for the unknown function $f(\mathbf{w},1)$ of the type: $f(\mathbf{w},1) = \hat{f}(\mathbf{w},1) \pm c_{K,B,||\cdot||_\infty}(1-\alpha)/\tau_{|K|}$, for all $(\mathbf{w},1) \in A$.

Recall that if the random field $\{X(\mathbf{t})\}$ is Gaussian, then the third-order cumulant function (and therefore the bispectrum as well) vanishes. Thus, the just obtained $(1-\alpha)100\%$ uniform confidence band for the bispectrum affords us the ability to perform a simple and easy test of the hypothesis

H_0: *"The random field $\{X(\mathbf{t})\}$ is Gaussian."*

The test rejects the normal hypothesis H_0 at approximate level α if the uniform band $\hat{f}(\mathbf{w},1) \pm c_{K,B,||\cdot||_\infty}(1-\alpha)/\tau_{|K|}$ does not fully (i.e., for all $(\mathbf{w},1) \in A$) "contain" the graph of the identically zero function. In other words, the test rejects the normal hypothesis if

$$0 \notin [\hat{f}(\mathbf{w},1) - c_{K,B,||\cdot||_\infty}(1-\alpha)/\tau_{|K|}, \hat{f}(\mathbf{w},1) + c_{K,B,||\cdot||_\infty}(1-\alpha)/\tau_{|K|}]$$

for at least one point $(\mathbf{w},1) \in A$.

Equivalently, if one constructs the band $0 \pm c_{K,B,||\cdot||_\infty}(1-\alpha)/\tau_{|K|}$ around the identically zero function, the test rejects the normal hypothesis if the graph of the estimator $\hat{f}(\mathbf{w},1)$ is not fully contained in this band for all $(\mathbf{w},1) \in A$; that is, the test rejects H_0 if $|\hat{f}(\mathbf{w},1)| > c_{K,B,||\cdot||_\infty}(1-\alpha)/\tau_{|K|}$ for at least one point $(\mathbf{w},1) \in A$.

6.7 Conclusions

In this chapter, the setup of data arising from a marked point process was addressed. In particular, the data are of the type $[\mathbf{t}_1, X(\mathbf{t}_1)], [\mathbf{t}_2, X(\mathbf{t}_2)], \ldots,$ $[\mathbf{t}_{N(K)}, X(\mathbf{t}_{N(K)})]$, where $\mathbf{t}_1, \mathbf{t}_2, \ldots, \mathbf{t}_{N(K)}$ are the $\mathbf{t}$-points generated by a homogeneous, Poisson process N (and ordered by some fashion) inside the compact, convex set K that constitutes our "observation region." The random field $\{X(\mathbf{t})\}$ generating the X-"marks" is assumed to be stationary and independent of the point process N.

The subsampling methodology was extended to cover this type of data, and the asymptotic validity of subsampling-based inference was demonstrated under very weak conditions. In addition, a mixing condition that is sufficient for our purposes was identified using strong mixing coefficients analogous to the ones defined in Chapter 5. The subsampling distribution is defined by an integral, and different ways consistently to approximate the integral by a sum were discussed.

It was described how to find confidence intervals for real-valued parameters and confidence regions for vector-valued parameters. The issue of variance estimation via subsampling was also addressed, and a number of examples showing the usefulness of subsampling with marked point process data were given.

7
Confidence Sets for General Parameters

7.1 Introduction

Let $X_1, \ldots, X_n$ denote a realization of a stationary time series. Suppose the infinite dimensional distribution of the infinite sequence is denoted P. The problem we consider is inference for a parameter $\theta(P)$. The focus of the present chapter is the case when the parameter space Θ is a metric space. The reason for considering such generality is to be able to consider the case when the parameter of interest is an unknown function, such as the marginal distribution of the process or the spectral distribution function of the process. Here, we need to extend the arguments of previous chapters to cover the more general case.

Recall that in Section 2.5 we already considered subsampling for more general parameters, such as a construction of uniform confidence bands for an entire function. Essentially, the problem was reduced to the simple case of estimating a distribution function on the real line by considering an appropriate real-valued function (such as a norm) of a function-valued process. The goal in this chapter is to achieve a stronger result, by estimating the distribution of an estimator taking values in an abstract space, such as a function space. In fact, the focus in this chapter will be to estimating a sampling distribution in a metric space, which may even be nonseparable.

In order to accomplish this task, we first develop a general result on the closeness of the empirical measure based on triangular arrays of dependent random variables, where the random variables take values in a (possibly nonseparable) metric space. In Section 7.2, we present such a result, which may be viewed as a generalization of the classical result of Varadarajan (1958), who considered a sequence of i.i.d. variables in a separable metric space. The threefold generalization to triangular arrays, to dependent variables, and to nonseparable metric spaces are all required for the statistical applications.

In Section 7.3, we apply the result of Section 7.2 to obtain a general result for subsampling. In Section 7.4, this result is applied to the special case of estimating the distribution of the empirical process of a stationary time series. The argument is seen to be quite simple and direct. Comparable results based on the (moving blocks) bootstrap are much more involved and they rely on heavier assumptions. In Section 7.5, the result is immediately applied to estimating the distribution of the spectral process (where no bootstrap counterpart has been established). Much of this chapter is extracted from Politis, Romano, and Wolf (1999).

7.2 A Basic Theorem for the Empirical Measure

Throughout this section, S denotes a (possibly nonseparable) metric space, equipped with a metric d. S is endowed with a σ-field **A**, which will be assumed large enough to contain all closed balls, but perhaps not as large as the Borel σ-field. The general problem considered concerns the closeness of the empirical distribution of S-valued observations to the underlying law. The observations are assumed stationary (though this can be generalized) and weakly dependent.

Weak dependence is quantified in terms of strong mixing coefficients. Specifically, let $\{X_t, t \in T\}$ denote a collection of random variables defined on some common probability space, and assume T is some subset of the integers. Let $\mathcal{F}_k$ denote the σ-field generated by $\{X_t, t \leq k\}$ and let $\mathcal{G}_j$ denote the σ-field generated by $\{X_t, t \geq j\}$. Define Rosenblatt's α-mixing coefficients by

$$\alpha_X(j) = \sup\{|P(AB) - P(A)P(B)| : A \in \mathcal{F}_k, B \in \mathcal{G}_{k+j}, k = 1, 2, \ldots\}. \tag{7.1}$$

Closeness of measures is described in terms of a metric metrizing weak convergence. Specifically, the Bounded-Lipschitz metric ρ_L is defined as follows. Let **L** be the class of **A**-measurable functions f satisfying

$$|f(x) - f(y)| \leq d(x, y) \text{ and } \sup_{x \in S} |f(x)| \leq 1.$$

For a probability Q and f in **L**, let the notation Q_f denote

$$Q_f = E_Q f = \int_S f(x) Q(dx).$$

For probability laws Q_1 and Q_2, define

$$\rho_L(Q_1, Q_2) = \sup\{|Q_1 f - Q_2 f| : f \in \mathbf{L}\}. \tag{7.2}$$

7.2 A Basic Theorem for the Empirical Measure

Convergence of $\rho_L(Q_n, Q)$ to zero and Q concentrating on a separable set implies weak convergence; see Pollard (1984, p.74).

The following theorem is a generalization of a classical result of Varadarajan (1958), who considered the case of i.i.d. observations in a separable metric space. Beran, LeCam, and Millar (1987) and Bickel and Millar (1992) extended his result to triangular arrays of i.i.d. variables in possibly nonseparable metric spaces. The result here covers the dependent case, which is needed in the next section for inference with time series.

Theorem 7.2.1. *Let $Y_{n,1}, \ldots, Y_{n,j_n}$ be S-valued, stationary observations with strong mixing sequence $\alpha_n(\cdot)$. Assume $j_n \to \infty$ as $n \to \infty$. Denote by Q_n the marginal distribution of $Y_{n,1}$, and let $\hat{Q}_n$ denote the empirical measure of the $Y_{n,t}$, $1 \le t \le j_n$. Assume $\{Q_n\}$ is δ-tight; that is, for every $\epsilon > 0$, there exists a compact set K and $\delta_n \downarrow 0$ so that $Q_n(K^{\delta_n}) > 1 - \epsilon$ for all n, where $K^\delta = \{x \in S : d(x, K) < \delta\}$. Assume the mixing coefficients satisfy $\sum_{t=k}^{j_n} \alpha_n(t)/j_n \to 0$ as $n \to \infty$.*

Then, $\rho_L(\hat{Q}_n, Q_n) \to 0$ in probability.

Remark 7.2.1. The issue of measurability cannot be ignored because $\rho_L(\hat{Q}_n, Q_n)$, being the sup of an uncountable collection of random variables, need not even be measurable. Instead of worrying about whether measurability holds, the proof shows the result is true if we interpret convergence in probability to mean convergence in outer probability (see van der Vaart and Wellner, 1996, Section 1.9).

Remark 7.2.2. The result can clearly be generalized to nonstationary observations if Q_n is replaced by the expectation of $\hat{Q}_n$.

Remark 7.2.3. The result holds for any metric metrizing weak convergence (by a subsequence argument).

Remark 7.2.4. The argument can be strengthened to yield an almost sure result when all the variables are defined on a common probability space.

Remark 7.2.5. The tightness assumption cannot be removed, even in the i.i.d. case; a counterexample is given in Beran, LeCam, and Millar (1987).

Proof of Theorem 7.2.1. For ease of notation, we assume $j_n = n$. Now, for any real-valued measurable function f defined on S which is uniformly bounded by one, $\hat{Q}_n f - Q_n f \to 0$ in probability. Indeed, $\hat{Q}_n f - Q_n f$ has mean zero and variance

$$\sigma_n^2(f) = n^{-1} Q_n f(f - Q_n f) + 2n^{-1} \sum_{t=1}^{n-1} (1 - \frac{t}{n}) Cov[f(Y_{n,1}), f(Y_{n,1+t})]$$

162   7. Confidence Sets for General Parameters

$$\leq n^{-1} + 8n^{-1} \sum_{t=1}^{n} \alpha_n(t) \equiv b_n \to 0, \tag{7.3}$$

by the standard strong mixing inequality for uniformly bounded variables (see Lemma A.0.2). The difficulty in establishing the theorem lies in showing this convergence is uniform over an uncountable collection of functions f.

Now, to show

$$\sup_{f \in \mathbf{L}} |\hat{Q}_n f - Q_n f| \to 0 \text{ in probability,}$$

it suffices to show

$$\sup_{f \in \mathbf{L_n}} |\hat{Q}_n f - Q_n f| \to 0 \text{ in probability,}$$

where $\mathbf{L_n}$ are the functions of the form $f(x)I(x \in K^{\delta_n})$, f is a function in $\mathbf{L}$ and δ_n is as in the statement of the theorem. To appreciate why, first note that $Q_n(K^{\delta_n}) \geq 1 - \epsilon$ by δ-tightness. Second, by the above variance calculation for general f, $\hat{Q}_n(K^{\delta_n})$ has variance bounded by b_n. So, by Chebychev,

$$\begin{aligned} Prob\,\{1 - \hat{Q}_n(K^{\delta_n}) \geq 2\epsilon\} &= Prob\,\{\hat{Q}_n(K^{\delta_n}) - Q_n(K^{\delta_n}) \\ &\leq 1 - Q_n(K^{\delta_n}) - 2\epsilon\} \\ &\leq Prob\,\{\hat{Q}_n(K^{\delta_n}) - Q_n(K^{\delta_n}) \leq -\epsilon\} \\ &\leq Prob\,\{|\hat{Q}_n(K^{\delta_n}) - Q_n(K^{\delta_n})| \geq \epsilon\} \\ &\leq Var\,[\hat{Q}_n(K^{\delta_n})]/\epsilon^2 \leq b_n/\epsilon^2 \to 0. \end{aligned}$$

So,

$$\sup_{f \in \mathbf{L}} |\hat{Q}_n f - Q_n f| \leq \sup_{f \in \mathbf{L_n}} |\hat{Q}_n f - Q_n f| + \sup_{f \in \mathbf{H_n}} |\hat{Q}_n f - Q_n f|,$$

where $\mathbf{H_n}$ is the collection of functions $\{f(\cdot)[1 - I(\cdot \in K^{\delta_n})] : f \in \mathbf{L}\}$. But, the last term

$$\sup_{f \in \mathbf{H_n}} |\hat{Q}_n f - Q_n f| \leq \sup_{f \in \mathbf{H_n}} |\hat{Q}_n f| + \sup_{f \in \mathbf{H_n}} |Q_n f|$$

is small because

$$\sup_{f \in \mathbf{H_n}} |Q_n f| \leq \epsilon$$

and

$$\sup_{f \in \mathbf{H_n}} |\hat{Q}_n f| \leq 1 - \hat{Q}_n(K^{\delta_n}) \leq 2\epsilon \text{ with probability tending to one,}$$

by the above. Hence, it is sufficient to show $\sup_{f \in \mathbf{L_n}} |\hat{Q}_n f - Q_n f|$ tends to zero in probability.

7.2 A Basic Theorem for the Empirical Measure

Fix $\epsilon > 0$. Next, let $\{f_1, \ldots, f_{m_\epsilon}\}$ be an ϵ-net (where the metric is sup norm) for the collection of functions $\mathbf{L_K} = \{f(\cdot)I(\cdot \in K)\}$. That is, if $f \in \mathbf{L}$, there is an i such that

$$\sup_{x \in K} |f(x) - f_i(x)| < \epsilon.$$

Note that the number m_ϵ of approximating functions in the ϵ-net is finite by the Arzela–Ascoli Theorem (see Dudley, 1989, Theorem 2.4.7). At this point, we cannot assume the approximating functions f_i are bounded-Lipschitz on all of S, because the Arzela–Ascoli theorem applies to the functions restricted to the compact set K. However, by the Kirszbraun–McShane extension theorem (see Theorems 6.1.1 and 11.2.3 of Dudley, 1989), the functions f_i can be assumed to be in $\mathbf{L}$. Now, if $x \in K^{\delta_n}$, then there exists $\tilde{x} \in K$ so that $d(x, \tilde{x}) < \delta_n$. Then, for any $f \in \mathbf{L}$,

$$|f(x) - f(\tilde{x})| \leq d(x, \tilde{x}) < \delta_n,$$

where we have used the fact that f is Lipschitz. This inequality is also true for the approximating functions f_i. So,

$$f(x) \leq f(\tilde{x}) + \delta_n \leq f_i(\tilde{x}) + \epsilon + \delta_n \leq f_i(x) + \epsilon + 2\delta_n.$$

Hence, for any $f \in \mathbf{L_n}$, there exists an $i \leq m_\epsilon$ satisfying

$$\sup_{x \in K^{\delta_n}} |f(x) - f_i(x)| \leq \epsilon + 2\delta_n. \tag{7.4}$$

Given an f, let $\tilde{f}$ denote the approximating function f_i satisfying (7.4). Then,

$$\sup_{f \in \mathbf{L_n}} |\hat{Q}_n f - Q_n f| \leq \sup_{f \in \mathbf{L_n}} |(\hat{Q}_n - Q_n)(f - \tilde{f})| + \max_{1 \leq i \leq m_\epsilon} |(\hat{Q}_n - Q_n) f_i|$$

$$\leq 2\epsilon + 4\delta_n + \max_{1 \leq i \leq m_\epsilon} |(\hat{Q}_n - Q_n) f_i|.$$

To show the left side tends to 0 in probability, fix any $\eta > 0$ and let $\epsilon = \eta/8$. Then,

$$Prob\{ \sup_{f \in \mathbf{L_n}} |\hat{Q}_n f - Q_n f| > \eta\} \leq Prob\{ \sup_{f \in \mathbf{L_n}} |(\hat{Q}_n - Q_n)(f - \tilde{f})| > \eta/2\}$$

$$+ Prob\{ \max_{1 \leq i \leq m_\epsilon} |(\hat{Q}_n - Q_n) f_i| > \eta/2\}$$

$$\leq Prob\{ \max_{1 \leq i \leq m_\epsilon} |(\hat{Q}_n - Q_n) f_i| > \eta/2\},$$

as soon as $2\epsilon + 4\delta_n \equiv \eta/4 + 4\delta_n$ is less than $\eta/2$, or equivalently when $\eta > 16\delta_n$. Finally, the last term tends to zero because it can be bounded by $m_\epsilon 4 b_n^2 / \eta^2$, where b_n is defined in (7.3). ∎

7.3 A General Theorem on Subsampling

In this section, we consider the problem of constructing asymptotically valid confidence sets for a parameter of a (strictly) stationary time series $\{X_t, t = 0, \pm 1, \pm 2, \ldots\}$, whose joint distribution will be denoted P. The variables X_t are all defined on some common probability space $(\Omega_1, \mathcal{F}_1, \mu_1)$ and take values in some general measure space $(\Omega_2, \mathcal{F}_2)$, though Ω_2 is usually assumed to be the real line. Hence, the joint law of the X_t variables, denoted by P, is a probability on the product space that is the countable product of Ω_2 endowed with the product σ-field.

Attention focuses on a parameter $\theta(P)$ that takes values in a parameter space Θ. At this point, nothing is assumed about Θ. The goal is to construct an asymptotically valid confidence set for $\theta(P)$ based on $X_1, \ldots, X_n$. The whole motivation of the present chapter is to present a general result for the case when Θ is quite large, such as an infinite-dimensional function space.

Suppose an estimator, $\hat{\theta}_n = \hat{\theta}_n(X_1, \ldots, X_n)$, of $\theta(P)$ is given. In order to construct a confidence set (such as a confidence band, region, or interval, depending on the context) for $\theta(P)$, some knowledge of the sampling distribution of $\hat{\theta}_n(X_1, \ldots, X_n)$ is required. More generally, let $R_n(X_1, \ldots, X_n; \theta(P))$ be a root, which is just some function of $X_1, \ldots, X_n$, $\theta(P)$ and n, taking values in a metric space S (endowed with a σ-field $\mathbf{A}$). The term root is now generalized from Chapter 1 so that the root may take values in a space S. For example, in the case where Θ is a linear space, we might take

$$R_n(X_1, \ldots, X_n; \theta(P)) = \tau_n[\hat{\theta}_n(X_1, \ldots, X_n) - \theta(P)], \qquad (7.5)$$

in which case $S = \Theta$; here, $\{\tau_n\}$ is just some normalizing sequence in anticipation of our asymptotic results. Alternatively, if Θ is a normed linear space with norm denoted $\|\cdot\|$, we might take

$$R_n(X_1, \ldots, X_n; \theta(P)) = \tau_n \|\hat{\theta}_n(X_1, \ldots, X_n) - \theta(P)\|,$$

so that S is the real line and possibly distinct from Θ. In any case, the idea is that if the sampling distribution of $R_n(X_1, \ldots, X_n; \theta(P))$ under P where known, this information could be used to obtain a confidence set for $\theta(P)$.

To fix ideas, consider the following examples. In all these examples, assume the time series consists of stationary real-valued observations. If P denotes the joint distribution of the process, let P_1 denote the marginal distribution of X_1. Let $\theta(P) = P_1$, a parameter that takes values in the space of distribution functions. Several choices for S exist, one being $D[-\infty, \infty]$ equipped with the uniform metric. Here, we could take $\hat{\theta}_n$ to be the empirical distribution function of the data. Alternatively, one might be interested in a simple real-valued functional of P_1, but our theory is intentionally general enough to handle general parameters. Certainly, if we can handle inference for P_1, we should be able to handle smooth functionals of P_1. A more important example in the context of modeling time series is the spectral distribution function, which again can be assumed to take values in a suitable function space. Note that in both of these examples, a function space equipped with the supremum norm is nonseparable. These examples will be developed in the next two sections.

Let $J_n(P)$ denote the law of $R_n(X_1, \ldots, X_n; \theta(P))$, regarded as a random element of a metric space S. We are implicitly assuming S is endowed with an appropriate σ-field so that R_n is measurable. The subsampling approximation to $J_n(P)$, denoted $L_{n,b}$, is the empirical distribution of the $n - b + 1$ values of $R_b(X_t, \ldots, X_{t+b-1}; \hat{\theta}_n)$ as t ranges from 1 to $n - b + 1$. The main assumption we will need is the following:

Assumption 7.3.1. *Assume $J_n(P)$ converges weakly to a limit law $J(P)$ which concentrates on a separable subset of S.*

As a preliminary step to analyzing the subsampling distribution $L_{n,b}$, we first analyze $H_{n,b}$, that is defined to be the empirical distribution of the $n - b + 1$ values of $R_b(X_t, \ldots, X_{t+b-1}; \theta(P))$.

Proposition 7.3.1. *Let $X_1, \ldots, X_n$ be a stationary time series with α-mixing sequence $\alpha_X(\cdot)$. Assume $\alpha_X(t) \to 0$ as $t \to \infty$. Also, assume $b/n \to 0$ and $b \to \infty$ as $n \to \infty$.*

Then, $\rho_L(H_{n,b}, J_n(P)) \to 0$ in probability.

Proof. Apply Theorem 7.2.1 with $Y_{n,t} = R_b(X_t, \ldots, X_{t+b-1}; \theta(P))$ and $j_n = n - b + 1$. Note that $H_{n,b}$ is the empirical distribution of S-valued stationary observations with exact distribution $J_b(P)$. So, Q_n is the distribution of $Y_{n,1}$ and the tightness assumption follows from Assumption 7.3.1 because $Q_n = J_b(P)$. Also, the mixing sequence of the n-th row $Y_{n,1}, \ldots, Y_{n,j_n}$ satisfies $\alpha_n(t) \leq \alpha_X(t - b)$ if $t - b \geq 0$. Bound $\alpha_n(t)$ by 1 otherwise. Then,

$$\sum_{t=1}^{j_n} \alpha_n(t)/j_n \leq \frac{1}{n - b + 1}[b + \sum_{t=b+1}^{n-b+1} \alpha_X(t - b)]$$

$$\leq \frac{b}{n - b + 1} + \frac{1}{n - b + 1} \sum_{t=1}^{n-b+1} \alpha_X(t) \to 0,$$

by assumptions on $\alpha_X(\cdot)$ and b. ∎

The previous result cannot in general be used for inference because the construction of $H_{n,b}$ involves the unknown $\theta(P)$. In order to obtain a result for the subsampling law $L_{n,b}$, we specialize a little. In particular, assume R_n takes the form (7.5) so that $\Theta = S$ and S is a normed linear space with norm $\|\cdot\|$ and $d(x, y) = \|x - y\|$ if x and y are in S.

166   7. Confidence Sets for General Parameters

Theorem 7.3.1. *Under the assumptions of Proposition 7.3.1 and the additional assumption that $\tau_b/\tau_n \to 0$ as $n \to \infty$, we have*

$$\rho_L(L_{n,b}, J_n(P)) \to 0 \text{ in probability.}$$

Proof. By the triangle inequality, it suffices to show $\rho_L(L_{n,b}, H_{n,b}) \to 0$ in probability. Make the same identifications as in the proof of Proposition 7.3.1. Since $L_{n,b}$ is the empirical distribution of the values

$$\tau_b[\hat{\theta}_b(X_t, \ldots, X_{t+b-1}) - \hat{\theta}_n] = \tau_b[\hat{\theta}_b(X_t, \ldots, X_{t+b-1}) - \theta(P)] + \tau_b[\theta(P) - \hat{\theta}_n],$$

we see that $L_{n,b}$ is just $H_{n,b}$ shifted by $\tau_b[\theta(P) - \hat{\theta}_n]$. Now, if $f \in \mathbf{L}$ is a (measurable) bounded Lipschitz function,

$$|L_{n,b}f - H_{n,b}f| \leq \frac{1}{n-b+1} \sum_{t=1}^{n-b+1} d(\hat{\theta}_{n,b,t}, \hat{\theta}_{n,b,t} - \tau_b[\theta(P) - \hat{\theta}_n])$$

$$= \|\tau_b[\theta(P) - \hat{\theta}_n]\|,$$

where $\hat{\theta}_{n,b,t} = \tau_b[\hat{\theta}_b(X_t, \ldots, X_{t+b-1}) - \theta(P)]$. So,

$$\rho_L(L_{n,b}, H_{n,b}) \leq \|\tau_b[\theta(P) - \hat{\theta}_n]\| = \frac{\tau_b}{\tau_n}\|\tau_n[\hat{\theta}_n - \theta(P)]\| \to 0 \text{ in probability}$$

by Assumption 7.3.1. ∎

7.4 Subsampling the Empirical Process

One of the great successes of Efron's (1979) i.i.d. bootstrap is that it can be used to approximate the distribution of the empirical process accurately, and hence certain functionals of the empirical process. There has been considerable interest in obtaining similar results for dependent data using the moving blocks bootstrap of Künsch (1989) and Liu and Singh (1992) (for example, see Bühlmann, 1994, and Naik-Nimbalkar and Rajarshi, 1994). Here, we obtain similar results based on subsampling using much simpler arguments. The aforementioned results rely on intricate chaining arguments and exponential inequalities. Our argument completely avoids such calculations if the original process of interest converges (which it is known to), whereas convergence of the moving blocks bootstrap process appears to require these arguments to be generalized and repeated. In fact, our simple arguments weaken the assumptions made on the underlying marginal distribution of the process, the choice of block size, and the mixing coefficients.

As before, let $X_1, \ldots, X_n$ denote a stretch of a stationary process, whose entire joint distribution is denoted P. Let P_1 denote the marginal distribution of X_1 and let $\hat{P}_n$ denote the empirical measure. Consider the empirical process $Z_n(\cdot)$ indexed by a class of functions $f \in \mathbf{F}$, defined by

$$Z_n(f) = n^{1/2}[\hat{P}_n f - P_1 f].$$

7.4 Subsampling the Empirical Process

Here, if the X_t take values in a space Ω_2, the functions f are assumed to be real valued with domain Ω_2. Regard Z_n as a random element of the metric space $L_\infty(\mathbf{F})$, the metric space of real-valued bounded functions on $\mathbf{F}$ with sup norm denoted $\|\cdot\|_\infty$. The goal is to approximate the distribution of Z_n, which we will denote by $J_n(P)$ in agreement with the general notation of Section 7.3.

Assumption 7.4.1. *Assume the law of Z_n converges weakly to a limiting process Z that concentrates on a separable subset of $L_\infty(\mathbf{F})$.*

Remark 7.4.1. To avoid measurability problems, we simply assume $\mathbf{F}$ to be a permissible class of functions, as in Pollard (1984). Also note that one needs to endow $L_\infty(\mathbf{F})$ with an appropriate σ-field or, alternatively, understand that weak convergence to be in the sense of Hoffman–Jørgensen; the latter approach is fully developed in Part 1 of van der Vaart and Wellner (1996). All our assumptions are wrapped in the assumption that Z_n converges weakly. Arcones and Yu (1994) have given a sufficient condition for this assumption to hold. They assume $\mathbf{F}$ is a VC graph class with envelope function F satisfying $PF^p < \infty$ for some $p > 2$ and β-mixing coefficients satisfying

$$\sum_{t=1}^\infty [\beta(t)]^{\frac{p-2}{p}} < \infty$$

and $t^{p/(p-2)} \log(t) \beta(t) \to 0$ as $t \to \infty$. Then, Z is a mean zero Gaussian process with

$$Cov[Z(f), Z(g)] = \sum_{j=-\infty}^\infty Cov[f(X_0), f(X_j)].$$

In the special case of the real line when $\mathbf{F}$ is the usual class of intervals, Deo (1973) has obtained a sufficient α-mixing condition and Yoshihara (1975) has considered the multidimensional case. Since no best sufficient mixing condition exists for the weak convergence of Z_n, we simply take Assumption 7.4.1 as given.

As in Section 7.3, the subsampling approximation to $J_n(P)$ is $L_{n,b}$, the empirical distribution of the $(n - b + 1)$ values $Z_{n,1}, \ldots, Z_{n,n-b+1}$, where

$$Z_{n,t}(f) = b^{1/2}[\hat{P}_{n,t}f - \hat{P}_n f],$$

and $\hat{P}_{n,t}$ is the empirical measure based on $X_t, \ldots, X_{t+b-1}$.

Theorem 7.4.1. *Assume Assumption 7.4.1 and that the $\{X_t\}$ process is α-mixing. Also assume that $b \to \infty$ and $b/n \to 0$ as $n \to \infty$.*

Then, $\rho_L(J_n(P), L_{n,b}) \to 0$ in probability.

Proof. Apply Theorem 7.3.1 with $R_n = n^{1/2}[\hat{P}_n(\cdot) - P_1(\cdot)]$, so that $S = \Theta = L_\infty(\mathbf{F})$. Here, $\tau_n = n^{1/2}$ and Assumption 7.4.1 is Assumption 7.3.1 specialized to the empirical process. ∎

Remark 7.4.2. We have aimed for a simple result by elementary but general methods. Even so, the assumptions are remarkably weak. In the case of the moving blocks bootstrap, Naik-Nimbalkar and Rajarshi (1994) assume the block size b of order n^p for some $p \in (0, 1/4)$, but they obtain an almost sure convergence result (in the special case of the real line). We could also obtain an almost sure convergence result simply by strengthening our arguments a little, though the statistical uses for the stronger result are not clear enough to warrant doing so at this time. Bühlmann (1994) assumes $b = n^p$ for $p \in (0, 1/2)$. These papers assume the X_t's are real and vector valued, respectively, and that the marginal distributions are continuous, which we do not need. The method utilized here applies to the case of the empirical process indexed by a general class of functions on an arbitrary space. Our mixing assumption is, of course, quite weak, with the real mixing assumption wrapped up in the verification of Assumption 7.4.1. The mixing conditions of the aforementioned works are stronger.

Remark 7.4.3. Needless to say, it follows that any continuous functional of the empirical process has an unknown distribution that can be consistently estimated by its subsampling counterpart. By looking at the supremum of the empirical process, asymptotically valid confidence bands for the unknown measure ensue.

Remark 7.4.4. Our result can actually be used to prove the corresponding result for the moving blocks bootstrap. By exploiting the linear structure of the empirical process, one sees that the moving blocks distribution can be obtained from the subsampling distribution by an appropriate normalized convolution operation. Since the subsampling distribution is an approximate Gaussian process, so must be a normalized convolution. This approach would simplify the arguments of Naik-Nimbalkar and Rajarshi (1994) and Bühlmann (1994).

7.5 Subsampling the Spectral Measure

The general theory in Sections 7.2 and 7.3 was motivated by the problem of approximating the distribution of the spectral process, which we define below. Let $F(\cdot)$ denote the spectral distribution function of a real-valued stationary time series, assumed to have a finite second moment. Here, $\theta = F(\cdot)$. Borrowing notation from Dahlhaus (1985), let $I_n(\lambda)$ denote the periodogram with tapered data, defined by

$$I_n(\lambda) = [2\pi H_{n,2}(0)]^{-1} d_n(\lambda) d_n(-\lambda),$$

where

$$d_n(\lambda) = \sum_{t=1}^{n} h[t/(n+1)] X_t \exp[-i\lambda t]$$

and

$$H_{n,k}(\lambda) = \sum_{t=1}^{n} h^k[t/(n+1)] \exp[-i\lambda t].$$

The data taper h is assumed of bounded variation and square integrable on $[0,1]$. Let $\hat{F}_n(\cdot)$ be the corresponding integrated periodogram given by

$$\hat{F}_n(\lambda) = \frac{2\pi}{n} \sum_{0 < 2\pi s/n \leq \lambda} I_n(2\pi s/n).$$

Take $\tau_n = n^{1/2}$ and regard $S_n(\cdot) = n^{1/2}[\hat{F}_n(\cdot) - F(\cdot)]$ as a random element of $D[0,\pi]$ endowed with the sup norm $\|\cdot\|$. Under suitable weak dependence conditions, the process $S_n(\cdot)$ converges weakly to a mean zero Gaussian process $S(\cdot)$ with covariance

$$Cov[S(\lambda), S(\mu)] = 2\pi G(\min\{\lambda,\mu\}) + 2\pi F_4(\lambda,\mu),$$

where

$$G(\lambda) = \int_0^\lambda f^2(\beta) d\beta$$

and

$$F_4(\lambda,\mu) = \int_0^\lambda \int_0^\mu f_4(\alpha, -\alpha, -\beta) d\alpha\, d\beta;$$

here, f is the spectral density and f_4 is the fourth-order cumulant spectrum (see, e.g., Brillinger, 1975). For various sets of conditions for this weak convergence to hold, see Anderson (1993), Brillinger (1975), and Dahlhaus (1985). Since the limit distribution is that of the supremum of a certain Gaussian process whose covariance structure depends on intricate fourth order properties of the underlying stationary process, analytical approximations to this limit law would be difficult to obtain, but Anderson (1993) makes progress in some special cases. In summary, the weak convergence of the process S_n has been well-studied and holds quite generally under weak dependence. As in Section 7.4, we assume this convergence as fundamental.

Assumption 7.5.1. *Assume the law of S_n converges weakly to a limiting process S whose paths concentrate on a separable subset of $D[0,\pi]$.*

Letting $J_n(P)$ denote the law of S_n, we immediately have the following result.

Theorem 7.5.1. *Assume Assumption 7.5.1 and that the $\{X_t\}$ process is α-mixing. Also assume that $b \to \infty$ and $b/n \to 0$ as $n \to \infty$.*

Then, $\rho_L(J_n(P), L_{n,b}) \to 0$ in probability.

Remark 7.5.1. The above arguments apply to the case where θ is the standardized spectral distribution function. Indeed, consider the process $Z_n(\cdot) = n^{1/2}[\hat{F}_n(\cdot)/\hat{F}_n(\pi) - F(\cdot)/F(\pi)]$. The weak convergence properties of Z_n can be deduced from that of Y_n, so that Assumption 7.3.1 holds here as well.

Remark 7.5.2. Remark 7.4.3 applies here as well. Thus, we can get asymptotically valid approximations to the distribution of the supremum of the spectral process, yielding a confidence band with asymptotic coverage probability equal to the nominal level. Actually, one needs to know a little more to claim the limiting coverage probability, namely that the limiting distribution in Assumption 7.5.1 is continuous. But, if the limit process is Gaussian with mean zero and continuous sample paths, as it is here, this continuity property follows by a general result of Tsirel'son (1975). Some simulation results of uniform confidence bands by subsampling for this example are presented in Politis, Romano, and You (1993).

Remark 7.5.3. In fact, the argument can be generalized to get uniform confidence bands for the spectral density itself, which is a harder problem. Here, Assumption 7.3.1 must be weakened so that $\tau_n \|\hat{\theta}_n - \theta(P)\| - c_n$ is assumed to have a limit distribution for some sequence $\{c_n\}$. This assumption holds for spectral density estimates (see Woodroofe and Van Ness, 1967).

7.6 Conclusions

In this chapter, the problem of estimating the distribution of an estimator taking values in a fairly abstract space was considered. The goal was to be able to be able to approximate the distribution of an estimator of a parameter that is not just real valued, such as an entire function. A basic weak convergence result, which can be viewed as a generalization of Varadarajan (1958) and may be of independent interest, was presented. This result was then applied to subsampling. Subsampling the empirical process and subsampling the spectral process were considered. The results were presented in the context of stationary observations. If the observations are assumed strictly i.i.d., the results and methods can be applied to subsampling over all subsamples of a given size (instead of those preserving time ordering) with little change in the arguments.

Part II

Extensions, Practical Issues, and Applications

8
Subsampling with Unknown Convergence Rate

8.1 Introduction

Let $X_1, \ldots, X_n$ be an observed stretch of a (strictly) stationary, strong mixing sequence of random variables $\{X_t, t \in Z\}$ taking values in an arbitrary sample space S; the probability measure generating the observations is denoted by P. The strong mixing condition means that the sequence $\alpha_X(k) = \sup_{A,B} |P(A \cap B) - P(A)P(B)|$ tends to zero as k tends to infinity, where A and B are events in the σ-algebras generated by $\{X_t, t < 0\}$ and $\{X_t, t \geq k\}$, respectively; the case where $X_1, \ldots, X_n$ are independent, identically distributed (i.i.d.) is an important special case of the general scenario.

In previous chapters, a general subsampling methodology has been put forth for the construction of large-sample confidence regions for a general unknown parameter $\theta(P)$ under very minimal conditions. To construct confidence regions for $\theta(P)$, we require an approximation to the sampling distribution (under P) of a standardized statistic $\hat{\theta}_n = \hat{\theta}_n(X_1, \ldots, X_n)$ that is consistent for $\theta(P)$ at some *known* rate τ_n; note that, in general, the rate of convergence τ_n depends on P as well, although this dependence will not be explicitly denoted. The subsampling methodology hinges on approximating the sampling distribution of $\hat{\theta}_n$ by the empirical distribution of restandardized subsample replicates of $\hat{\theta}_b$, where the subsample size b is a positive integer smaller than the original sample size n.

To review, assume there is a nondegenerate asymptotic distribution for the normalized statistic $\tau_n(\hat{\theta}_n - \theta(P))$; that is, there is a $J(x, P)$, continuous in x, such that

$$J_n(x, P) \equiv Prob_P\{\tau_n(\hat{\theta}_n - \theta(P)) \leq x\} \xrightarrow[n \to \infty]{} J(x, P) \quad (8.1)$$

for any real number x (and by continuity of the limit, this convergence is uniform in x as well). Then, in the i.i.d. case, the subsampling methodology was shown to 'work' provided that

$$b \xrightarrow[n \to \infty]{} \infty \quad (8.2)$$

and

$$\frac{b}{n} \xrightarrow[n \to \infty]{} 0. \quad (8.3)$$

In the general case of stationary observations, it was seen sufficient to have the underlying sequence to be strong mixing. Actually, the further assumption $\tau_b/\tau_n \to 0$ as $n \to \infty$ must be satisfied as well, but this is trivially satisfied in the case where $\tau_n = n^\beta$ for some $\beta > 0$, or even if $\tau_n = n^\beta L(n)$ where $L(\cdot)$ is a normalized slowly varying function; see Section 8.2 for the definition of the notion of a normalized slowly varying function. In other words, subsampling yields confidence regions for $\theta(P)$ of asymptotically correct coverage under the very weak assumption (8.1), provided care is taken so that (8.2), (8.3), and possibly a mixing condition, are satisfied as well.

Although existence of an asymptotic distribution is almost a *sine qua non* condition for the purposes of approximating the large-sample sampling distribution of the statistic $\hat{\theta}_n$, and some form of weak dependence condition (e.g., mixing) is required for consistency of estimation as the sample size increases, it may be possible that the convergence rate $\tau_.$ (as a function of n) is unknown because P is unknown and because we might be unwilling to even assume that P belongs in a particular class of measures that share a particular type of rate $\tau_.$. In that case, it is difficult to see how subsampling could be used for construction of confidence regions; however, subsampling would readily give some information on the shape of $J(x, P)$—as a function of x—which can be a helpful diagnostic tool (cf. Sherman, 1992, Sherman and Carlstein, 1996).

The aim of this chapter is actually to show that it is possible to drop the hypothesis concerning the explicit knowledge of the convergence rate in order to use subsampling for the construction of nonparametric confidence regions. In particular, we will first use the subsampling methodology to derive a consistent estimator of the rate $\tau_.$, and we will then use the estimated rate to get asymptotically correct confidence regions based on subsampling. The underlying idea is that it is possible to construct the subsampling distribution of $\hat{\theta}_n$ itself (rather than that of $\tau_n \hat{\theta}_n$); the speed by which it degenerates to a Dirac measure as $n \to \infty$ is closely related to τ_b. Constructing several subsampling distributions for different choices of b gives valuable information on the shape of τ_n as a function of n.

Note that, in contrast to Chapter 10, in this chapter we are not concerned with second-order accuracy properties that usually require strong distributional assumptions. Rather, we opt for a minimalist approach and show that first-order consistency is attainable under the minimal assumption that there is a (continuous and strictly increasing) large-sample distribution of the statistic in question whose form may be unknown. The results presented here are based on the paper by Bertail, Politis, and Romano, (1999), and indicate that it is not necessary to know *a priori* the shape of the convergence rate τ (as a function of n) in order to apply the subsampling methodology. Indeed, the subsampling methodology may provide the only foreseeable way to construct first-order accurate confidence regions for the unknown parameter $\theta(P)$ without imposing any assumptions on P whatsoever. However, see also Sherman and Carlstein (1997) for a different subsampling approach in the same context. Their construction is clever in that it completely avoids estimating the rate, but it produces unnecessarily long or inefficient intervals, and so will not be considered here.

The following examples help illustrate why it may be necessary to estimate the convergence rate before subsampling.

Example 8.1.1 (Nonnormal limit distribution). Consider the functional $\theta(P) = (E_P X)^2$ and its empirical counterpart $\hat{\theta}_n = \left(n^{-1}\sum_{i=1}^{n} X_i\right)^2$, where the X_i's are i.i.d. with variance σ^2. If $E_P X \neq 0$, $n^{1/2}(\hat{\theta}_n - \theta(P))$ converges weakly to $N(0, 4\theta(P)\sigma^2)$ and the usual bootstrap works. If $E_P X = 0$, $n(\hat{\theta}_n - \theta(P))$ converges weakly to $\sigma^2 \chi_1^2$ and the usual bootstrap fails; see, for example, Datta (1995), where a modified bootstrap is proposed to remedy the situation. Note that subsampling (or bootstraping with smaller resampling size) would work in this case as well, *provided* the rate of convergence were known, that is, provided it were known whether or not $E_P X = 0$. For a related discussion of a similar phenomenon in the context of U-statistics, see Example 2.3.1.

Example 8.1.2 (Extreme of n i.i.d. observations). Another interesting example is the case of the extreme order statistic. It is well known that the convergence rate of the extreme depends on the domain of attraction of the underlying distribution and that the usual bootstrap fails because of a certain lack of uniformity (Bickel and Freedman, 1981). Moreover since the underlying distribution is in general unknown, the convergence rate is unknown, and we cannot construct the subsampling distribution, unless we have some extra information on P.

Example 8.1.3 (Mean of a time series with long range memory). Suppose $\bar{X}_n = n^{-1}\sum_{t=1}^{n} X_t$ is the sample mean, and let $\theta(P) = EX_1$ be the true mean. Suppose that the sequence $\{X_t\}$ is strong mixing, but that the mixing coefficients decrease to zero slowly. Specifically, suppose that $\sum_{k=1}^{\infty} |Cov(X_1, X_k)| = \infty$; then the variance of $\bar{X}_n$ is not of order n^{-1}. Suppose that actually $\sum_{k=1}^{n} Cov(X_1, X_k) \sim n^{2\beta}$, and therefore $Var(\bar{X}_n) \sim \sigma_\infty^2 n^{2\beta-1}$, for some $0 < \beta < 1/2$, and $\sigma_\infty^2 > 0$. Assuming $E|X_1|^{2+\delta} < \infty$, for some $\delta > 0$, it follows from Rosenblatt (1984) that $n^{\frac{1}{2}-\beta}(\bar{X}_n - \theta(P))$ has an asymptotic normal $N(0, \sigma_\infty^2)$ distribution. To use either the normal limit distribution or its subsampling approximation, the exponent β must be estimated from the data (see, for example, Künsch, 1989). In fact, in the context of long-range dependent data, the limiting distribution for the sample mean need not be normal at all and can be quite complicated; see Theorem 3.1 of Beran (1994). For example, suppose $X_t = G(Z_t)$, where $\{Z_t\}$ is a stationary Gaussian process with long range correlations. Then, the convergence rate of the sample mean depends on the rate of decay of the correlations of the process $\{Z_t\}$, as well as the so-called Hermite rank of the function G. Moreover, the limiting process may be nonnormal (even if G is such that X_t has a finite mean and variance). In both the normal and nonnormal cases, the convergence rate however is of the form n to a power. Further complications arise if both long range dependence in the $\{X_t\}$ process and heavy tails in the distribution of X_t are present.

Remark 8.1.1. Hall, Lahiri, and Jing (1996) considered the formally analogous problem of the mean of a long range dependent time series (with finite second moment). They showed that subsampling is consistent for estimation of the studentized/self-normalized sample mean, that is, a result formally analogous to the results that we develop in this chapter, and also Chapter 11; notably, see Theorem 11.3.1. Incidentally, note that the "sampling window" method attributed to Hall and Jing (1996) is nothing else but the subsampling method for time series and random fields developed by Politis and Romano (1994b) and described in Chapters 2 and 3.

Example 8.1.4 (Nonparametric regression). Suppose bivariate data (Y_i, X_i), $i = 1, \ldots, n$ are available and we are interested in estimating the conditional expectation $E(Y|X = x)$, for some x, using some nonparametric regression technique. Many such smoothing methods are currently available, e.g., lowess, supsmu, AVAS, and ACE, and are included as ready-to-use functions in statistical software such as S-Plus. One can employ the classical bootstrap with resample size n to get a measure of accuracy of the estimated $E(Y|X = x)$; as a matter of fact, it is hard to avoid using the bootstrap in this case (Efron, 1994). Nevertheless, for robustness purposes (and in particular because the classical bootstrap would typically require strong regularity conditions in order to be valid here), one might want to use subsampling (or bootstrap with resample size $b < n$) instead; to do this, it is unavoidable that the rate of convergence of the estimated $E(Y|X = x)$ to its true value should be calculated or estimated.

Example 8.1.5 (Mean of a heavy-tailed distribution). Consider the problem of constructing a confidence interval for the mean of an unknown distribution having possibly infinite variance. This is a classic example where the bootstrap is known to fail, as previously discussed in Section 1.3. Here, the rate of convergence of the sample mean as an estimator of the population mean depends on the tail behavior of the distribution, which is assumed unknown. See Chapter 11 for a detailed discussion.

Example 8.1.6 (The autoregressive parameter). Consider the problem of constructing a confidence interval for the parameter ρ in the autoregressive model

$$X_t = \mu + \rho X_{t-1} + \epsilon_t,$$

where $\{\epsilon_t\}$ is a strictly stationary white noise innovation sequence and ρ is assumed to be in $(-1, 1]$. The ordinary least squares estimator converges at rate $n^{1/2}$ if $|\rho| < 1$; for $\rho = 1$ and $\mu = 0$, it converges at rate n; for $\rho = 1$ and $\mu \neq 0$, it converges at rate $n^{3/2}$. Again, the convergence rate depends on unknown parameters. A detailed discussion of this problem is deferred to Chapter 12.

The structure of the remainder of the chapter is as follows: Section 8.2 presents the basic idea of rate estimation using subsampling, and the proposed estimators are shown to be asymptotically consistent at an appropriate rate. Section 8.3 shows how the estimated rate can be employed in the construction of asymptotically correct confidence regions based on another round of subsampling. Section 8.4 summarizes our conclusions.

8.2 Estimation of the Convergence Rate

8.2.1 Convergence Rate Estimation: Univariate Parameter Case

In this subsection, let us make the simplifying assumption that $\hat{\theta}_n$ and $\theta(P)$ are real valued, and that $J(x, P)$ is continuous in x on the whole real line. Although the case of i.i.d. data is a special case of the strong mixing case, the construction of the subsampling distribution can take advantage of the i.i.d. structure, when such a structure exists; of course, if one is unsure regarding the independence assumption, it is safer (or more robust) to operate under the general strong mixing assumption.

  i. **General case (strong mixing data).** Define Y_a to be the subsequence $(X_a, X_{a+1}, \ldots, X_{a+b-1})$, for $a = 1, \ldots, q$, and $q = n - b + 1$; note that Y_a consists of b consecutive observations from the $X_1, \ldots, X_n$ sequence, and the order of the observations is preserved.

178   8. Subsampling with Unknown Convergence Rate

ii. Special case (i.i.d. data). Let $Y_1, \ldots, Y_q$ be equal to the $q = \binom{n}{b}$ subsets of size b chosen from $\{X_1, \ldots, X_n\}$, and then ordered in any fashion; here the subsets Y_a consist of unordered observations, and the statistic $\hat{\theta}_n$ is usually assumed to be a symmetric function of its arguments $X_1, \ldots, X_n$.

In either case, let $\hat{\theta}_{n,b,a}$ be the value of the statistic $\hat{\theta}_b$ applied to the subsample Y_a. The subsampling approximation to the sampling distribution of the root $\tau_n(\hat{\theta}_n - \theta(P))$, based on a block size b, is defined by

$$L_{n,b}(x \mid \tau.) \equiv q^{-1} \sum_{a=1}^{q} 1\{\tau_b(\hat{\theta}_{n,b,a} - \hat{\theta}_n) \leq x\}. \tag{8.4}$$

Note that, unlike in previous chapters, we make the dependence of the subsampling approximation on the rate $\tau.$ explicit in the notation. In both the strong mixing and i.i.d. cases, it was shown in previous chapters that if (8.1), (8.2), (8.3) hold, then

$$\sup_x |L_{n,b}(x \mid \tau.) - J(x, P)| = o_P(1) \tag{8.5}$$

as n tends to infinity.

Denote by $L_{n,b}(x \mid 1)$ the subsampling approximation to the sampling distribution of the root $(\hat{\theta}_n - \theta(P))$.

Given a distribution G on the real line and a number $t \in (0, 1)$, we will let $G^{-1}(t)$ denote the quantile transformation, that is,

$$G^{-1}(t) = \inf \{x : G(x) \geq t\},$$

which reduces to the regular inverse of the function G if G happens to be invertible. Note that we have

$$L_{n,b}(x \tau_b^{-1} \mid 1) = L_{n,b}(x \mid \tau.) \tag{8.6}$$

and thus it is easy to see that

$$L_{n,b}^{-1}(t \mid \tau.) = \tau_b \, L_{n,b}^{-1}(t \mid 1). \tag{8.7}$$

The following lemma will be required. It is quite similar to Lemma 1.2.1. For the purposes of the lemma, define $k_0 = \sup\{x : J(x, P) = 0\}$ and $k_1 = \inf\{x : J(x, P) = 1\}$.

Lemma 8.2.1. *Assume that $J(x, P)$ is continuous and strictly increasing on (k_0, k_1) as a function of x. If (8.5) is true as n tends to infinity, then*

$$L_{n,b}^{-1}(t \mid \tau.) = J^{-1}(t, P) + o_P(1)$$

for any $t \in (0, 1)$.

Proof. Let $\epsilon > 0$. By (8.5), we have that

$$Prob_P\{|L_{n,b}(x \mid \tau.) - J(x, P)| < \epsilon\} \to 1 \text{ uniformly in } x.$$

So let $z = L_{n,b}^{-1}(t - \epsilon \mid \tau.)$; thus with probability tending to one,
$$L_{n,b}(z \mid \tau.) \geq t - \epsilon \to J(z, P) \geq t - 2\epsilon \to z \geq J^{-1}(t - 2\epsilon, P).$$
Similarly, let $y = J^{-1}(t, P)$; thus with probability tending to one,
$$J(y, P) \geq t \Rightarrow L_{n,b}(y \mid \tau.) \geq t - \epsilon \Rightarrow y \geq L_{n,b}^{-1}(t - \epsilon \mid \tau.).$$
Hence, for any t and any $\epsilon > 0$, we have that
$$J^{-1}(t - 2\epsilon, P) \leq L_{n,b}^{-1}(t - \epsilon \mid \tau.) \leq J^{-1}(t, P)$$
with probability tending to one. Note that the assumptions imply that $J^{-1}(t - \epsilon, P) \to J^{-1}(t, P)$ as $t \to 0$. Thus, we can let $\epsilon \to 0^+$ to conclude that $L_{n,b}^{-1}(t \mid \tau.) = J^{-1}(t, P) + o_P(1)$. ∎

As a consequence of Lemma 8.2.1 and equation (8.7) we have that
$$\tau_b \, L_{n,b}^{-1}(t \mid 1) = J^{-1}(t, P) + o_P(1)$$
or equivalently
$$L_{n,b}^{-1}(t \mid 1) = \tau_b^{-1} J^{-1}(t, P) + o_P(\tau_b^{-1}). \tag{8.8}$$

It is observed that $L_{n,b}^{-1}(t \mid 1)$ is approximately proportional to τ_b^{-1}; thus, if we construct several subsampling distributions for different sizes b, we should be able to estimate the convergence rate.

More precisely, for any point $t > J(0, P)$ we have
$$\log\left(L_{n,b}^{-1}(t \mid 1)\right) = \log\left(J^{-1}(t, P)\right) - \log(\tau_b) + o_P(1). \tag{8.9}$$
Consequently, if we choose $b_1 \neq b_2$, we have then
$$\log\left(L_{n,b_2}^{-1}(t \mid 1)\right) - \log\left(L_{n,b_1}^{-1}(t \mid 1)\right) = \log(\frac{\tau_{b_1}}{\tau_{b_2}}) + o_P(1);$$
it follows that with two subsampling distributions, it is possible to estimate the shape of the function $\tau.$ up to some slowly varying function. If τ_n is of the form $\tau_n = n^\beta$ where β is an unknown constant, then
$$\log(\tfrac{b_1}{b_2})^{-1} \left(\log\left(L_{n,b_2}^{-1}(t \mid 1)\right) - \log\left(L_{n,b_1}^{-1}(t \mid 1)\right)\right) = \beta + o_P(\log(\tfrac{b_1}{b_2})^{-1}). \tag{8.10}$$

It is easy to see that the mean over several points t_j, $j = 1, \ldots, J$ of the left-hand side of (8.10) is a consistent estimator of β, provided that we choose b_1 and b_2 such that $\log(\tfrac{b_1}{b_2}) \xrightarrow[n \to \infty]{} \infty$; for instance take $b_i = n^{\beta_i}$, where $1 > \beta_1 > \beta_2 > 0$ are some constants. More generally, we can construct I subsampling distributions based on subsample sizes b_i, $i = 1, \ldots, I$; then equation (8.9) evaluated at some points $t_j > J(0, P)$, $j = 1, \ldots, J$, yields
$$y_{i,j} \equiv \log\left(L_{n,b_i}^{-1}(t_j \mid 1)\right) = a_j - \beta \log(b_i) + u_{i,j} \tag{8.11}$$

where $a_j \equiv \log\left(J^{-1}(t_j, P)\right)$ and $u_{i,j} = o_P(1)$, $i = 1, \ldots, I$ and $j = 1, \ldots, J$. The above equation may be interpreted as an analysis of covariance setup, in which case the following estimator of β suggests itself, namely

$$\beta_{I,J} \equiv -\frac{\sum_{i=1}^{I}(y_{i,.} - \overline{y})(\log(b_i) - \overline{\log})}{\sum_{i=1}^{I}(\log(b_i) - \overline{\log})^2}, \tag{8.12}$$

where

$$y_{i,.} = J^{-1}\sum_{j=1}^{J} y_{i,j} = J^{-1}\sum_{j=1}^{J} \log\left(L_{n,b_i}^{-1}(t_j \mid 1)\right),$$

$$\overline{y} = (IJ)^{-1}\sum_{i=1}^{I}\sum_{j=1}^{J} y_{i,j},$$

and

$$\overline{\log} = I^{-1}\sum_{i=1}^{I} \log(b_i).$$

Notice that if $I = 2$ and $J = 1$, then

$$\beta_{2,1} = \log(\tfrac{b_1}{b_2})^{-1}\left(\log\left(L_{n,b_2}^{-1}(t \mid 1)\right) - \log\left(L_{n,b_1}^{-1}(t \mid 1)\right)\right)$$

is the estimator suggested by (8.10). The following theorem asymptotically validates the use of the estimator $\beta_{I,J}$ as an estimator of β.

Theorem 8.2.1. *Assume that (8.1) holds for $\tau_n = n^\beta$, for some $\beta > 0$, and some $J(x, P)$ continuous and strictly increasing on (k_0, k_1) as a function of x. Let $b_i = \text{const.} \, n^{\gamma_i}$, $1 > \gamma_1 > \cdots > \gamma_I > 0$, and consider some points $t_j > J(0, P)$, $j = 1, \ldots J$.*

Then, $\beta_{I,J} = \beta + o_P((\log n)^{-1})$.

Proof. First note that equations (8.2) and (8.3) follow from our assumption on the b_i's, hence equation (8.5)—with any of the b_i's used as subsample size—is seen to hold as well from the results in Chapters 2 and 3.

Now note that, under (8.5), Lemma 8.2.1 validates the ANOVA equation (8.11), and in particular the fact that $u_{i,j} = o_P(1)$; therefore (8.12) yields

$$\beta_{I,J} = \beta - \frac{\sum_{i=1}^{I}(u_{i,.} - \overline{u})(\log(b_i) - \overline{\log})}{\sum_{i=1}^{I}(\log(b_i) - \overline{\log})^2},$$

where $\overline{u} = (IJ)^{-1}\sum_{i=1}^{I}\sum_{j=1}^{J} u_{i,j}$. Also note that under the stated assumptions on b_i we have $\overline{\log} = A\log n$, and $\sum_{i=1}^{I}(\log(b_i) - \overline{\log})^2 = B(\log n)^2$, for some A, B constants; therefore, for fixed I and J, it is seen that $\beta_{I,J} = \beta + o_P((\log n)^{-1})$. ∎

Remark 8.2.1. In the i.i.d. case, it may be the case that the bootstrap with resample size equal to b provides an asymptotically consistent estimator of the distribution $J(P)$ under conditions (8.2) and (8.3); this can be checked (Bickel et al., 1997), or enforced by letting $0.5 > \gamma_1 > \cdots > \gamma_I > 0$ in Theorem 8.2.1 (see Section 2.3). In such a case, this bootstrap distribution estimator could equally be used in place of the subsampling distribution estimator $L_{n,b}$ for the purposes of rate estimation; however, there does not seem to be an advantage in doing this.

Remark 8.2.2. Under stronger conditions, the rate of convergence of $\beta_{I,J}$ to β may be faster than $(\log n)^{-1}$. For example, it may be assumed not only that (8.1) holds, but that the convergence to the limit law $J(x, P)$ occurs at some rate, e.g., $J_n(x, P) = J(x, P) + O_P(n^{-1/2})$, or some other asymptotic expansion (see, for example, Barbe and Bertail, 1995, or Bertail, 1997).

Remark 8.2.3. If τ_n is a more complicated function, equation (8.9) suggests estimating the shape of $h(n) = log(\tau_n)$, under the constraint that $h(\cdot)$ is an increasing function equal to $+\infty$ at $+\infty$, using some nonparametric technique. In many interesting situations, τ_n is a regular varying function of index β, say $\tau_n = n^\beta L(n)$ where L is a normalized slowly varying function that is such that $L(1) = 0$ and for any $\lambda > 0$, $\lim_{x\to\infty} \frac{L(\lambda x)}{L(x)} = 1$ (see Bingham, Goldie, and Teugels, 1987). By the Karamata representation theorem, there exists $\epsilon(\cdot), \epsilon(u) \xrightarrow[u\to\infty]{} 0$ such that $L(n) = \exp \int_1^n u^{-1}\epsilon(u)du$, and (8.9) may be written

$$\log\left(L_{n,b}^{-1}(t\mid 1)\right) = \log\left(J^{-1}(t, P)\right) - \beta \log(b) + \int_1^b u^{-1}\epsilon(u)du + o(1),$$

which may be seen, for a fixed t, as a partial spline regression model (see Engle et al., 1986). However, in that case we necessarily need more than two subsampling distributions to estimate the slowly varying function part. Since we only want to obtain a consistent estimator of the convergence rate, that is, a $\hat\tau_n$ such that $\hat\tau_n/\tau_n \to 1$ in probability, we only need a consistent estimator of $L(\cdot)$, that is, a $\hat L(n)$ such that $\hat L(n)/L(n) \to 1$ in probability, and an estimator $\hat\beta$ such that $\hat\beta = \beta + o_P((\log n)^{-1})$. We conjecture that this goal may be achieved by constructing $I = \log(n)$ subsampling distributions with $b_i = n^{\gamma_i}, i = 1, \ldots, I$, where $0 < \gamma_1 < \cdots < \gamma_I < 1$. However, it is likely that a huge sample size is needed to obtain an accurate estimator of the slowly varying function L; therefore, we focus on the more feasible task of estimating β under the assumption of $\tau_n = n^\beta$ from now on.

Note that $L_{n,b}(0 \mid \tau_\cdot) = L_{n,b}(0 \mid 1) = J(0, P) + o_P(1)$; that is, $J(0, P)$ may be estimated without knowing the rate τ_n; thus, choosing the t_j in order to apply Theorem 8.2.1 will not be a problem in practice. Nevertheless, the requirement $t_j > J(0, P)$ seems a bit cumbersome. We now show that, by looking at differences of quantiles, this requirement can be circumvented.

Let t_{2j}, for $j = 1, \ldots, J$, be some points in $(0.5, 1)$, and let t_{2j-1}, for $j = 1, \ldots, J$, be some points in $(0, 0.5)$; as before, let b_i be some different subsample sizes, for $i = 1, \ldots, I$. By equation (8.8) it follows that

$$y_{i,j} \equiv \log\left(L_{n,b_i}^{-1}(t_{2j} \mid 1) - L_{n,b_i}^{-1}(t_{2j-1} \mid 1)\right) = a_j - \beta \log(b_i) + u_{i,j} \quad (8.13)$$

where $a_j \equiv \log\left(J^{-1}(t_{2j}, P) - J^{-1}(t_{2j-1}, P)\right)$ and $u_{i,j} = o_P(1)$, $i = 1, \ldots, I$ and $j = 1, \ldots, J$. Here as well we can use the estimator $\beta_{I,J}$ as defined in (8.12), where now

$$y_{i,\cdot} = J^{-1} \sum_{j=1}^{J} y_{i,j} = J^{-1} \sum_{j=1}^{J} \log\left(L_{n,b_i}^{-1}(t_{2j} \mid 1) - L_{n,b_i}^{-1}(t_{2j-1} \mid 1)\right),$$

$\bar{y} = (IJ)^{-1} \sum_{i=1}^{I} \sum_{j=1}^{J} y_{i,j}$ and $\overline{\log} = I^{-1} \sum_{i=1}^{I} \log(b_i)$. The following theorem asymptotically validates the use of the new estimator $\beta_{I,J}$ as an estimator of β.

Theorem 8.2.2. *Assume that (8.1) holds for $\tau_n = n^\beta$, for some $\beta > 0$ and some $J(x, P)$ continuous and strictly increasing on (k_0, k_1) as a function of x. Let $b_i = \text{const.} \, n^{\gamma_i}$, $1 > \gamma_1 > \cdots > \gamma_I > 0$, and consider some points $t_{2j} \in (0.5, 1)$, and $t_{2j-1} \in (0, 0.5)$, for $j = 1, \ldots J$. Then $\beta_{I,J} = \beta + o_P((\log n)^{-1})$.*

Proof. The proof is similar to the proof of Theorem 8.2.1 and is omitted. ∎

For example, suppose that $J = 1$ and that $t_1 = 0.25$ and $t_2 = 0.75$; then $y_{i,j}$ as given in (8.13) is the logarithm of the inter-quartile range of the subsampling distribution $L_{n,b_i}(x \mid 1)$, and a_j as given in (8.13) is the logarithm of the inter-quartile range of the limit distribution $J(x, P)$. Nevertheless, it is recommended to take $J > 1$, that is, to look at many differences of quantiles, in order that $\beta_{I,J}$ becomes more accurate.

In the following subsection, a different way to side-step the cumbersome requirement $t_j > J(0, P)$ of Theorem 8.2.1 is proposed in the more general setting of a vector-valued parameter θ.

8.2.2 Convergence Rate Estimation: Multivariate Parameter Case

Now assume that θ takes values in a normed linear space Θ with norm $\|\cdot\|$, and that $\hat\theta_n = \hat\theta_n(X_1, \ldots, X_n)$ is an estimator consistent for $\theta(P)$ at rate $\tau_n = n^\beta$, where β may be unknown. A confidence region for $\theta(P)$ can be constructed if an approximation to the sampling distribution of $\tau_n \|\hat\theta_n - \theta(P)\|$ is available; for example, Θ might be a function space, and $\|\cdot\|$ might be the sup-norm in which case the confidence region has the form of a uniform confidence band for the unknown function $\theta(P)$ (cf. Politis, Romano, and You, 1993).

8.2 Estimation of the Convergence Rate

Consequently, let us denote

$$J_{n,\|\cdot\|} \equiv Prob_P(\tau_n\|\hat{\theta}_n - \theta(P)\| \leq x) \tag{8.14}$$

and we will assume as before that

$$J_{n,\|\cdot\|}(x, P) \xrightarrow[n\to\infty]{} J_{\|\cdot\|}(x, P) \tag{8.15}$$

for some $J_{\|\cdot\|}(x, P)$ continuous in x.

The subsample values $\hat{\theta}_{n,b,a}$ are defined in a fashion analogous to the construction leading to rate-equation (8.2.1) and the subsampling approximation to the sampling distribution of the root $\tau_n\|\hat{\theta}_n - \theta(P)\|$ is given by

$$L_{n,b,\|\cdot\|}(x \mid \tau.) = q^{-1}\sum_{a=1}^{q} 1\{\tau_b\|\hat{\theta}_{n,b,a} - \hat{\theta}_n\| \leq x\}. \tag{8.16}$$

Politis, Romano, and You (1993) showed that if (8.15), (8.2), (8.3) hold, then

$$\sup_x |L_{n,b,\|\cdot\|}(x \mid \tau.) - J_{\|\cdot\|}(x, P)| = o_P(1) \tag{8.17}$$

as n tends to infinity. The following analogue of Lemma 8.2.1 will be useful here. For the lemma, we let

$$\tilde{k}_0 = \sup\{x : J_{\|\cdot\|}(x, P) = 0\} \text{ and } \tilde{k}_1 = \inf\{x : J_{\|\cdot\|}(x, P) = 1\}.$$

Lemma 8.2.2. *Assume that $J_{\|\cdot\|}(x, P)$ is continuous and strictly increasing on $(\tilde{k}_0, \tilde{k}_1)$ as a function of x. If (8.17) is true as n tends to infinity, then*

$$L_{n,b,\|\cdot\|}^{-1}(t \mid \tau.) = J_{\|\cdot\|}^{-1}(t, P) + o_P(1)$$

for any $t \in (0, 1)$.

Proof. Similar to the proof of Lemma 8.2.1. ∎

Now let

$$y_{i,j} \equiv \log\left(L_{n,b_i,\|\cdot\|}^{-1}(t_j \mid 1)\right) = a_j - \beta \log(b_i) + u_{i,j} \tag{8.18}$$

where $a_j \equiv \log\left(J_{\|\cdot\|}^{-1}(t_j, P)\right)$ and $u_{i,j} = o_P(1)$, $i = 1, \ldots, I$ and $j = 1, \ldots, J$. Now we can use $\beta_{I,J}$ exactly as given in definition (8.12)—but now using the $y_{i,j}$s as given above in (8.18)—to get a consistent estimator of β.

Theorem 8.2.3. *Assume that (8.15) holds for $\tau_n = n^\beta$, for some $\beta > 0$ and some $J_{\|\cdot\|}(x, P)$ continuous and strictly increasing on $(\tilde{k}_0, \tilde{k}_1)$ as a function of x. Let $b_i = \text{constant} \cdot n^{\gamma_i}$, $1 > \gamma_1 > \cdots > \gamma_I > 0$, and consider some points $t_j > 0$, $j = 1, \ldots J$. Then $\beta_{I,J} = \beta + o_P((\log n)^{-1})$.*

Proof. Similar to the proof of 8.2.1 using Lemma 8.2.2 and the results of Politis, Romano, and You (1993). ∎

Although Theorem 8.2.3 is quite similar to Theorem 8.2.1, note that the cumbersome condition $t_j > J(0, P)$ has been replaced by the more natural condition $t_j > 0$; the reason is that $J_{\|\cdot\|}(0, P) = 0$ because of the continuity of $J_{\|\cdot\|}(x, P)$. Theorem 8.2.3 might of course be used even for real-valued parameters; see also Theorem 8.3.2 in what follows.

8.3 Subsampling with Estimated Convergence Rate

Having shown that it is possible to consistently estimate the rate of convergence $\tau_{.}$ in general situations, we will now focus on our original goal, namely, the construction of confidence regions for the unknown parameter $\theta(P)$.

The obvious strategy in order to use subsampling in the case of unknown rate of convergence $\tau_{.}$ is to estimate $\tau_{.}$, and use the estimated $\tau_{.}$ in constructing the subsampling distribution. As Theorems 8.3.1 and 8.3.2 below demonstrate, this plug-in strategy gives valid results, meaning that subsampling with an estimated rate of convergence yields confidence regions for $\theta(P)$ of asymptotically correct coverage.

The following theorem establishes the asymptotic validity of subsampling with an estimated rate of convergence in the case of real-valued $\hat{\theta}_n$ and θ.

Theorem 8.3.1. *Assume that (8.1) holds for $\tau_n = n^\beta$, for some $\beta > 0$, and some $J(x, P)$ continuous in x; also assume conditions (8.2) and (8.3). Let $\hat{\beta} = \beta + o_P((\log n)^{-1})$, and put $\hat{\tau}_n = n^{\hat{\beta}}$. Then,*

$$\sup_x |L_{n,b}(x \mid \hat{\tau}_{.}) - J(x, P)| = o_P(1). \tag{8.19}$$

Let $t \in (0,1)$, and let $c_n(t) = L_{n,b}^{-1}(t \mid \hat{\tau}_{.})$ be the t-th quantile of the subsampling distribution $L_{n,b}(x \mid \hat{\tau}_{.})$. Then,

$$\text{Prob}_P\{\hat{\tau}_n(\hat{\theta}_n - \theta(P)) \leq c_n(t)\} \xrightarrow[n \to \infty]{} t. \tag{8.20}$$

Thus the asymptotic coverage probability of the interval $[\hat{\theta}_n - \hat{\tau}_n^{-1} c_n(t), \infty)$ is the nominal level t.

Proof. Let x be a real number and note that

$$L_{n,b}(x \mid \hat{\tau}_{.}) \equiv q^{-1} \sum_{a=1}^{q} 1\{b^{\hat{\beta}}(\hat{\theta}_{n,b,a} - \hat{\theta}_n) \leq x\}$$

$$= q^{-1} \sum_{a=1}^{q} 1\{b^{\hat{\beta}}(\hat{\theta}_{n,b,a} - \theta(P)) - b^{\hat{\beta}}(\hat{\theta}_n - \theta(P)) \leq x\}.$$

8.3 Subsampling with Estimated Convergence Rate

Put $U_n(x) = q^{-1} \sum_{a=1}^{q} 1\{b^{\hat{\beta}}(\hat{\theta}_{n,b,a} - \theta(P)) \le x\}$ and, for some $\epsilon > 0$, $E_n = \{b^{\hat{\beta}}|\hat{\theta}_n - \theta(P)| \le \epsilon\}$. Since $\hat{\beta} = \beta + o_P((\log n)^{-1})$, it follows that $n^{\hat{\beta}}/n^{\beta} \to 1$, in probability; thus, $n^{\hat{\beta}} = n^{\beta}(1+o_P(1))$. Similarly, $b^{\hat{\beta}}/b^{\beta} \to 1$, in probability, and $b^{\hat{\beta}} = b^{\beta}(1+o_P(1))$.

As in the proofs of Theorems 2.2.1 and 3.2.1, it follows that (8.2) and (8.3) imply that $P(E_n) \xrightarrow[n\to\infty]{} 1$; hence, with probability tending to one,

$$U_n(x-\epsilon) \le L_{n,b}(x \mid \hat{\tau}.) \le U_n(x+\epsilon).$$

Now it suffices to show that $U_n(x)$ converges to $J(x, P)$ in probability; for in that case we could let $\epsilon \to 0$ in the above inequality to conclude that $L_{n,b}(x \mid \hat{\tau}.) \to J(x, P)$ in probability. The uniform convergence (8.19) would then be a consequence of Polya's theorem, since $J(x, P)$ is continuous.

Note that from the results of Theorems 2.2.1 and 3.2.1 again we know that, for every x, $V_n(x) \to J(x, P)$ in probability, where

$$V_n(x) = q^{-1} \sum_{a=1}^{q} 1\{b^{\beta}(\hat{\theta}_{n,b,a} - \theta(P)) \le x\}.$$

We now claim that the already established convergence $V_n(x) \to J(x, P)$ implies the desired convergence $U_n(x) \to J(x, P)$ in probability. To appreciate why, note that, for any $\epsilon_1 > 0$,

$$U_n(x) = q^{-1} \sum_{a=1}^{q} 1\{b^{\beta}(\hat{\theta}_{n,b,a} - \theta(P)) \le x\frac{b^{\beta}}{b^{\hat{\beta}}}\} \le V_n(x + \epsilon_1)$$

where the above inequality holds with probability tending to one. A similar argument shows that $U_n(x) \ge V_n(x - \epsilon_1)$ with probability tending to one.

But from the results of Theorems 2.2.1 and 3.2.1 it also follows that $V_n(x+\epsilon_1) \to J(x+\epsilon_1, P)$ in probability, and that $V_n(x-\epsilon_1) \to J(x-\epsilon_1, P)$ in probability. Therefore, letting $\epsilon_1 \to 0$, we have that $U_n(x) \to J(x, P)$ in probability as required.

Finally note that the proof of (8.20) is very similar to the proof of Theorem 1 of Beran (1984) given result (8.19) and noting that $\hat{\tau}_n/\tau_n \to 1$ in probability, since $\hat{\beta} = \beta + o_P((\log n)^{-1})$. ∎

Remark 8.3.1. Note that the estimator $\hat{\beta}$ in Theorem 8.3.1 can be obtained by *any* method, as long as $\hat{\beta} = \beta + o_P((\log n)^{-1})$; for example, any of the rate estimation methods discussed so far in Theorems 8.2.1 or 8.2.2, could be used, provided that $J(x, P)$ is strictly increasing for x in its support (k_0, k_1) which is a reasonable assumption.

Remark 8.3.2. If a particular model can be assumed, then it might be more effective to use a model-specific estimator of the convergence rate. The model-specific rate estimators should be more accurate (if the model assumed is indeed true), and thus the subsampling distribution with estimated rate will be more accurate as a result. For instance, in Example 8.1.3

we only need to estimate β which is linked to the strong mixing coefficients of the model; see, for example, Beran (1994) or Robinson (1994) regarding estimation of the 'long-memory' parameter β. Nevertheless, it is reassuring to know that there are several model-free estimators (given by our Theorems 8.2.1, 8.2.2, and 8.2.3) that are accurate enough to be used in conjunction with confidence intervals based on subsampling.

The next theorem establishes the asymptotic validity of subsampling with an estimated rate of convergence in the general case where θ (and therefore $\hat{\theta}_n$ as well) takes values in a normed linear space Θ with norm $\|\cdot\|$. In the case Θ is the real line, Theorem 8.3.2 allows for the construction of symmetric confidence intervals for $\theta(P)$.

Theorem 8.3.2. *Assume that (8.15) holds for $\tau_n = n^\beta$, for some $\beta > 0$ and some $J_{\|\cdot\|}(x, P)$ continuous in x; also assume (8.2) and (8.3). Let $\hat{\beta} = \beta + o_P((\log n)^{-1})$, and put $\hat{\tau}_n = n^{\hat{\beta}}$. Then,*

$$\sup_x |L_{n,b,\|\cdot\|}(x \mid \hat{\tau}_.) - J_{\|\cdot\|}(x, P)| = o_P(1). \quad (8.21)$$

Let $t \in (0,1)$, and let $c_n(t) = L_{n,b,\|\cdot\|}^{-1}(t \mid \hat{\tau}_.)$ be the t-th quantile of the subsampling distribution $L_{n,b,\|\cdot\|}(x \mid \hat{\tau}_.)$. Then,

$$Prob_P\{\hat{\tau}_n \|\hat{\theta}_n - \theta(P)\| \leq c_n(t)\} \xrightarrow[n \to \infty]{} t. \quad (8.22)$$

Thus the asymptotic coverage probability under P of the confidence region $\{\theta : \hat{\tau}_n \|\hat{\theta}_n - \theta\| \leq c_n(t)\}$ is the nominal level t.

Proof. Similar to the proof of Theorem 8.3.1 using Theorem 8.2.3 in conjunction with the results of Politis, Romano, and You (1993). ∎

Remark 8.3.3. As long as $J_{\|\cdot\|}(x, P)$ is strictly increasing for x in its support $(\tilde{k}_0, \tilde{k}_1)$, we can use the estimator $\beta_{I,J}$ of Theorem 8.2.3 as the estimator $\hat{\beta}$ employed in Theorem 8.3.2.

Remark 8.3.4. Theorems 8.3.1 and 8.3.2 show that the construction of large-sample confidence regions based on subsampling can be performed in a fairly automatic and totally model-free fashion, in the sense that we do not have to adapt the resampling methodology to the model as it is usually the case for the bootstrap. Nonetheless, the subsampling methodology may yield poor approximations of the true distribution for finite sample sizes. Some ways of improving the accuracy of the subsampling distributions (for instance, through linear combinations of subsampling distributions) are proposed in Chapter 10.

8.4 Conclusions

In this chapter, a simple method for the estimation of the rate of convergence of a general statistic $T_n = T_n(X_1, \ldots, X_n)$ to a parameter θ of interest was proposed, and its large-sample consistency was shown in the case of i.i.d. or strong mixing data. In addition, it was shown that the general subsampling methodology for the construction of large-sample confidence regions that was introduced in Chapters 2 and 3 may be used even if the rate of convergence of $\hat{\theta}_n$ to θ is unknown, and an estimated rate (perhaps the one estimated by a first round of subsampling as proposed here) can be used instead without spoiling the large-sample validity. Notably, a very recent paper by Sherman and Carlstein (1997) proposes a different way of constructing confidence intervals in the case of unknown rate of convergence that bypasses the difficult issue of rate estimation; however, their intervals are larger by orders of magnitude as compared to the ones studied in this chapter. For some simulation results, the reader is referred to Bertail, Politis, and Romano (1999). For more concrete application of the problems considered in this chapter, the reader is referred to Chapter 11 and Chapter 12. In those chapters, some special structure is exploited, so that an alternative approach based on subsampling self-normalizing statistics is presented. The method discussed in this chapter applies more generally.

9
Choice of the Block Size

9.1 Introduction

The main practical problem in applying the subsampling method lies in choosing the block size b. This problem is shared by all blocking methods, such as, for example, the moving blocks bootstrap or Carlstein's (1986) variance estimator (see Sections 3.8 and 3.9). The asymptotic conditions, at least for first-order theory, are usually $b \to \infty$ and $b/n \to 0$ as $n \to \infty$. Although any choice of b satisfying these conditions will yield the required consistency of subsampling methods, the two asymptotic conditions do not give much guidance with respect to how to choose b when faced with a finite sample. The aim of this chapter is to provide some guidelines for this task. It turns out that the optimal choice of b depends on the purpose for which subsampling is used.

In Section 9.2, choice of the block size for variance estimation is discussed. Then, the applications of confidence interval construction and hypothesis testing are considered in Sections 9.3 and 9.4, respectively. Finally, Section 9.5 provides two simulation studies to shed some light on the small sample behavior of the subsampling method in general and our methods for choosing the block choice in particular. All tables appear at the end of the chapter.

9.2 Variance Estimation

9.2.1 Case of the Sample Mean

Let $\underline{X}_n = (X_1, \ldots, X_n)$ be an observed stretch of a stationary, strong mixing sequence of real-valued random variables $\{X_t, t \in \mathbb{Z}\}$ with mixing coefficients $\alpha_X(\cdot)$. We again consider the sample mean $\bar{X}_n = n^{-1} \sum_{t=1}^{n} X_t$ as an estimator of the true mean $\theta(P) = EX_t$; also let $R(k) = Cov(X_0, X_k)$ denote the autocovariance at lag k, and $f(w) = (2\pi)^{-1} \sum_{s=-\infty}^{\infty} R(s) \cos(ws)$ the spectral density of the $\{X_t\}$ process.

Note that, under suitable moment and mixing conditions, our Theorem 3.4.1 yields $\sqrt{n}(\bar{X}_n - \theta(P))/\sigma_n \xrightarrow{\mathcal{L}} N(0, 1)$, where $\sigma_n^2 = Var(\sqrt{n}\bar{X}_n) = R(0) + 2\sum_{s=1}^{n}(1 - s/n)R(s)$. In order to actually use this Central Limit theorem to set confidence intervals for $\theta(P)$, estimating the variance σ_n^2, or its asymptotic limit σ_∞^2 becomes crucial.

Many estimators of $\sigma_\infty^2 = \lim_{n\to\infty} Var(\sqrt{n}\bar{X}_n) = \sum_{s=-\infty}^{\infty} R(s)$ have been proposed in the literature; probably the most popular one is

$$\hat{\sigma}_{b_M,n}^2 = \frac{b_M}{Q} \sum_{i=1}^{Q} (\bar{X}_{i,b_M,h_L} - \bar{X}_n)^2, \qquad (9.1)$$

where $\bar{X}_{i,b_M,h_L} = b_M^{-1} \sum_{t=h_L(i-1)+1}^{h_L(i-1)+b_M} X_t$ is the mean of the block of consecutive data $\{X_{h_L(i-1)+1}, \ldots, X_{h_L(i-1)+b_M}\}$, the numbers h_L, b_M are integers depending on the sample size n, and $Q = \lfloor \frac{n-b_M}{h_L} \rfloor + 1$, with $\lfloor \cdot \rfloor$ being the integer part; b_M is the block's size here, and h_L is the amount of "lag" between the starting points of block i and block $i+1$ as in Chapter 5. If $h_L = b_M$, there is no overlap between block i and block $i+1$, whereas if $h_L = 1$, the overlap is the maximum possible. Q is the total number of such blocks that can be extracted from the data.

The estimator $\hat{\sigma}_{b_M,n}^2$ is very intuitive, and is found in the literature in many asymptotically equivalent variations and under many different names, including the following: Bartlett spectral density estimator (at the origin), moving blocks bootstrap and moving blocks jackknife, subsampling variance estimator, overlapping or nonoverlapping batch means estimator. See Politis and Romano (1995) for a list of references and more discussion on Bartlett's estimator in its different forms. In particular, if $h_L = b_M$, $\hat{\sigma}_{b_M,n}^2$ is *identical* to Carlstein's variance estimator $\hat{\sigma}_{n,CARL}^2$ (in the sample mean case and with subsample size $b = b_M$), while if $h_L = 1$, $\hat{\sigma}_{b_M,n}^2$ is *identical* to the subsampling variance estimator $\hat{\sigma}_{n,SUB}^2$, both of which were discussed in Chapter 3.

Under regularity conditions, $\hat{\sigma}_{b_M,n}^2$ is a consistent and asymptotically normal estimator. The regularity conditions are moment and mixing conditions and conditions on the design parameters; typically $b_M \to \infty$, but with $b_M/n \to 0$ and $h_L/b_M \to a \in [0, 1]$ as $n \to \infty$. Consistency is immediate, because the first two moments of $\hat{\sigma}^2_{b_M,n}$ can be asymptotically calculated to be

$$Bias(\hat{\sigma}^2_{b_M,n}) = E\hat{\sigma}^2_{b_M,n} - \sigma^2_\infty = -b_M^{-1} \sum_{k=-\infty}^{\infty} |k|R(k) + o(1/b_M) + O(b_M/n), \tag{9.2}$$

$$Var(\hat{\sigma}^2_{b_M,n}) = 2c\frac{b_M}{n}\sigma^4_\infty + o(b_M/n), \tag{9.3}$$

where $c = a(1 + 2\sum_{k=1}^{\lfloor 1/a \rfloor}(1-ka)^2)$, if $a > 0$, and $c = 2/3$, if $a = 0$. Since c is smallest for $a = 0$, the full-overlap case corresponding to $h_L = 1$ is recommended as mentioned in our Chapter 3; thus we set $h_L = 1$ in what follows. To see that in this case $\hat{\sigma}^2_{b_M,n}$ is nothing other than the subsampling variance estimator $\hat{\sigma}^2_{n,SUB}$ (using a subsample size $b = b_M$) of the variance of $\bar{X}_n$ note that, if $T_n = \bar{X}_n$ is the sample mean, then $\hat{\sigma}^2_{b,n}$ is the variance of a random variable with cumulative distribution function given by the subsampling distribution $L_{n,b}(x)$ of our Chapter 3.

At this point, a most important difficulty in this estimation setup should be mentioned, namely, the choice of b_M in practice. By equations (9.2) and (9.3) it follows that the choice

$$b_M \approx \left(\frac{(\sum_{k=-\infty}^{\infty}|k|R(k))^2}{c\sigma^4_\infty}\right)^{1/3} n^{1/3} \tag{9.4}$$

minimizes the asymptotic MSE (mean squared error) of the estimator $\hat{\sigma}^2_{b_M,n}$.

However, there are many unknown parameters on the right-hand side of (9.4) that complicate things. To start with, σ^2_∞ is unknown; as a matter of fact this is exactly what $\hat{\sigma}^2_{b_M,n}$ is estimating! One way out of this difficulty is to first use *some* choice of b_M to get a "pilot" estimator of σ^2_∞; the "pilot" estimator would then be used in the right-hand-side of (9.4) to get a better value of b_M. This "plug-in" method has been used in the literature of bandwidth selection of a nonparametric probability or spectral density function estimator and shown to work asymptotically, but not much can be said regarding finite samples.

Another complication is the presence of the term $\sum_{k=-\infty}^{\infty}|k|R(k)$ in this formula for the optimal choice of b_M. Recall that $\sigma^2_\infty \equiv 2\pi f(0)$, where $f(w) = (2\pi)^{-1}\sum_{s=-\infty}^{\infty} R(s)\cos(ws)$ is the spectral density function, and note that the quantity $\sum_{k=-\infty}^{\infty}|k|R(k)$ is related to the first derivative of $f(w)$ at $w=0$. Consequently, the estimation of $\sum_{k=-\infty}^{\infty}|k|R(k)$ is *harder*—involving yet another bandwidth choice—as compared to the estimation of $f(0)$ which was our original problem. Indeed, the inherent inaccuracy in estimating $\sum_{k=-\infty}^{\infty}|k|R(k)$ is sufficient to make us suspicious of the "plug-in" approach in practical, finite-sample applications; see also the discussion in Léger, Politis, and Romano (1992).

9.2 Variance Estimation

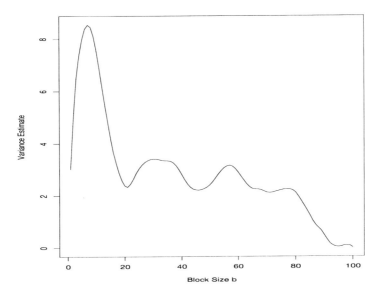

FIGURE 9.1. The subsampling variance estimator $\hat{\sigma}^2_{b_M,n}$ as a function of b_M for a particular data set of size $n = 100$ generated from a Gaussian MA(2) model.

To illustrate the difficulty of the practical problem of block size choice, consider Figure 9.1, which essentially shows a graph of the estimator $\hat{\sigma}^2_{b_M,n}$ as a function of b_M for a particular data set of size $n = 100$ generated from a simple moving average MA(2) model that is defined recursively via the equation

$$X_t = a_0 Z_t + a_1 Z_{t-1} + a_2 Z_{t-2}, \qquad (9.5)$$

where the Z_t's are i.i.d. normal $N(0,1)$ random variables, and $a_i = 1$, for $i = 0, 1,$ and 2.

The graph makes the great sensitivity of $\hat{\sigma}^2_{b_M,n}$ on the choice of b_M apparent; for a good choice of b_M (in the neighborhood of $b_M = 10$), $\hat{\sigma}^2_{b_M,n}$ is very accurate and close to the true value of $Var(\sqrt{100}\bar{X}_{100}) \simeq 9$, but for b_M away from the value 10, $\hat{\sigma}^2_{b_M,n}$ can be in error by an order of magnitude. Thus, the choice of block size b_M in practice is a delicate issue totally analogous to the issue of bandwidth choice in nonparametric spectral density estimation; as a matter of fact, insights from the experience in bandwidth choice for "optimal" smoothing can be very helpful in choosing the block size b_M in practice (see, Politis and Romano, 1995, for more details).

Coming back to the MA(2) simulation, it is apparent that the practitioner faced with that data set would not know in general that the data actually come from a normal MA(2) model such as the one given by equation (9.5) (with unknown a_i coefficients in general); but even if this MA(2) model information *were* somehow available, the mean squared error of estimator $\hat{\sigma}^2_{b_M,n}$ (with optimal choice of block size b_M) would *still* remain of order $n^{-1/3}$. Nevertheless, in this finite-parameter case, σ^2_∞ is actually estimable at $\sqrt{n}$-consistency rate by directly estimating the three nonzero autocovariances $R(0), R(1), R(2)$ and plugging them in the formula $\sigma^2_\infty = R(0) + 2R(1) + 2R(2)$.

The reason for this unsatisfactory performance of estimator $\hat{\sigma}^2_{b_M,n}$ in a simple model such as the MA(2) is its rather large bias. In Section, 10.5.5 we address this issue and show how we can bias-correct the estimator $\hat{\sigma}^2_{b_M,n}$ and kill two birds with one stone: (a) produce a very accurate variance estimator that is actually $\sqrt{n}$-consistent in the moving average case discussed above but whose applicability is *not* limited to the moving average class of models, and (b) provide a simple, empirical way for selecting the MSE-optimal block size for this new bias-corrected estimator.

9.2.2 General Case

Note that, up to the moment, not much is known in terms of block size choice for a general statistic $\theta(P)$. Nevertheless, the sample mean example is of prime importance because many nonlinear statistics from time series can be approximated by linear ones in direct analogy to the discussion on differentiability (Fréchet, etc.) in Section 1.6 (see, for example, Künsch, 1989, Politis and Romano, 1992a, or Bühlmann and Künsch, 1995). To elaborate, suppose we are dealing with an estimator $\hat{\theta}_n$ calculated from our real-valued series $X_1, \ldots, X_n$, and estimating the parameter $\theta(P)$ at rate $\sqrt{n}$. Let us assume for simplicity that $\theta(P)$ is a feature of the l-dimensional marginal distribution of the process $\{X_t\}$, where l is some (fixed) positive integer.

Observe that one can construct a new multivariate series $\{X_t^{(l)}\}$ by blocking, that is, letting $X_t^{(l)} = (X_t, \ldots, X_{t+l-1})$ for $t = 0, \pm 1, \pm 2, \ldots$. Then, for many statistics of interest, a function $G_l(\cdot)$ can be found such that

$$\hat{\theta}_n = \frac{1}{n-l+1} \sum_{t=1}^{n-l+1} W_t + o_P(1/\sqrt{n}), \qquad (9.6)$$

where $W_t = G_l(X_t^{(l)})$ for $t = 1, 2, \ldots, n-l+1$. Therefore, the problem is (approximately) reduced to a sample mean problem for the new sequence $\{W_t\}$.

As an immediate example, assume that $EX_t = 0$, and consider the sample autocovariance $\hat{R}(l)$ at some fixed lag l. Then obviously

$$\hat{R}(l) = \frac{1}{n-l+1} \sum_{t=1}^{n-l+1} W_t + o_P(1/\sqrt{n}),$$

where $W_t = X_t X_{t+l}$; some more examples can be found in Politis and Romano (1992a) and Bühlmann and Künsch (1995).

In many cases of interest, one can take l to be equal to one and $G_l(\cdot)$ to be equal to the influence function $g_F(\cdot)$ (see Section 1.6). Note that $g_F(\cdot)$ depends on the underlying distribution, but it can be replaced by a consistent estimate $g_{\hat{F}_n}(\cdot)$. For example, if $\theta(F)$ is the median of X_1, then $g_F(X_t) = \text{sign}(X_t - \theta(F))$, which can be estimated by $g_{\hat{F}_n}(X_t) = \text{sign}(X_t - \hat{\theta}_n)$, where $\hat{\theta}_n$ denotes the sample median. Using the estimated influence function then yields

$$\hat{\theta}_n = \frac{1}{n}\sum_{t=1}^{n} g_{\hat{F}_n}(X_t) + o_P(1/\sqrt{n}), \qquad (9.7)$$

and again the problem is (approximately) reduced to variance estimation for a sample mean. This approach has been discussed in detail by Bühlmann and Künsch (1994) in the context of block size selection for the moving blocks bootstrap. They make use of the fact that the moving blocks variance estimator is asymptotically equivalent to the Bartlett variance estimator, where the block size b plays the role of the inverse of the bandwidth. Hence, methods for choosing the bandwidth for kernel estimators can be employed to pick the block size. These results directly apply to subsampling as well due to the equivalence of the moving blocks and subsampling variance estimators for the sample mean.

9.3 Estimation of a Distribution Function

A more important, but also more ambitious, goal than estimating the variance of an estimator is to construct confidence intervals or regions for unknown parameters. To this end, the sampling distribution of the properly standardized estimator has to be estimated. For the construction of one-sided or equal-tailed two-sided confidence intervals, one needs to estimate a one-sided distribution function. For symmetric two-sided confidence intervals and multidimensional confidence regions, one needs to estimate a two-sided distribution function; for example, see Sections 3.2 and 3.3 for a discussion of the estimation of one-sided and two-sided distribution functions.

Hall, Horowitz, and Jing (1996) showed that for the moving blocks bootstrap, in the context of stationary observations, the optimal asymptotic rate for b depends critically on the application. Indeed, it is $n^{1/3}$, $n^{1/4}$, and $n^{1/5}$ for bias or variance estimation, estimation of a one-sided distribution function, and estimation of a two-sided distribution function, respectively. For this reason the condition of Lahiri (1992) on the block size for the moving blocks bootstrap, namely $b = o(n^{1/4})$, seems to be overly restrictive (see Remark 4.4.3).

In the previous section, we showed that the optimal block size choice for the subsampling variance estimator of the sample mean of dependent data is proportional to $n^{1/3}$. In Chapter 10, we also give the optimal rates for one-sided and two-sided subsampling distributions in the context of i.i.d. or strong mixing data.

Nevertheless, although the optimal rates are known to us, the proportionality constants associated with those rates are typically hard to pin-point and a practical problem ensues. To address this problem, Hall et al. (1996) present a a data-driven rule for selecting the block size. The first step is to find an optimal block size for time series of smaller length than the original, say $m < n$. This is done by applying the moving blocks method to all $n - m + 1$ subseries of length m and comparing the results to the outcome of applying the moving blocks to the entire series of length n. Denote the thus found optimal block size for a series of length m by $\hat{b}_m$. The optimal block size for the original series is then estimated by $\hat{b}_n = (n/m)^{1/k} \hat{b}_m$, where $k = 3$, 4, or 5 depending on context.

It is not difficult to modify this rule for the subsampling method. Unfortunately, simulation studies did not give promising results. For example, from Subsection 9.5.1 and Table 9.1 it seems clear that for an AR(1) process with positive autoregressive parameter ρ the optimal block size increases with ρ. However, the distribution of the chosen block size with the Hall, Lahiri, and Jing (1996) method remained basically unchanged when we varied ρ between 0.2 and 0.9. This was true for sample sizes of $n = 128$ and $n = 256$.

In this section, we will present two approaches for choosing a block size for the subsampling method. The first one uses a calibration idea dating back to Loh (1987). The crux is to use subsampling on a large number of pseudo sequences generated by a standard bootstrap. It is therefore restricted to scenarios where standard bootstrap method would also work. These basically encompass asymptotically linear statistics and parameters depending smoothly on a finite-dimensional marginal distribution of the underlying probability mechanism.

Since a significant part of the appeal of the subsampling method stems from the fact that it works under weak conditions, where bootstrap methods may fail, we also need an alternative approach that can be used whenever subsampling applies. The idea of this alternative approach will be to minimize the variability of confidence interval width over a range of block sizes.

We do not claim any asymptotic optimality for either approach. The aim is rather to provide sensible answers for small to moderate sample sizes. The effect of different choices of b diminishes as the sample size increases.

9.3.1 Calibration Method

One can think of the accuracy of an approximate or asymptotic confidence procedure—such as normal, bootstrap, or subsampling methods—in terms of its *calibration* (Loh, 1987). Suppose we use the procedure to construct a confidence interval with nominal confidence level $1 - \alpha$. We can denote the actual confidence level by $1 - \lambda$. Here, α is known to us while λ typically is not. An asymptotic method only ensures that $1 - \lambda$ will tend to $1 - \alpha$ as the sample size tends to infinity. For a finite sample size, the two levels generally will not be the same. If we knew the calibration function $h : 1 - \alpha \to 1 - \lambda$, we could construct a confidence region with exactly the desired coverage. First we would let $1 - \lambda$ be the desired level and then construct an interval with nominal level $1 - \alpha$ such that $h(1 - \alpha) = 1 - \lambda$. For example, if $h(0.98) = 0.95$, then a confidence interval with nominal level 98% would be an actual 95% confidence interval.

Fortunately, the calibration function $h(\cdot)$ can be estimated. One way of doing this would be to assume an approximating, parametric model with a known parameter θ_0. By then using a Monte Carlo approach a natural estimate $\hat{h}(\cdot)$ would be easy to find: one would generate many pseudo sequences, compute a $1 - \alpha$ interval for each sequence, and take the proportion of intervals that contain θ_0. Instead, we will opt for a model-free approach by using a bootstrap distribution P_n^* as the pseudo data-generating mechanism. For independent observations, we can use Efron's bootstrap. In the context of dependent data, we need to employ a bootstrap suitable for time series. The moving blocks bootstrap (Künsch, 1989, and Liu and Singh, 1992; see Section 3.9) lends itself to the task.

The application of the calibration scheme to the subsampling method can be done conditional on a *reasonable* block size. This means that a sensible block size b is fixed and one calibrates the subsampling intervals using this particular block size. Thereby, the problem of finding the "optimal" block size is eliminated. In some scenarios, one has a pretty good idea what a reasonable block size will be, either from prior experience or related simulation studies. Otherwise, see Remarks 9.3.2 and 9.3.3 below. To describe the calibration technique more formally, see the following algorithm.

Algorithm 9.3.1 (Calibration by adjusting the confidence level).

1. Generate K pseudo sequences $X_1^{*k}, \ldots, X_n^{*k}$, according to a suitable bootstrap distribution P_n^*.
   For each sequence, $k = 1, \ldots, K$,

   a. Compute an $1 - \alpha$ level confidence interval $\text{CI}_{1-\alpha}^k$, for a grid of values of α in the neighborhood of λ.

2. For each α compute $\hat{h}(1 - \alpha) = \#\{\hat{\theta}_n \in \text{CI}_{1-\alpha}^k\}/K$.
3. Interpolate $\hat{h}(\cdot)$ between the grid values.
4. Find the value of α satisfying $\hat{h}(1 - \alpha) = 1 - \lambda$.
5. Construct a confidence interval with nominal confidence level $1 - \alpha$.

Remark 9.3.1. In case the moving blocks bootstrap is employed in Step 1, a block size b_{MB} is required. The choice of this block size has a second order effect and is therefore not very important. However, if an

196   9. Choice of the Block Size

automatic selection method is preferred, a "nested bootstrap" can be used. That means that one would use the moving blocks bootstrap in both Steps 1 and 1a of the above algorithm with the same block size $b_M B$, limiting the grid of α values to $\alpha = \lambda$. Repeating this algorithm for a number of b_{MB} values, one then would select the value b_{MB} which yields estimated coverage closest to $1 - \lambda$.

Remark 9.3.2. If the calibration scheme is used to calibrate the subsampling method, one needs to start out with a reasonable block size b. In situations where one does not know what a reasonable block size is, the following idea can be used. In the same way as the actual confidence level can be regarded as function of the nominal confidence level (conditional on a fixed block size), it can be considered as a function of the block size (conditional on a fixed nominal level). Fixing the nominal level at the desired level, that is, choosing $\alpha = \lambda$, one can therefore estimate the block calibration function $g : b \to 1 - \lambda$, using an analogous calibration algorithm:

Algorithm 9.3.2 (Calibration by adjusting the block size).

1*. Generate K pseudo sequences $X_1^{*k}, \ldots, X_n^{*k}$, according to a moving blocks bootstrap distribution P_n^*.
For each sequence, $k = 1, \ldots, K$,

    1a*. Compute an $1 - \lambda$ level confidence interval CI_b^k, for a selection of block sizes b.

2*. For each b compute $\hat{g}(b) = \#\{\hat{\theta}_n \in \text{CI}_b^k\}/K$.

A reasonable block size will then satisfy $\hat{g}(b) \approx 1 - \lambda$.

Obviously, if at this step already a block size b with $\hat{g}(b) = 1 - \lambda$ is found, no further calibration is needed. Using this particular b, one then constructs a confidence interval with nominal level $1 - \alpha = 1 - \lambda$.

Remark 9.3.3. An alternative and maybe more straightforward approach to find a reasonable block size is to use the minimum volatility method of Subsection 9.3.2.

Remark 9.3.4. Two-sided equal-tailed intervals should always be computed as the intersection of two separately calibrated one-sided intervals. Particularly if the sampling distribution of $\hat{\theta}_n$ is asymmetric, the amount of calibration needed in the lower tail can be different from the one needed in the upper tail.

Remark 9.3.5. The motivation for using a bootstrap distribution P_n^* in Step 1 rather than a fixed parametric model is to match the true data-generating mechanism, at least in an asymptotic sense. An alternative approach would be the use of a nested sequence of parametric models, a so-called *sieve*, say $\{\mathbf{P}_n\}$. Of the class $\mathbf{P}_n$, one would then employ an estimator $\hat{P}_n$ (for example, obtained by maximum likelihood) in Step 1.

If the class $\mathbf{P}_n$ is allowed to become "richer" in an appropriate way as n increases, then the true data-generating mechanism can also be matched asymptotically. For example, if the observed sequence is a univariate, stationary time series, one could use for $\mathbf{P}_n$ the class of $AR(p_n)$ models, where $p_n \to \infty$ and $p_n/n \to 0$ as $n \to \infty$. (For a reference on sieves, see Shen and Wong, 1994.)

Remark 9.3.6. The algorithm is stated in terms of constructing confidence intervals for univariate parameters. It works analogously for confidence regions or confidence bands for multivariate or functional parameters, respectively.

For an illustration of how one would use the calibration method, see Figure 9.2 for an artificial example. Suppose one wants to construct a 95% confidence interval, which corresponds to $\lambda = 0.05$. The calibration function $h(\cdot)$ is estimated at the discrete points $0.90, 0.91, \ldots, 0.99$, and linearly interpolated inbetween. The estimate then says that one should construct a confidence region using a nominal level around 0.978.

Since the calibration is based on the estimated calibration function $\hat{h}(\cdot)$ rather than the true function $h(\cdot)$, calibrated intervals still are not exact. It can easily be shown that calibrating an asymptotically correct procedure, such as subsampling confidence intervals, results in an asymptotically correct procedure again. In the context of i.i.d. random variables it is known that calibrated confidence intervals have better asymptotic properties than uncalibrated intervals (e.g., Efron and Tibshirani, 1993). In technical terms, calibrated confidence intervals for i.i.d. data are generally second-order correct versus first-order correct. Corresponding results for dependent data are work in progress and have not yet been obtained. Simulation studies support the conjecture that also in the context of dependent data calibrated intervals generally give better results; see Section 9.5.

The problem of the calibration method is that it relies on a suitable bootstrap distribution P_n^*, Efron's bootstrap or the moving blocks, to estimate the calibration function $h(\cdot)$. This will give good results if P_n^* is a reliable approximation of the underlying probability mechanism; that is, if bootstrapping would also result in asymptotically correct confidence regions. Clearly, an alternative method is needed that can be used in cases where Efron's or the moving blocks bootstrap are inconsistent.

9.3.2 Minimum Volatility Method

In this subsection, another approach is presented that will work under more general conditions than the calibration technique, that is, whenever subsampling applies.

The method is of heuristic nature and we do not claim any optimality properties. It is based on the fact that for the subsampling method to be

9. Choice of the Block Size

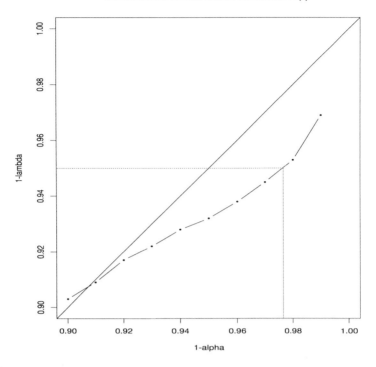

FIGURE 9.2. An artificial example. Estimates of $h(1 - \alpha)$ for a grid of α values $\{0.01, 0.02, \ldots, 0.1\}$. For a confidence interval with desired confidence level 0.95 we should construct an interval with nominal level 0.978.

consistent, the block size b needs to tend to infinity with the sample size n but at a smaller rate, satisfying $b/n \to 0$. Indeed, for b too close to n all subsample statistics ($\hat{\theta}_{n,b,i}$ or $\hat{\theta}_{n,b,t}$) will be almost equal to $\hat{\theta}_n$, resulting in the subsampling distribution being too tight and in undercoverage of subsampling confidence intervals. Lahiri (1998) makes this intuition precise by proving, in the context of mean-like statistics, that for $b/n \to 1$, the subsampling approximation collapses to a point mass at zero. On the other hand, if b is too small, the intervals can undercover or overcover depending on the state of nature. For theoretical examples, see Section 2.2 and for some simulation studies, Section 9.5. This leaves a number of b-values in the "right range" where we would expect almost correct results, at least for large sample sizes. Hence, in this range, the confidence intervals should be "stable" when considered as a function of the block size. This idea is exploited by computing subsampling intervals for a large number of block sizes b and then looking for a region where the intervals do not change very much. Within this region, an interval is then picked according to some arbitrary criterion.

While this method can be carried out by "visual inspection," it is desirable to also have some automatic selection procedure, at the very least when simulation studies are to be carried out. The procedure we propose is based on minimizing a running standard deviation. Assume one computes subsampling intervals for block sizes b in the range of b_{small} to b_{big}. The endpoints of the confidence intervals should change in smooth fashion, as b changes. This might be somewhat violated if we use a stochastic approximation, such as (2.12), for moderate or large sample sizes. In that case, it seems sensible to enforce some smoothness by applying a running mean to the endpoints of the intervals. Finally, a running standard deviation applied to the endpoints determines the volatility around a specific b value, and the value of b associated with the smallest volatility is chosen. Here is a more formal description of the algorithm.

Algorithm 9.3.3 (Minimizing confidence interval volatility).

1. For $b = b_{small}$ to $b = b_{big}$ compute a subsampling interval for $\theta(P)$ at the desired confidence level, resulting in endpoints $I_{b,low}$ and $I_{b,up}$.
2. If a stochastic approximation such as (2.12) was used in Step 1, smooth the lower and upper endpoints separately, using a running mean of span m. This means replace $I_{b,low}$ by the average of the values $\{I_{b-m,low}, I_{b-m+1,low}, \ldots, I_{b+m,low}\}$ and do the analogue for $I_{b,up}$.
3. For each b, compute a volatility index VI_b as the standard deviation of the interval endpoints in a neighborhood of b. More specifically, for a small integer k, let VI_b be equal to the standard deviation of the endpoints $\{I_{b-k,low}, \ldots, I_{b+k,low}\}$ plus the standard deviation of the endpoints $\{I_{b-k,up}, \ldots, I_{b+k,up}\}$.
4. Pick the value b^* corresponding to the smallest volatility index and report $[I_{b^*,low}, I_{b^*,up}]$ as the final confidence interval.

Remark 9.3.7. It is obvious how to modify the algorithm for one-sided confidence intervals.

Remark 9.3.8. The range of b-values, determined by b_{small} and b_{big}, which is included in the minimization algorithm is not very important, as long as it is not too narrow. In the context of data-dependent choice of block size of Sections 2.7 and 3.6, we can think of b_{small} as corresponding to j_n and of b_{big} as corresponding to k_n. Of course, the dependence on n has been suppressed in our notation here.

Remark 9.3.9. To make the algorithm more computationally efficient, it might be desirable to skip a number of b values in a regular fashion. For example, include only every other or every third b between b_{small} and b_{big}.

Remark 9.3.10. The algorithm contains two model parameters, m and k. Simulation studies have shown that the algorithm is very insensitive to the choice of either parameter. We typically employ $m = 2$ or $m = 3$ and also $k = 2$ or $k = 3$.

Remark 9.3.11. The algorithm is stated for the construction of confidence intervals. It can be easily modified if one wants to construct confidence regions or bands instead, by minimizing the volatility of some appropriate measure of size. For example, if a confidence region is needed, one can use volume or width of a one-dimensional projection.

Remark 9.3.12. For an illustration of the algorithm, see Sections 11.5 and 12.4.

9.4 Hypothesis Testing

As was mentioned in Sections 2.6 and 3.5, in some instances a hypothesis testing problem does not translate into constructing a confidence region, since the null hypothesis does not translate into a null hypothesis on a parameter θ. Alternative testing procedures based on subsampling, which do not require the existence of a suitable parameter, were presented in above two sections. Of course, in applying these procedures, one also has to choose a block size b. It is natural to ask whether analogues of the methods proposed in the previous section can be used.

9.4.1 Calibration Method

The calibration method of Subsection 9.3.1 involves checking whether an estimated parameter is contained in confidence regions computed from pseudo samples. Clearly, this renders it inapplicable for our current purpose. A modification is feasible, however, by generating pseudo samples that impose the null hypothesis and thereby estimating the true (or actual) level of the test. Suppose a test is carried out at nominal significance level α. Denote the actual significance level by λ. Again, one can think of λ as an unknown function $h(\alpha)$, the calibration function. Given an estimate $\hat{h}(\cdot)$ and a desired actual level λ, one should carry out the test using the nominal level α which satisfies $\hat{h}(\alpha) = \lambda$. The modifications of Algorithm 9.3.1 are given as follows.

Algorithm 9.4.1 (Calibration by adjusting the significance level).

1. Generate K pseudo sequences $X_1^{*k}, \ldots, X_n^{*k}$, according to a suitable distribution P_n^* that imposes the null hypothesis.
   For each sequence, $k = 1, \ldots, K$,

   1a. Carry out a test at significance level α, for a grid of values of α in the neighborhood of λ.

2. For each α compute $\hat{h}(\alpha) = \{\#$ times the test rejects at level $\alpha\}/K$.
3. Interpolate $\hat{h}(\cdot)$ between the grid values.
4. Find the value of α satisfying $\hat{h}(\alpha) = \lambda$.
5. Carry out a test at nominal significance level α.

Remark 9.4.1. Remark 9.3.2 applies in spirit; that is, an analogous algorithm can be used to find a reasonable block size b, which is employed in above algorithm. By generating pseudo data from P_n^*, one can estimate the calibration function $g : b \to \lambda$. A reasonable block size will then satisfy $\hat{g}(b) \approx \lambda$, where λ is the desired significance level. Again, if at this step already a block size b with $\hat{g}(b) = \lambda$ can be found, no further calibration is needed. With this particular b, the test is carried out at nominal level $\alpha = \lambda$.

Remark 9.4.2. The appropriate mechanism P_n^* will depend on the testing problem at hand. In certain applications, it could be a bootstrap distribution that imposes the null hypothesis. In other applications, it may be a (semi-)parametric model. Recall that the null hypothesis states that the underlying distribution P belongs to a certain class of distributions $\mathbf{P_0}$. If, under the null, P can be consistently estimated, we can let P_n^* be the member of $\mathbf{P_0}$ that is "closest" to the sample $X_1, \ldots, X_n$. For example, P_n^* might be obtained by a constrained maximum likelihood or a minimum distance estimator.

In some applications, the consistent estimation of the true model, even under the null only, might be cumbersome or even impossible. This renders the calibration method inapplicable. A more general approach is then provided by the minimum volatility method.

9.4.2 Minimum Volatility Method

While Algorithm 9.3.3 involves the construction of confidence intervals, an equivalent description can be given which does not. Since a subsampling interval is nothing but the point estimate $\hat{\theta}_n$ plus or minus scaled quantiles of an appropriate subsampling distribution such as $L_{n,b}$ or $L_{n,b,|\cdot|}$, it would be equivalent to minimize the volatility of these quantiles as b ranges from b_{small} to b_{big}. It is exactly this idea that has its analogue when subsampling is used for hypothesis testing.

Algorithm 9.4.2 (Minimizing volatility of critical values).

1. For $b = b_{small}$ to $b = b_{big}$ compute a subsampling quantile $g_{n,b}(1 - \alpha)$, as defined in (2.15) or (3.38), for the desired significance level α.
2. If a stochastic approximation was used in Step 1, smooth the quantiles, using a running mean of span m. This means replace $g_{n,b}(1 - \alpha)$ by the average of the values $\{g_{n,b-m}(1 - \alpha), \ldots, g_{n,b+m}(1 - \alpha)\}$.

3. For each b, compute a volatility index VI_b as the standard deviation of the quantiles in a neighborhood of b. More specifically, for a small integer k, let VI_b be equal to the standard deviation of the values $\{g_{n,b-k}(1-\alpha), \ldots, g_{n,b+k}(1-\alpha)\}$.
4. Pick the value b^* corresponding to the smallest volatility index and use $g_{n,b^*}(1-\alpha)$ as the critical value of the test.

Remarks 9.3.9 and 9.3.10 apply as well.

Remark 9.4.3. The analogy to the minimum volatility method for construction of confidence intervals is actually only valid under the null hypothesis, since only then T_n converges to a nondegenerate limiting distribution. In the notation of Sections 2.6 and 3.5, under the null and the regularity conditions stated, $g_{n,b}(1-\alpha)$ converges to $g(1-\alpha, P)$ in probability.

However, under the alternative, $g_{n,b}(1-\alpha)$ will typically tend to infinity in probability. Indeed, assume that for $P \in \mathbf{P_1}$, $\tau_n(t_n(X_1, \ldots, X_n) - t(P))$ converges to a continuous law $G_1(P)$; this assumption will quite often hold. Hence, $g_{n,b}(1-\alpha)$ will be approximately equal to $g_1(1-\alpha, P) + \tau_b \, t(P)$; recall that we assume $t(P) > 0$ under the alternative. For example, in regular cases, $\tau_n = n^{1/2}$ and so $g_{n,b}(1-\alpha)$ will tend to increase in a square root fashion as a function of the block size b. Since the test rejects the null if and only if $T_n > g_{n,b}(1-\alpha)$, in finite samples the power tends to decrease with the block size. Therefore, it is important to not choose b_{big} very big in Step 2 of the algorithm. This is in contrast to applying the minimum volatility method for the construction of confidence intervals.

To exaggerate the difference, consider the (inconsistent) choice $b = n$. In that case, it is easy to see that a subsampling confidence interval for $\theta(P)$ will be given by the single point $\hat{\theta}_n$ only. Hence, if the null hypothesis translates into a null hypothesis on a parameter and one constructs a confidence interval to do the test, the null would typically be rejected with probability one in finite samples. On the other hand, if the direct testing approach is used, it is easy to see that $g_{n,b}(1-\alpha) = T_n$, and the null would never be rejected.

9.5 Two Simulation Studies

The goal of this section is to examine the small sample performance of the subsampling method via simulation studies. Focus is on the applications of the univariate mean and multivariate linear regression. As data-generating mechanisms, both stationary and nonstationary (or heteroscedastic) processes will be employed. Note that some further, context-related simulation studies can be found in Chapters 11, 12, and 13.

9.5.1 Univariate Mean

The first simulation study is concerned with the univariate mean. We assess finite sample performance of nominal 95% two-sided confidence intervals. Performance is measured by estimated true coverage probability.

As the data generating process, a simple AR(1) model is used given by

$$X_t = \rho X_{t-1} + \epsilon_t,$$

where $\{\epsilon_t\}$ is white noise. The closer the AR(1) parameter ρ is to one in absolute value, the stronger is the dependence of the X_t sequence.

As a first question, we are interested in the effect of the block size; that is, how does coverage probability change as b is changed? Then, the two automatic methods for dealing with the block size of Section 9.3 are compared. Also, both equal-tailed and symmetric intervals are considered.

To address the first question, we compute confidence intervals for different block sizes b, keeping the sample size n fixed, and estimate coverage probability as the proportion of intervals that contain the true mean. For our simulations we use $n = 256$ and $b = 4, 8, 16, 32,$ and 64. As far as the innovation sequence for our AR(1) model is concerned, the ϵ_t are chosen to be i.i.d. from either a standard normal distribution or an exponential distribution with parameter one, shifted to have expected value zero. Hence, the true mean is always zero. As far as the AR(1) parameter is concerned, the values $\rho = 0.2, 0.5, 0.8, 0.95,$ and -0.5 are considered. The simulation results are based on 1000 random samples for each scenario, and they are presented in Table 9.1.

As to be expected, performance gets worse as ρ increases in absolute value. Also, the best fixed block size clearly depends on ρ. Larger block sizes are needed to best capture stronger dependency structures. Note that the intervals tend to undercover, except for the negative AR(1) parameter $\rho = -0.5$ where we find overcoverage for small block sizes. Symmetric intervals perform slightly better than equal-tailed intervals.

Next, the two methods for dealing with the block size in an automatic way are compared. The results are also contained in Table 9.1. Some remarks are in order.

The calibration method is based on Algorithm 9.3.1:

1. The moving blocks bootstrap was utilized to generate the pseudo sequences in Step 1. Using a nested bootstrap approach for a variety of block sizes, we found the following b_{MB} to be satisfactory block sizes: $b_{MB} = 20$ for $|\rho| \leq 0.8$ and $b_{MB} = 40$ for $\rho = 0.95$.
2. To find reasonable block sizes for the subsampling method, the block calibration function $g(\cdot)$ was estimated (see Remark 9.3.2). The so chosen block sizes were as follows: $b = 10$ for $\rho = 0.2$, $b = 15$ for $|\rho| = 0.5$, $b = 20$ for $\rho = 0.8$, and $b = 40$ for $\rho = 0.95$.
3. A practical issue is the number K of bootstrap samples that are generated in order to estimate the calibration function $h(\cdot)$. For any *real*

application, one should choose $K = 1000$ at the least. However, this turned out the be computationally too expensive in the context of a *simulation study*, and we settled for the number $K = 200$.

The minimum volatility method is based on Algorithm 9.3.3. Throughout, we used $b_{small} = 4$, $b_{big} = 40$, and $k = 2$. Since no stochastic simulation was employed in Step 1, Step 2 and thus the choice of m were not relevant.

Not surprisingly, the calibration method gives better results and typically outperforms the best fixed block size. This is due to the fact that it adjusts the nominal coverage level to correct for over- or undercoverage. Note, however, that especially when the dependence is strong these adjustments are not quite sufficient and the intervals still undercover. On the other hand, the minimum volatility method compares favorably to the best fixed block size. This is encouraging, since in applications where bootstrap methods are inconsistent, we cannot use the calibration method. Therefore, it is very valuable to have an automatic method that generally does almost as well as the typically unknown optimal fixed block size.

As with fixed block sizes, symmetric intervals perform somewhat better than equal-tailed intervals for automatic block size selection also.

9.5.2 *Linear Regression*

The second simulation study is concerned with multivariate least squares linear regression. We assess finite sample performance of confidence intervals for a (single) regression parameter. Again, performance is measured in terms of coverage probability of two-sided nominal 95% intervals. For comparison, also some more standard kernel approaches, based on asymptotic normality, are included. There, a confidence interval is given by the estimator plus or minus 1.96 times its standard error, which is computed using a kernel technique. Andrews (1991) compared various kernels applied to covariance estimation in multivariate linear regression. He found that the so-called quadratic spectral (QS) kernel has certain asymptotic optimality properties, which were then substantiated in a Monte Carlo study. In a follow-up paper, Andrews and Monahan (1992) suggested prewhitened kernel estimators as an improvement over regular kernel estimators, at least when coverage probabilities are of main interest. Again, the prewhitened QS (QS-PW) kernel seems favored over other kernels. For all kernel methods, an automatic bandwidth selection, as proposed by Andrews (1991), is employed in our simulations.

The same basic model as in Andrews (1991) is used, given by

$$y_t = x_t'\beta + \epsilon_t, \tag{9.8}$$

where $x_{t,1} = 1$ and $x_t = (1, \dot{x}_t)'$ and β_0 are 5×1 vectors. Throughout, we are concerned with constructing confidence intervals for the regression parameter β_2. Without loss of generality, we set β equal to zero. As far as the data generating processes are concerned, four models are employed. Three of them were suggested by Andrews (1991). In the first model, AR(1)-HOMO, errors and regressors are independent AR(1) processes.

AR(1)-HOMO: $\dot{x}_{t,j} = \rho \dot{x}_{t-1,j} + \nu_{t,j}$ and $\epsilon_t = \rho \epsilon_{t-1} + \nu_t^\epsilon$.

Here, and for the following models, the $\{\nu_{t,j}\}$ and $\{\nu_t^\epsilon\}$ are mutually independent white noise processes.

The second model, AR(1)-HET1, is a variation of the first one in the sense that multiplicative heteroskedasticity is overlaid on the errors. To be more specific, let

AR(1)-HET1: $\dot{x}_{t,j} = \rho \dot{x}_{t-1,j} + \nu_{t,j}$, $\tilde{\epsilon}_t = \rho \tilde{\epsilon}_{t-1} + \nu_t^\epsilon$ and $\epsilon_t = |x_{t,2}| \tilde{\epsilon}_t$.

In Andrews' notation, HET stands for some kind of heteroskedasticity, although one should note that for this process the ϵ_t are only conditionally heteroskedastic.

Next, a similar process is considered with the difference that the errors are unconditionally heteroskedastic, displaying a "seasonal" pattern.

AR(1)-SEASON: $\dot{x}_{t,j} = \rho \dot{x}_{t-1,j} + \nu_{t,j}$ and $\epsilon_t = \rho \epsilon_{t-1} + a_t \nu_t^\epsilon$.

$\{a_t\}$ is the infinite repetition of the sequence $\{1, 1, 1, 2, 3, 1, 1, 1, 1, 2, 4, 6\}$.

In the final process, both the errors and the regressors are independent MA(1) processes.

MA(1)-HOMO: $\dot{x}_{t,j} = \nu_{t,j} + \theta \nu_{t-1,j}$ and $\epsilon_t = \nu_t^\epsilon + \theta \nu_{t-1}^\epsilon$.

For all four data-generating mechanisms, $\{\nu_{t,j}\}$ and $\{\nu_t^\epsilon\}$ are independent i.i.d. innovation sequences, having either a standard normal distribution or a centered exponential distribution with variance one. The values considered for the parameters ρ and θ are 0.2, 0.5, 0.8, 0.95, and -0.5. The sample size employed is $n = 256$. For each scenario, 1000 random samples were generated. Some remarks concerning the block size choice methods of Section 9.3 are in order.

The calibration method is based on Algorithm 9.3.1:

1. The moving blocks bootstrap was utilized to generate the pseudo sequences in Step 1. Using a nested bootstrap approach for a variety of block sizes, we found the following b_{MB} to be satisfactory block sizes: $b_{MB} = 20$ for both the AR(1) models with parameters $|\rho| \leq 0.8$; $b_{MB} = 40$ for these models with parameter $\rho = 0.95$; $b_{MB} = 20$ for the MA(1)-HOMO model across all parameters considered.
2. To find reasonable block sizes for the subsampling method, the block calibration function $g(\cdot)$ was estimated (see Remark 9.3.2). The so chosen block sizes were as follows: For the AR(1)-HET model, we used $b = 10$ throughout; for the AR(1)-HOMO and AR(1)-SEASON models, we used $b = 20$ for $|\rho| \leq 0.8$ and $b = 40$ for $|\rho| = 0.95$; for the MA(1)-HOMO model, we used $b = 20$ throughout.

3. A practical issue is the number K of bootstrap samples that are generated in order to estimate the calibration function $h(\cdot)$. For any *real application*, one should choose $K = 1000$ at the least. However, this turned out the be computationally too expensive in the context of a *simulation study*, and we settled for the number $K = 200$.

The minimum volatility method is based on Algorithm 9.3.3. Throughout, we used $b_{small} = 8$, $b_{big} = 50$, and $k = 2$. Since no stochastic simulation was employed in Step 1, Step 2 and thus the choice of m were not relevant.

The results of the simulations are presented in Tables 9.1–9.5.

For the AR(1) models, it can be seen that the coverage of the kernel intervals gets progressively worse as ρ increases in absolute value and can be as low as 63% for the AR(1)-HET1 model. As an aside, we do not find superior performance of the prewhitened intervals. On the the other hand, the calibrated subsampling intervals exhibit pretty constant coverage probability, except for the extreme case $\rho = 0.95$, where it can drop as low as 84%. Subsampling intervals based on the minimum volatility method are somewhere in between in terms of performance, at least when symmetric intervals are used; equal-tailed intervals perform rather poorly.

For the MA(1)-HOMO model, all intervals perform satisfactorily, no matter what the value of the parameter θ. This is not too surprising since an MA(1) sequence with i.i.d. innovations is 1-dependent, that is, observations separated by more than one time unit are always independent.

Finally, note that the kernel intervals seem to be affected slightly by the distribution of the innovation sequences, with exponential innovations yielding somewhat worse performance throughout. The same cannot be said for the subsampling intervals, which appear rather robust in this respect.

9.6 Conclusions

The goal of this chapter was to propose some methods to choose the block size b in practice. The applications of variance estimation, estimation of a distribution function, and hypothesis testing were discussed.

In the section on variance estimation, focus was on the case of the sample mean. It was pointed out that the subsampling variance estimator is asymptotically equivalent to the well-known Bartlett kernel estimator, with the block size b playing the role of the inverse of the bandwidth for the Bartlett kernel. Hence, any of the methods used to pick the bandwidth of a kernel estimator can be employed to choose the block size. This approach carries over to "mean-like" statistics by linearization. The reader is also referred to Section 10.5.5, where an improved subsampling variance estimator, based on extrapolation, will be proposed whose corresponding optimal block size is perhaps easier to identify.

9.6 Conclusions

For the estimation of a distribution function and hypothesis testing, two methods were proposed, namely, the calibration method and the minimum volatility method; note that either method varies somewhat depending on the application (distribution function versus testing). The idea of the calibration method is to start out with a reasonable block size and then to adjust the confidence or significance level by estimating the calibration function, which maps nominal level to actual level. The estimation is done by generating pseudo data from an appropriate mechanism, such as a bootstrap distribution. The downside of this approach is that it is not always possible to find such a mechanism. Examples are situations where subsampling yields asymptotically correct inference but bootstrap methods or (semi-)parametric models do not.

On the other hand, the minimum volatility method can be employed whenever subsampling applies and it is therefore more robust. The method is based on the fact that subsampling will behave poorly for block sizes b very small or very large compared to the "optimal" choice, but will be "stable" in an appropriate range of b; for hypothesis testing, this was seen to be only true under the null hypothesis, though. The method tries to locate this range by first computing intervals or quantiles for a large number of block sizes b and then applying a running standard deviation to interval endpoints or quantiles (considered as functions of b). Note that an alternative use of this approach is to yield a reasonable block size b to be employed by the calibration method in the context of confidence interval construction.

While we believe that both methods give reasonable results—a belief that is supported by simulation studies in this and in other chapters—the issue of choosing the block size is far from settled. Further research will have to be devoted to this difficult problem.

Finally, the small sample behavior of subsampling was examined with the help of some simulation studies. The cases of the univariate mean and multivariate linear regression were considered, employing both stationary and nonstationary (or heteroskedastic) data generating mechanisms. Note that some other, context-dependent simulation studies can be found in Chapters 11–13. Of interest was also to judge the effect of the method for choosing the block size, calibration method versus minimum volatility method. The results were encouraging and, as expected, somewhat better for the calibration method. However, the minimum volatility method did almost as well as the best fixed block size, which will be unknown in practical applications, in the context of the univariate mean; and it outperformed kernel methods, based on using asymptotic normality, in the context of linear regression. Since the minimum volatility method can be used for more complicated statistics, where the calibration method or traditional inference fail, it is reassuring to know that it gives good results in cases where its performance can be compared to that of competitors.

9.7 Tables

TABLE 9.1. Mean, AR(1) model, n = 256. Estimated coverage probabilities of various nominal 95% confidence intervals for the univariate mean. The estimates are based on 1000 replications for each scenario.

Parameter	Interval	$b=4$	$b=8$	$b=16$	$b=32$	$b=64$	Calib.	Min. Vol.
		Gaussian innovations						
$\rho = 0.2$	Equal-tailed	0.92	0.92	0.90	0.88	0.80	0.92	0.90
$\rho = 0.2$	Symmetric	0.92	0.92	0.91	0.89	0.84	0.94	0.92
$\rho = 0.5$	Equal-tailed	0.86	0.89	0.88	0.86	0.79	0.92	0.90
$\rho = 0.5$	Symmetric	0.87	0.90	0.89	0.88	0.81	0.94	0.92
$\rho = 0.8$	Equal-tailed	0.74	0.82	0.85	0.85	0.78	0.86	0.85
$\rho = 0.8$	Symmetric	0.74	0.84	0.87	0.87	0.81	0.91	0.87
$\rho = 0.95$	Equal-tailed	0.41	0.51	0.63	0.69	0.66	0.69	0.70
$\rho = 0.95$	Symmetric	0.41	0.53	0.64	0.73	0.71	0.75	0.74
$\rho = -0.5$	Equal-tailed	0.96	0.95	0.93	0.91	0.83	0.94	0.91
$\rho = -0.5$	Symmetric	0.97	0.95	0.94	0.92	0.86	0.94	0.92
		Exponential innovations						
$\rho = 0.2$	Equal-tailed	0.90	0.91	0.91	0.87	0.81	0.91	0.89
$\rho = 0.2$	Symmetric	0.92	0.92	0.92	0.89	0.84	0.93	0.91
$\rho = 0.5$	Equal-tailed	0.86	0.89	0.89	0.87	0.78	0.89	0.89
$\rho = 0.5$	Symmetric	0.88	0.90	0.90	0.88	0.81	0.93	0.91
$\rho = 0.8$	Equal-tailed	0.70	0.79	0.84	0.84	0.77	0.86	0.87
$\rho = 0.8$	Symmetric	0.70	0.81	0.86	0.86	0.81	0.91	0.90
$\rho = 0.95$	Equal-tailed	0.44	0.56	0.64	0.71	0.68	0.72	0.68
$\rho = 0.95$	Symmetric	0.44	0.57	0.67	0.74	0.71	0.77	0.71
$\rho = -0.5$	Equal-tailed	0.93	0.93	0.92	0.90	0.83	0.92	0.91
$\rho = -0.5$	Symmetric	0.97	0.95	0.93	0.91	0.85	0.94	0.92

TABLE 9.2. Regression, AR(1)-HOMO model, n = 256. Estimated coverage probabilities of various nominal 95% confidence intervals the regression parameter β_2. The estimates are based on 1000 replications for each scenario.

Parameter	Interval	Calibration	Min. Vol.	QS	QS-PW
	Gaussian innovations				
$\rho = 0.2$	Equal-tailed	0.96	0.92		
$\rho = 0.2$	Symmetric	0.96	0.94	0.92	0.94
$\rho = 0.5$	Equal-tailed	0.95	0.90		
$\rho = 0.5$	Symmetric	0.95	0.94	0.93	0.94
$\rho = 0.8$	Equal-tailed	0.89	0.82		
$\rho = 0.8$	Symmetric	0.93	0.91	0.89	0.89
$\rho = 0.95$	Equal-tailed	0.69	0.63		
$\rho = 0.95$	Symmetric	0.88	0.80	0.76	0.74
$\rho = -0.5$	Equal-tailed	0.94	0.91		
$\rho = -0.5$	Symmetric	0.94	0.94	0.93	0.95
	Exponential innovations				
Parameter	Interval	Calibration	Min. Vol.	QS	QS-PW
$\rho = 0.2$	Equal-tailed	0.95	0.95		
$\rho = 0.2$	Symmetric	0.96	0.96	0.93	0.93
$\rho = 0.5$	Equal-tailed	0.97	0.91		
$\rho = 0.5$	Symmetric	0.94	0.95	0.92	0.93
$\rho = 0.8$	Equal-tailed	0.90	0.84		
$\rho = 0.8$	Symmetric	0.93	0.92	0.86	0.85
$\rho = 0.95$	Equal-tailed	0.73	0.60		
$\rho = 0.95$	Symmetric	0.84	0.78	0.73	0.71
$\rho = -0.5$	Equal-tailed	0.96	0.93		
$\rho = -0.5$	Symmetric	0.96	0.95	0.93	0.94

TABLE 9.3. Regression, AR(1)-HET model, n = 256. Estimated coverage probabilities of various nominal 95% confidence intervals the regression parameter β_2. The estimates are based on 1000 replications for each scenario.

		Gaussian innovations			
Parameter	Interval	Calibration	Min. Vol.	QS	QS-PW
$\rho = 0.2$	Equal-tailed	0.96	0.91		
$\rho = 0.2$	Symmetric	0.95	0.94	0.93	0.93
$\rho = 0.5$	Equal-tailed	0.97	0.86		
$\rho = 0.5$	Symmetric	0.95	0.92	0.90	0.91
$\rho = 0.8$	Equal-tailed	0.96	0.79		
$\rho = 0.8$	Symmetric	0.94	0.87	0.86	0.83
$\rho = 0.95$	Equal-tailed	0.85	0.70		
$\rho = 0.95$	Symmetric	0.84	0.77	0.67	0.64
$\rho = -0.5$	Equal-tailed	0.92	0.88		
$\rho = -0.5$	Symmetric	0.92	0.91	0.90	0.91
		Exponential innovations			
Parameter		Calibration	Min. Vol.	QS	QS-PW
$\rho = 0.2$	Equal-tailed	0.94	0.78		
$\rho = 0.2$	Symmetric	0.93	0.88	0.89	0.88
$\rho = 0.5$	Equal-tailed	0.91	0.80		
$\rho = 0.5$	Symmetric	0.93	0.87	0.89	0.88
$\rho = 0.8$	Equal-tailed	0.96	0.77		
$\rho = 0.8$	Symmetric	0.94	0.84	0.83	0.81
$\rho = 0.95$	Equal-tailed	0.89	0.70		
$\rho = 0.95$	Symmetric	0.89	0.75	0.66	0.63
$\rho = -0.5$	Equal-tailed	0.95	0.84		
$\rho = -0.5$	Symmetric	0.93	0.88	0.88	0.89

TABLE 9.4. Regression, AR(1)-SEASON model, n = 256. Estimated coverage probabilities of various nominal 95% confidence intervals the regression parameter β_2. The estimates are based on 1000 replications for each scenario.

Parameter	Interval	Calibration	Min. Vol.	QS	QS-PW
		Gaussian innovations			
$\rho = 0.2$	Equal-tailed	0.95	0.89		
$\rho = 0.2$	Symmetric	0.96	0.95	0.92	0.91
$\rho = 0.5$	Equal-tailed	0.93	0.88		
$\rho = 0.5$	Symmetric	0.94	0.94	0.88	0.85
$\rho = 0.8$	Equal-tailed	0.90	0.83		
$\rho = 0.8$	Symmetric	0.93	0.92	0.86	0.82
$\rho = 0.95$	Equal-tailed	0.79	0.65		
$\rho = 0.95$	Symmetric	0.92	0.82	0.75	0.72
$\rho = -0.5$	Equal-tailed	0.93	0.87		
$\rho = -0.5$	Symmetric	0.94	0.94	0.90	0.91
		Exponential innovations			
Parameter	Interval	Calibration	Min. Vol.	QS	QS-PW
$\rho = 0.2$	Equal-tailed	0.95	0.92		
$\rho = 0.2$	Symmetric	0.95	0.96	0.90	0.90
$\rho = 0.5$	Equal-tailed	0.97	0.92		
$\rho = 0.5$	Symmetric	0.97	0.96	0.88	0.86
$\rho = 0.8$	Equal-tailed	0.96	0.87		
$\rho = 0.8$	Symmetric	0.96	0.95	0.85	0.81
$\rho = 0.95$	Equal-tailed	0.77	0.70		
$\rho = 0.95$	Symmetric	0.92	0.83	0.73	0.71
$\rho = -0.5$	Equal-tailed	0.96	0.91		
$\rho = -0.5$	Symmetric	0.97	0.96	0.88	0.88

TABLE 9.5. Regression, MA(1)-HOMO model, n = 256. Estimated coverage probabilities of various nominal 95% confidence intervals the regression parameter β_2. The estimates are based on 1000 replications for each scenario.

Parameter	Interval	Calibration	Min. Vol.	QS	QS-PW
\multicolumn{6}{c}{**Gaussian innovations**}
$\theta = 0.2$	Equal-tailed	0.94	0.92		
$\theta = 0.2$	Symmetric	0.95	0.94	0.94	0.93
$\theta = 0.5$	Equal-tailed	0.96	0.92		
$\theta = 0.5$	Symmetric	0.95	0.93	0.93	0.94
$\theta = 0.8$	Equal-tailed	0.95	0.92		
$\theta = 0.8$	Symmetric	0.95	0.94	0.93	0.95
$\theta = 0.95$	Equal-tailed	0.95	0.92		
$\theta = 0.95$	Symmetric	0.95	0.94	0.93	0.95
$\theta = -0.5$	Equal-tailed	0.95	0.93		
$\theta = -0.5$	Symmetric	0.94	0.95	0.94	0.94

Parameter	Interval	Calibration	Min. Vol.	QS	QS-PW
\multicolumn{6}{c}{**Exponential innovations**}
$\theta = 0.2$	Equal-tailed	0.95	0.93		
$\theta = 0.2$	Symmetric	0.95	0.95	0.92	0.92
$\theta = 0.5$	Equal-tailed	0.96	0.93		
$\theta = 0.5$	Symmetric	0.96	0.94	0.93	0.94
$\theta = 0.8$	Equal-tailed	0.97	0.93		
$\theta = 0.8$	Symmetric	0.96	0.95	0.92	0.92
$\theta = 0.95$	Equal-tailed	0.97	0.93		
$\theta = 0.95$	Symmetric	0.96	0.95	0.92	0.92
$\theta = -0.5$	Equal-tailed	0.96	0.94		
$\theta = -0.5$	Symmetric	0.95	0.95	0.93	0.93

10
Extrapolation, Interpolation, and Higher-Order Accuracy

10.1 Introduction

In this chapter, we consider $\underline{X}_n = (X_1, \ldots, X_n)$ to be an observed stretch of a stationary, strong mixing sequence of real-valued random variables $\{X_t, t \in \mathbb{Z}\}$. The probability measure generating the observations is again denoted by P. As mentioned in Appendix A, the strong mixing condition amounts to $\alpha_X(k) = \sup_{A,B} |P(A \cap B) - P(A)P(B)| \to 0$ as k tends to infinity, where A and B are events in the σ-algebras generated by $\{X_t, t < 0\}$ and $\{X_t, t \geq k\}$, respectively. The case where $X_1, \ldots, X_n$ are independent, identically distributed (i.i.d.) will be treated here as an important special case where $\alpha_X(k) = 0$ for all $k > 0$.

In the previous chapters, a general subsampling methodology has been put forth for the construction of large-sample confidence regions for a general unknown parameter $\theta(P)$ under minimal conditions. The subsampling method hinges on estimating the sampling distribution of a statistic $\hat{\theta}_n = \hat{\theta}_n(\underline{X}_n)$ that converges weakly to the parameter $\theta(P)$ at some rate τ_n.

In this chapter, we make the simplifying assumption that $\hat{\theta}_n$ and $\theta(P)$ are real valued. To obtain asymptotically pivotal (or at least, scale-free) statistics, a standardization (also known as "studentization" when the norming is data-based and random) is often required. Since we will specifically discuss the influence of the studentization, we introduce a statistic $\hat{\sigma}_n = \hat{\sigma}_n(\underline{X}_n)$ converging in probability to some constant $\sigma > 0$. Typically, $\sigma^2 = \sigma^2(P)$ stands for the asymptotic variance of $\tau_n \hat{\theta}_n$, but this is not necessarily always the case. Without loss of generality, the unstudentized case corresponds to letting $\hat{\sigma}_n = 1$.

Under the assumption that there is a nondegenerate asymptotic distribution for the centered, "studentized" statistic $\tau_n \hat{\sigma}_n^{-1}(\hat{\theta}_n - \theta(P))$, and in particular if there is a distribution $J^*(x, P)$, *continuous* in x, such that

$$J_n^*(x, P) \equiv Prob_P\{\tau_n \hat{\sigma}_n^{-1}(\hat{\theta}_n - \theta(P)) \leq x\} \to J^*(x, P) \quad (10.1)$$

as $n \to \infty$, for any real number x, the subsampling methodology was shown in Subsections 2.5.1 and 3.2.3 to "work" asymptotically provided that the integer subsample size b satisfies

$$b \to \infty \quad (10.2)$$

and

$$\max(\frac{b}{n}, \frac{\tau_b}{\tau_n}) \to 0 \quad (10.3)$$

as $n \to \infty$. In other words, subsampling "works" in that it may be used to yield confidence regions for $\theta(P)$ of asymptotically correct coverage under the very weak condition (10.1), provided care is taken (by proper choice of the subsample size b), so that (10.2) and (10.3) are satisfied as well. In fact, the condition τ_b/τ_n can even be removed, as was done in Corollary 2.2.1.

Although i.i.d. data can be seen as a special case of stationary strong mixing data, the construction of the subsampling distribution can take advantage of the i.i.d. structure when such a structure exists. Of course, if one is unsure regarding the independence assumption, it is safer (or more robust) to operate under the general strong mixing assumption.

i. **General case (strong mixing data).** Define Y_i to be the subsequence $(X_i, X_{i+1}, \ldots, X_{i+b-1})$, for $i = 1, \ldots, q$, and $q = n-b+1$. Note that Y_i consists of b consecutive observations from the $X_1, \ldots, X_n$ sequence, and the order of the observations is preserved.

ii. **Special case (i.i.d. data).** Let $Y_1, \ldots, Y_q$ be equal to the $q = \binom{n}{b}$ subsets of size b chosen from $\{X_1, \ldots, X_n\}$, and then ordered in any fashion; here the subsets Y_i consist of unordered observations.

In either case, let $\hat{\theta}_{n,b,i}$ and $\hat{\sigma}_{n,b,i}$ be the values of statistics $\hat{\theta}_b$ and $\hat{\sigma}_b$ calculated from the subsample Y_i. The subsampling distribution of the root $\tau_n \hat{\sigma}_n^{-1}(\hat{\theta}_n - \theta(P))$, based on a subsample size b, is defined as usual by

$$L_{n,b}^*(x) \equiv q^{-1} \sum_{i=1}^{q} 1\{\tau_b \hat{\sigma}_{n,b,i}^{-1}(\hat{\theta}_{n,b,i} - \hat{\theta}_n) \leq x\}. \quad (10.4)$$

In Chapters 2 and 3, it was shown that if (10.1) is true, and if b is chosen in a way that (10.2) and (10.3) hold, then

$$\sup_x |L_{n,b}^*(x) - J^*(x, P)| = o_P(1). \quad (10.5)$$

Equation (10.5) can be used to construct confidence intervals for $\theta(P)$ of asymptotically correct coverage using the quantiles of $L^*_{n,b}(\cdot)$ as approximations to the corresponding quantiles of $J^*(\cdot, P)$. Note also that (10.1) and (10.5) imply that

$$\sup_x |L^*_{n,b}(x) - J^*_n(x, P)| = o_P(1) \qquad (10.6)$$

as well.

Nevertheless, the rate of convergence of $L^*_{n,b}(x)$ to $J^*(x, P)$ is of some concern to us in order to see how large the sample size should be such that the approximation of $J^*(x, P)$ (or of $J^*_n(x, P)$) by $L^*_{n,b}(x)$ be reasonably accurate. In the sample mean case, the benchmark for comparison is provided by the Berry–Esseen theorem, which states that the normal approximation is in error by $O_P(1/\sqrt{n})$ from $J^*_n(x, P)$. Any approximation that outperforms the normal approximation is called *higher-order accurate*.

The subsampling distribution in this case turns out to be a relatively low-accuracy approximation to the true sampling distribution $J^*_n(x, P)$, and is actually worse than the asymptotic normal distribution (with estimated variance). This phenomenon may be explained by the fact that the Berry–Esseen bound for the subsampling distribution based on subsamples of size b gives an error whose first term is $O(1/\sqrt{b})$ and hence cannot become $o_P(1/\sqrt{n})$ (not even $O_P(1/\sqrt{n})$) with b chosen in a way that (10.2) and (10.3) hold.

In Section 10.2, some background on the notions of extrapolation and interpolation is given, and a reference to the notion of the generalized jackknife is made. In Section 10.3 we discuss the use of the finite population correction factor in the subsampling distribution for i.i.d. data, and show that the extrapolation of two or more (finite-population-corrected) subsampling distributions has the potential of exhibiting higher-order accuracy under some regularity conditions; the same is true for an appropriately constructed (robust) interpolation. Notably, both the robust interpolation as well as the extrapolation of two (or more) subsampling distributions retain their asymptotic consistency/validity even if the regularity conditions required for higher-order accuracy break down. Extrapolation for general statistics is covered in Section 10.4. In Section 10.5, a partial generalization of the interpolation/extrapolation ideas is offered in the case of dependent sequences satisfying a strong mixing assumption; interestingly, a finite population correction factor enters in the picture here as well (though with less dramatic consequences). Subsection 10.5.5 presents an improved, bias-corrected, variance estimator for the sample mean of a time series; the bias-corrected estimator is actually nothing other than the extrapolation of two subsampling variance estimators. Finally, Section 10.6 addresses the issue of moderate deviations in subsampling distribution estimation.

10.2 Background

In Bertail (1997), it was proved that the subsampling distribution admits, for suitable b (typically ensuring that (10.3) holds), the same Edgeworth expansion as $J_n^*(x, P)$—when such an expansion exists—but in powers of b instead of n; that is, if

$$J_n^*(x, P) = J^*(x, P) + f_1(n)^{-1} p(x, P) + o(f_1(n)^{-1} p(x, P))$$

for some increasing function $f_1(\cdot)$, and some smooth function $p(x, P)$, then

$$L_{n,b}^*(x) = J^*(x, P) + f_1(b)^{-1} p(x, P) + o_P(f_1(b)^{-1} p(x, P)).$$

This result has a straightforward consequence when there exists a standardization $\hat{\sigma}_n$ such that the asymptotic distribution $J(P)$ is known and pivotal, that is, independent of P. In particular, if the rate of the first term in the Edgeworth expansion $f_1(n)$ is known (typically $f_1(n) = n^{1/2}$ in the regular case), then it is possible to improve the subsampling distribution by considering a linear combination of that distribution with the asymptotic distribution:

$$L_{n,b}^{*,int}(x) = \left(1 - \frac{f_1(b)}{f_1(n)}\right) J^*(x) + \frac{f_1(b)}{f_1(n)} L_{n,b}^*(x).$$

This type of linear (convex) combination with positive coefficients may be seen as an *interpolation* in that $L_{n,b}^{*,int}(x)$ is an intermediate point on the straight line segment joining $L_{n,b}^*(x)$ to the asymptotic $J^*(x)$, in the same way that sample size n is an intermediate point between sample sizes b and ∞. To see this, visualize the graph of $L_{n,b}^*(x)$ as a function of b, with x held fixed. Note the ordering $b < n < \infty$ and recall that we are interested in obtaining an estimate of the ordinate (sampling distribution) at sample size n (based on the ordinates at sample sizes b and ∞). This interpolation idea was first considered in Booth and Hall (1993) in the particular case of the sample mean of i.i.d. data, where they show that the error of estimator $L_{n,b}^{*,int}(x)$ can be as low as $O_P(n^{-5/6})$.

To explain the notion of interpolation in more detail, note that, in the sample mean case, the limit distribution $J^*(\cdot, P)$ is the normal that exhibits zero skewness. At the same time, the finite-sample distribution $J_n^*(\cdot, P)$ has skewness approximately $c_\gamma/\sqrt{n}$, for some constant c_γ. Note that this skewness is generally nonzero (albeit tending to zero). Similarly, the subsampling distribution $L_{n,b}^*(\cdot)$ has skewness approximately $c_\gamma/\sqrt{b}$, which is further from zero as compared to $c_\gamma/\sqrt{n}$. Now the interpolation $L_{n,b}^{*,int}(\cdot)$ has for skewness an intermediate value between $c_\gamma/\sqrt{b}$ and zero. Knowing that $c_\gamma/\sqrt{n}$ is also an intermediate value between $c_\gamma/\sqrt{b}$ and zero, the coefficients in the linear combination yielding $L_{n,b}^{*,int}(\cdot)$ can be chosen so that the skewness of $L_{n,b}^{*,int}(\cdot)$ matches the skewness of $J_n^*(\cdot, P)$—hence the property of higher order accuracy achieved by the interpolation $L_{n,b}^{*,int}(\cdot)$.

Furthering the discussion on the sample mean of i.i.d. data, note that another way to reduce the skewness of $L_{n,b}^*(\cdot)$ and have it match the skewness of $J_n^*(\cdot, P)$ is to consider a k-fold convolution of $L_{n,b}^*(\cdot)$ with itself, where $k \approx n/b$. Note that, in effect, convolving $L_{n,b}^*(\cdot)$ with itself amounts to applying a normalizing transformation. A different normalizing transformation was used by Tu (1992) with the same purpose, namely, showing higher-order accuracy of the (transformed) subsampling distribution.

Note that interpolation and the related "normalizing" methods rely on the fact that the form of the limit distribution $J^*(\cdot, P)$ is assumed known. Nevertheless, the generality of the subsampling methodology lies in the fact that $J^*(x, P)$ does not have to be known in order for subsampling to work. Therefore, it is of interest to seek improvements upon the subsampling distribution estimators that do not explicitly involve $J^*(x, P)$. In the present chapter, we will explore a notion of *extrapolation* similar to the notion of Richardson extrapolation considered by Bickel and Yahav (1988), Politis and Romano (1995), and Bickel, Götze, and van Zwet (1997); our exposition relies heavily on unpublished results of Bertail and Politis (1996).

Taking a linear combination of two (or more) subsampling distribution estimators with different subsample sizes (say b_1 and b_2, with $b_2 < b_1 < n$), an improved (extrapolated) subsampling distribution estimator can be constructed. Again for some fixed x, visualize the graph of $L_{n,b}^*(x)$ as a function of b. The notion of extrapolation can now be understood by realizing that the extrapolated subsampling distribution estimator is a point on the straight line passing through $L_{n,b_1}^*(x)$ and $L_{n,b_2}^*(x)$ but lying outside the line segment joining those two points (and closer to the larger subsample size, i.e., b_1 here), again reflecting the ordering $b_2 < b_1 < n$ of the (sub)sample sizes, and where the unknown is still the ordinate (sampling distribution) at sample size n.

As noticed by Bertail (1997), the interpolation idea used to obtain this second-order correct approximation is very similar to the idea of bias reduction: it is easy to see that the interpolated $L_{n,b}^{*,int}(x)$ is the generalized jackknife (cf. Gray, Schucany, and Watkins, 1972) involving the two non-second order correct estimators $L_{n,b}^*(x)$ and $J^*(x)$; that is,

$$L_{n,b}^{*,int}(x) = \frac{\det\begin{pmatrix} L_{n,b}^*(x) & f_1(b_n)^{-1} - f_1(n)^{-1} \\ J^*(x) & -f_1(n)^{-1} \end{pmatrix}}{\det\begin{pmatrix} 1 & f_1(b_n)^{-1} - f_1(n)^{-1} \\ 1 & -f_1(n)^{-1} \end{pmatrix}}.$$

In what follows, we will show that the generalized jackknife involving the two non-second order correct estimators $L_{n,b_1}^*(x)$ and $L_{n,b_2}^*(x)$ can be used to provide the linear combination effecting the aforementioned extrapolation of subsampling distribution estimators. We will also attempt to quantify the improvement achieved by the extrapolation.

It is clear that, for these results to be useful, an adequate standardization is needed for the statistic in question to be asymptotically pivotal. Were the asymptotic distribution unknown or difficult to handle, it would be appealing if extrapolation (or generalized jackknife) of different subsampling distributions (corresponding to different subsample sizes) also yielded a second-order correct distribution. In fact, this result does not hold directly as noted in Bickel, Götze, and van Zwet (1997). If no adequate standardization for the statistics is available, there is actually little hope for second-order correctness to hold.

10.3   I.I.D. Data: The Sample Mean

10.3.1   Finite Population Correction

Having observed the i.i.d. sample $(X_1, \ldots, X_n)$, we can compute a statistic $\hat{\theta}_n$. Now note that obtaining the subsamples $Y_1, \ldots, Y_q$ amounts to drawing samples of size b without replacement from the finite population $(X_1, \ldots, X_n)$. It is then of no surprise that the finite population correction for the sampling variance might be useful.

To fix ideas, consider the sample mean $\hat{\theta}_n = \bar{X}_n = n^{-1} \sum_{i=1}^n X_i$, and note that $Var(n^{1/2} \bar{X}_n) = \sigma^2 \equiv Var X_1$. Nevertheless,

$$Var(b^{1/2}(\bar{X}_{n,b,1} - \bar{X}_n)) = (1 - \frac{b}{n})\sigma^2 \neq \sigma^2,$$

where $\bar{X}_{n,b,1}$ is the sample mean of subsample Y_1, which is drawn at random but without replacement from the finite population $(X_1, \ldots, X_n)$. Therefore, the finite population correction factor

$$f = \frac{b}{n}$$

presents itself. Although $b/n \to 0$ asymptotically, it may be better, in view of potential second-order properties, to take this factor into account in constructing the subsampling distribution, and thus to consider

$$\tilde{L}^*_{n,b}(x) \equiv q^{-1} \sum_{i=1}^q 1\{\tau_b(1-f)^{-1/2} \hat{\sigma}^{-1}_{n,b,i}(\hat{\theta}_{n,b,i} - \hat{\theta}_n) \leq x\} = L^*_{n,b}(x(1-f)^{1/2}),$$

with $\tau_n = \sqrt{n}$. (The factor $(1-f)^{1/2} = [(n-b)/n]^{1/2}$ is clearly similar to $(\tau_n - \tau_b)/\tau_n$, which in the present context is just $(n^{1/2} - b^{1/2})/n^{1/2}$. This latter correction factor was considered in Corollary 2.2.1 and (2.11). Throughout this chapter, we utilize the correction factor $(1-f)^{1/2}$, but the other one could be analyzed by similar methods.) Thus, a more appropriate normalization factor in the subsampling distribution may be τ_r (instead of τ_b), where r is defined by

$$r = b/(1-f);$$

see also Shao and Tu (1995) for related ideas in the case of delete-d jackknife. So, we define the corrected subsampling distribution as

$$\tilde{L}_{n,b}^*(x) = q^{-1} \sum_{i=1}^{q} 1\{\tau_r \hat{\sigma}_{n,b,i}^{-1}(\hat{\theta}_{n,b,i} - \hat{\theta}_n) \le x\}.$$

Clearly, the factor $(1-f)$ has no first-order asymptotic effect on the subsampling distribution which remains consistent under very weak assumptions provided that $f \to 0$. However, the $(1-f)$ factor is of great importance for second-order properties, as will be shown in the next subsection.

10.3.2 The Studentized Sample Mean

Consider the problem of estimating the mean $\theta(P) = EX_1$ by the sample mean $\hat{\theta}_n = \bar{X}_n = n^{-1} \sum_{i=1}^{n} X_i$, and define

$$\hat{\sigma}_n^2 = (n-1)^{-1} \sum_{i=1}^{n} (X_i - \bar{X}_n)^2$$

as usual. Assuming $EX_1^4 < \infty$ and the usual Cramér condition, Bhattacharya and Ghosh (1978) give the Edgeworth expansion:

$$J_n^*(x, P) = P\left\{n^{1/2} \hat{\sigma}_n^{-1}(\bar{X}_n - \theta(P)) \le x\right\}$$

$$= \Phi(x) + n^{-1/2} p_1(x, P) \phi(x) + O(n^{-1}), \tag{10.7}$$

where

$$p_1(x, P) = \frac{k_3}{6}(2x^2 + 1)$$

and

$$k_3 = \frac{E(X_1 - EX_1)^3}{(E(X_1 - EX_1)^2)^{3/2}}$$

is the skewness of X_1; in the above, $\Phi(\cdot)$ and $\phi(\cdot)$ represent the standard normal distribution and density function, respectively.

Following Booth and Hall (1993), under the Cramér condition and assuming that $E|X_i|^{8+\eta} < \infty$, for some $\eta > 0$, we have from Babu and Singh (1985) an Edgeworth expansion for sampling without replacement from a finite population, with $b/n \xrightarrow[n \to \infty]{} 0$. So for any $\epsilon > 0$,

$$\tilde{L}_{n,b}^*(x) = \Phi(x) + b^{-1/2} p_1(x, P) \phi(x) + b^{-1} p_2(x, P) \phi(x) \tag{10.8}$$

$$- b^{1/2} n^{-1} \frac{1}{4} k_3 \phi(x) + O_P(b^{-1/2} n^{-1/2+\epsilon} + b^{3/2} n^{-2}) + o(b^{-1}),$$

where
$$p_2(x,P) = 12^{-1}k_4(x^3 - 3x) - 18^{-1}k_3^2(x^5 + 2x^3 - 3x) - 4^{-1}(x^3 + 3x)$$
and
$$k_4 = \frac{E(X_1 - EX_1)^4}{(E(X_1 - EX_1)^2)^2}.$$

Notice that for b such that $b/n \to 0$, the error of approximating $J_n^*(x,P)$ by $\widetilde{L}_{n,b}^*(x)$ cannot be made smaller than $O_P(n^{-1/2})$. Thus even with the finite population correction, the subsampling distribution will not be second-order correct.

If we do not take into account the finite population correction factor, then we have

$$L_{n,b}^*(x) = \widetilde{L}_{n,b}^*(x) + O_P(\frac{b}{n}). \tag{10.9}$$

Consequently, the best rate that we can achieve with distribution $L_{n,b}^*(x)$ is obtained for $b \propto n^{2/3}$, where the symbol $\propto$ is used to denote "proportional." Thus, the error of the approximation cannot be smaller than $O_P(n^{-1/3})$, as pointed out in Politis and Romano (1992c).

10.3.3 Estimation of a Two-Sided Distribution

Motivated by the improved accuracy of two-sided bootstrap distributions (see, for example, Hall, 1992), we now consider the accuracy of the two-sided subsampling distribution of the sample mean $\hat{\theta}_n = \bar{X}_n$. First note that equation (10.7) implies that

$$J_{n,|\cdot|}^*(x,P) = P\left\{n^{1/2}\hat{\sigma}_n^{-1}|\bar{X}_n - \theta(P)| \leq x\right\} =$$

$$= J_n^*(x,P) - J_n^*(-x,P) = \Phi(x) - \Phi(-x) + O(n^{-1}), \tag{10.10}$$

where the even symmetry of the functions $p_1(x,P)$ and $\phi(x)$ was used. Similarly, equation (10.8) implies that, for arbitrary $\epsilon > 0$,

$$\widetilde{L}_{n,b,|\cdot|}^*(x) = q^{-1}\sum_{i=1}^{q}\mathbf{1}\{\tau_r\hat{\sigma}_{n,b,i}^{-1}|\hat{\theta}_{n,b,i} - \hat{\theta}_n| \leq x\} = \widetilde{L}_{n,b}^*(x) - \widetilde{L}_{n,b}^*(-x) =$$

$$= \Phi(x) - \Phi(-x) + 2b^{-1}p_2(x,P)\phi(x) + O_P(b^{-1/2}n^{-1/2+\epsilon} + b^{3/2}n^{-2}) + o(b^{-1}); \tag{10.11}$$

in the above, $\tau_n = \sqrt{n}$, and $\hat{\sigma}_n^2 = (n-1)^{-1}\sum_{i=1}^{n}(X_i - \bar{X}_n)^2$ as before.

Note that equation (10.11) holds for any $\epsilon > 0$; so let us fix a number $\epsilon \in (0, 1/10)$. We now see that, as long as the regularity conditions ensuring (10.8) hold true, $\widetilde{L}_{n,b,|\cdot|}^*(x)$ considered as an approximation to

$J^*_{n,|\cdot|}(x,P)$ possesses an approximation error of $o_P(n^{-1/2})$, provided we take $b \propto n^\gamma$ for some $\gamma \in (1/2, 1)$. The optimal rate is actually obtained by letting $b \propto n^{4/5}$, in which case the rate of approximation of $J^*_{n,|\cdot|}(x,P)$ by $\widetilde{L}^*_{n,b,|\cdot|}(x)$ becomes $O_P(n^{-4/5})$. Despite this finding, $\widetilde{L}^*_{n,b,|\cdot|}(x)$ cannot be considered higher-order accurate as it is still inferior to the normal approximation, given by $\Phi(x) - \Phi(-x)$ here, that possesses an error of $O(n^{-1})$. However, the two-sided subsampling distribution estimation seems to be more accurate than the one-sided analogue.

It should also be stressed that the $O_P(n^{-4/5})$ rate of the two-sided subsampling distribution is achieved *only* if the finite population correction is used. If we do not take into account the finite population correction factor, then we have

$$L^*_{n,b,|\cdot|}(x) = q^{-1} \sum_{i=1}^{q} 1\{\tau_b \hat{\sigma}^{-1}_{n,b,i} |\hat{\theta}_{n,b,i} - \hat{\theta}_n| \leq x\} = \widetilde{L}^*_{n,b,|\cdot|}(x) + O_P\left(\frac{b}{n}\right). \tag{10.12}$$

It is now apparent that the rate of approximation of $J^*_{n,|\cdot|}(x,P)$ by $L^*_{n,b,|\cdot|}(x)$ can never become $o_P(1/\sqrt{n})$; it can, however, become of order $O_P(1/\sqrt{n})$, by letting $b \propto n^{1/2}$, thus still yielding an improved rate of approximation as compared to the one-sided (uncorrected) subsampling distribution $L^*_{n,b}(x)$; see the discussion related to equation (10.9).

10.3.4 Extrapolation

We now return to the problem of accurate estimation of the one-sided distribution of the sample mean $\hat{\theta}_n$ of i.i.d. data. The extrapolation of two finite-population corrected subsampling distributions (one with subsample size b_1 and the other with b_2) is given by

$$\widetilde{L}^{*,ext}_{(2)}(x) = \lambda_1 \widetilde{L}^*_{n,b_1}(x) + \lambda_2 \widetilde{L}^*_{n,b_2}(x),$$

where λ_1 and λ_2 are chosen to solve

$$\lambda_1 + \lambda_2 = 1$$
$$\lambda_1 f_1^{-1}(b_1) + \lambda_2 f_1^{-1}(b_2) = f_1^{-1}(n),$$

with $f_1(n) = \sqrt{n}$; see also Bickel and Yahav (1988), who used a similar extrapolation to reduce the computational costs in constructing bootstrap distributions. It is easy to see that this extrapolation is actually given by

$$\widetilde{L}^{*,ext}_{(2)}(x) = \frac{\det \begin{pmatrix} \widetilde{L}^*_{n,b_1}(x) & b_1^{-1/2} - n^{-1/2} \\ \widetilde{L}^*_{n,b_2}(x) & b_2^{-1/2} - n^{-1/2} \end{pmatrix}}{\det \begin{pmatrix} 1 & b_1^{-1/2} - n^{-1/2} \\ 1 & b_2^{-1/2} - n^{-1/2} \end{pmatrix}} \tag{10.13}$$

and may be interpreted as the generalized jackknife of two non-second-order correct distributions estimating the true finite sample distribution $J_n^*(x, P)$. Just as in the case of the jackknife, we may hope that this estimator is second-order correct. More precisely, if we choose b_1 and b_2 such that $\frac{b_i}{n} \xrightarrow{n\to\infty} 0$ (for $i = 1, 2$), and $\frac{b_2}{b_1} \xrightarrow{n\to\infty} C \in [0, 1)$, a straightforward calculation shows that

$$\tilde{L}_{(2)}^{*,ext}(x) = \Phi(x) + n^{-1/2} p_1(x, P)\phi(x)$$
$$- b_1^{-1/2} b_2^{-1/2} p_2(x, P)\phi(x) + b_1^{1/2}(1 + C^{1/2}) n^{-1} \frac{1}{4} k_3 \phi(x)$$
$$+ O_P(b_1^{-1/2} n^{-1/2+\epsilon} + b_1^{3/2} n^{-2}) + o(b_1^{1/2} n^{-1}).$$

It follows that

$$\tilde{L}_{(2)}^{*,ext}(x) - J_n^*(x, P) = -b_1^{-1/2} b_2^{-1/2} p_2(x, P)\phi(x) \qquad (10.14)$$

$$+ b_1^{1/2}(1 + C^{1/2}) n^{-1} \frac{1}{4} k_3 \phi(x) + O_P(b_1^{-1/2} n^{-1/2+\epsilon} + b_1^{3/2} n^{-2}) + o(b_1^{1/2} n^{-1}).$$

Minimizing the order of the right-hand side of (10.14) leads us to choose b_1 and b_2 such that $b_1 b_2^{1/2}$ will be proportional to n. This implies that the best choice is given by

$$b_1 = C_1 n^{2/3}, \quad b_2 = C_2 n^{2/3}$$

with $C_2 < C_1$. In this case, we have

$$\sup_x |\tilde{L}_{(2)}^{*,ext}(x) - J_n^*(x, P)| = O_P(n^{-2/3}), \qquad (10.15)$$

which shows that the extrapolation of the two finite-population-corrected subsampling distributions of the studentized mean is indeed second-order correct as hoped for. Notice that the order of the whole approximation is worse than the one that we obtain by considering the interpolation of one subsampling distribution with its asymptotic distribution when the latter is known (see Booth and Hall, 1993, who obtain an error of size $n^{-5/6}$). Thus, if the asymptotic distribution is known explicitly and interpolation is in fact feasible, then interpolation would be preferable to extrapolation. However, as mentioned earlier, extrapolation is more robust, as it remains first-order correct even under misspecification of the form of the asymptotic distribution.

It is very important to point out here that if we do not take into account the finite population correction factor, then the second-order validity of the extrapolated version of the two distributions fails; for instance, see Bickel, Götze, and van Zwet (1997, p. 17), where the importance of this correction factor was also recognized. Indeed in the absence of the finite population correction, if we let $L_{(2)}^{*,ext}(x)$ be the extrapolation of $L_{n,b_1}^*(x)$ and $L_{n,b_2}^*(x)$, then we have

$$L^{*,ext}_{(2)}(x) = \Phi(x) + n^{-1/2}p_1(x,P)\phi(x) + O_P\left(\frac{b_1}{n}\right)$$
$$+ O_P(b_1^{-1/2}n^{-1/2+\epsilon} + b_1^{3/2}n^{-2} + b_1^{-1/2}b_2^{-1/2} + b_1^{1/2}\,n^{-1});$$

once again, to obtain the second-order correctness, we would have to choose $n^{1/2} = o(b_1)$, and the loss induced by the sampling-without-replacement scheme (typically of order b_1/n) implies that second-order correctness cannot be attained. However, it is easy to see that the (uncorrected) extrapolation $L^{*,ext}_{(2)}(x)$ improves over only one subsampling distribution. If we choose $b_2 = Cb_1 = Cn^{1/2}$, $C < 1$, then we obtain an approximation correct up to $O_P(n^{-1/2})$, instead of the best rate $O_P(n^{-1/3})$ for a single subsampling distribution; see the discussion after equation (10.9).

Note that this $O_P(n^{-1/2})$ error rate for the (uncorrected) extrapolation $L^{*,ext}_{(2)}(x)$ also holds if we choose $\hat{\sigma}_n = 1$, that is, in the unstudentized case. It is obvious that this $O_P(n^{-1/2})$ rate cannot be improved upon since, in the unstudentized case, the two subsampling distributions (and thus their extrapolation as well) asymptotically approximate $\Phi(x/(\sigma+\xi))$ rather than $\Phi(x/\sigma)$, where $\xi = O_P(n^{-1/2})$. Thus, the order of magnitude of the error of the unstudentized extrapolated distribution will not be smaller than the discrepancy in the scaling, that is, $O_P(n^{-1/2})$.

Of course, in the case of the mean, interpolation and/or extrapolation of subsampling distributions can be thought to give no real gain, since we already know that the usual (studentized) bootstrap gives an approximation up to $O_P(n^{-1})$ (see, e.g., Hall, 1992). It is well known that the same is true for smooth functions of means; similarly, generalizations to Fréchet differentiable functionals may also be obtained using the results of Barbe and Bertail (1995).

Even though the usual bootstrap works in these mean-like cases, subsampling may still be useful for computational reasons, since constructing two subsampling distributions requires less simulations than constructing the usual bootstrap distribution (see Bickel and Yahav, 1988; Bertail, 1997). Moreover, subsampling for the mean will be valid under greater generality as compared to the bootstrap; for example, subsampling will work even in the infinite variance case; see the next subsection, as well as Chapter 11. Quite notably, the extrapolation of subsampling distributions shares this great generality of validity.

To elaborate on this important point, note that, in contrast to the interpolation schemes studied in Booth and Hall (1993) and Bertail (1997), the extrapolation does not depend on the asymptotic approximation and is thus more robust. Indeed, if the original assumptions break down, for example, if the assumptions leading to the Edgeworth expansion break down or even if the statistic $\hat{\theta}_n$ is not asymptotically Gaussian, then the interpolation is not applicable and the bootstrap fails (for example, see Chapter 11). On the other hand, the extrapolation of the corrected (with the finite population factor) or uncorrected subsampling distributions remains a consistent distribution estimator as long as equation (10.1) holds; the higher-order property of the (corrected) extrapolation comes as a "bonus" in case the regularity conditions ensuring the validity of the Edgeworth expansion (10.8) happen to be in effect.

We will discuss a general extrapolation procedure in Subsection 10.4.1. However, before going into that, we next give a robust way of doing the interpolation of subsampling distributions that does not explicitly use the form of the asymptotic distribution.

10.3.5 Robust Interpolation

As in Subsection 10.3.2, we again consider the problem of estimating the mean $\theta(P)$ of the i.i.d. sequence $X_1, X_2, \ldots, X_n$. Throughout the present subsection, we assume that the (centered) variable $X_1 - \theta(P)$ lies in the domain of attraction of a *symmetric* stable distribution with index of stability $1 < \alpha \leq 2$; in other words, we are in the setup of Chapter 11 with the skewness parameter β being zero. The value of α is generally unknown; the case $\alpha = 2$ corresponds to the case of a Gaussian limit law, while for $\alpha < 2$, we have a "heavy-tailed" distribution.

We will now employ a symmetrization idea to "have our cake and eat it too." Specifically, an interpolative distribution estimator will be contructed that is not only consistent in the fat-tailed case, but it is also higher-order accurate in the case where higher-order accurate estimation is possible (when regularity conditions are met). All this will be achieved without requiring knowledge of the parameter α.

To this effect, let $\tilde{L}_{n,b}^{*,flip}(x) = 1 - \tilde{L}_{n,b}^{*}(-x)$, for all x, and

$$\tilde{L}_{n,b}^{*,symm}(x) = (\tilde{L}_{n,b}^{*}(x) + \tilde{L}_{n,b}^{*,flip}(x))/2 = (\tilde{L}_{n,b}^{*}(x) + 1 - \tilde{L}_{n,b}^{*}(-x))/2.$$

Finally, construct the partially symmetrized interpolated subsampling distribution

$$\tilde{L}_{n,b}^{*,rob}(x) = (1 - \sqrt{b/n})\tilde{L}_{n,b}^{*,symm}(x) + \sqrt{b/n}\tilde{L}_{n,b}^{*}(x) \tag{10.16}$$

that should be compared with the definition of $L_{n,b}^{*,int}(x)$ previously. We now give the following result showing that $\tilde{L}_{n,b}^{*,rob}(x)$ justifies the title "robust interpolation."

Theorem 10.3.1. *Assume the distribution of $X_1 - \theta(P)$ lies in the domain of attraction of a symmetric stable distribution with index of stability $1 < \alpha \leq 2$. Also assume conditions (10.2) and (10.3). Then,*

(a) $\sup_x |\tilde{L}_{n,b}^{*,rob}(x) - J_n^*(x, P)| = o_P(1).$

If in addition it so happens that $\alpha = 2$, and the two Edgeworth expansions

(10.7) and (10.8) hold true (for some arbitrarily small constant $\epsilon > 0$), then

(b) $\sup_x |\tilde{L}_{n,b}^{*,rob}(x) - J_n^*(x, P)| = o_P(1/\sqrt{n})$

if $b \geq c_0 n^{1/2+\epsilon}$ as $n \to \infty$, for some constant $c_0 > 0$.

Proof. Part (a) is a consequence of Proposition 11.4.3. Regarding part (b), note that $p_1(x, P) = p_1(-x, P)$, and $-p_2(x, P) = p_2(-x, P)$, for all x, and that $p_2(x, P)\phi(x)$ is bounded; hence, under the assumed Edgeworth expansions and the condition $b \geq c_0 n^{1/2+\epsilon}$, it follows that

$$\tilde{L}_{n,b}^{*,rob}(x) = \Phi(x) + n^{-1/2} p_1(x, P)\phi(x) + O_P(b^{-1}) + O_P(\sqrt{b}/n) \quad (10.17)$$

and part (b) is proved. ∎

Finally, observe that equation (10.17) implies that the most accurate approximation by $\tilde{L}_{n,b}^{*,rob}(x)$ is achieved by taking $b \propto n^{2/3}$, yielding an approximation error of order $O_P(n^{-2/3})$. As is to be expected, this is slightly worse than the error of size $O_P(n^{-5/6})$ obtained by Booth and Hall (1993) in their (nonrobust) interpolation, but notably it is identical to the rate of approximation we achieved by extrapolation in equation (10.15).

10.4 I.I.D. Data: General Statistics

10.4.1 Extrapolation

In this section, we consider a general statistic $\hat{\theta}_n$ (studentized or standardized by an appropriate statistic $\hat{\sigma}_n > 0$), and assume that its studentized sampling distribution admits an Edgeworth expansion (not necessarily normal), i.e., that

$$J_n^*(x, P) \equiv P(\tau_n \hat{\sigma}_n^{-1}(\hat{\theta}_n - \theta(P)) \leq x) =$$

$$= J^*(x, P) + f_1(n)^{-1} p(x, P) + O(f_2(n)^{-1}), \quad (10.18)$$

for some increasing functions f_1 and f_2; here, $J^*(x, P)$ is assumed to be uniformly continuously differentiable and $p(x, P)$ is assumed to be uniformly continuous in x. Bertail (1997) gives very weak conditions ensuring that, if (10.18) holds, then

$$L_{n,b}^*(x) = J^*(x, P) + f_1(b)^{-1} p(x, P) + O_P(f_2(b)^{-1}). \quad (10.19)$$

In the usual case corresponding to $f_i(n) = n^{i/2}$, $i = 1, 2$, and without the correction factor, result (10.19) holds with $b = O(n^{1/2}/\log n)$, provided that the distribution of $\hat{\sigma}_n$ does not have a very fat tail. This condition is automatically satisfied either if $\hat{\sigma}_n = 1$ (the unstudentized case), or if

$\hat{\sigma}_n$ satisfies some exponential inequality, or if $\hat{\sigma}_n$ is a truncated estimator. Better choices of b may be possible in some specific situations, in particular if one takes into account the finite population correction factor.

For the sake of generality, we will now assume that (10.19) holds for some b satisfying (10.2) and (10.3), and some increasing functions f_1 and f_2. Now let $b_i = c_i^2 b$, where the c_i, $i = 1, 2$, are some positive constants different from each other, and consider the subsampling distributions L_{n,b_i}^* for $i = 1, 2$. The extrapolation of two subsampling distributions is defined by

$$L_{(2)}^{*,ext}(x) = \frac{\det \begin{pmatrix} L_{n,b_1}^*(x) & f_1(b_1)^{-1} - f_1(n)^{-1} \\ L_{n,b_2}^*(x) & f_1(b_2)^{-1} - f_1(n)^{-1} \end{pmatrix}}{\det \begin{pmatrix} 1 & f_1(b_1)^{-1} - f_1(n)^{-1} \\ 1 & f_1(b_2)^{-1} - f_1(n)^{-1} \end{pmatrix}}, \qquad (10.20)$$

and it is easy to see that (by construction), if (10.19) holds, then

$$L_{(2)}^{*,ext}(x) - J_n^*(x, P) = O(f_2^{-1}(b)).$$

Thus, the extrapolation $L_{(2)}^{*,ext}(x)$ yields a smaller error as compared to the $O(f_1^{-1}(b))$ rate attained by each individual subsampling distribution. Of course this result relies heavily on the assumption that a sequence $b = b(n)$ can be found so that (10.19) holds. For example, note that the results of the previous section for the sample mean case suggest that such a choice for b that ensures (10.19) cannot be found, unless the finite population correction is used throughout.

For this reason, Bertail and Politis (1996) conjecture that in a great number of situations, extrapolation in conjunction with a finite population correction, that is, using τ_r instead of τ_b in the construction of the subsampling distribution with subsample size b, where $r = b/(1 - f)$, will indeed yield second-order accuracy. For example, the aforementioned higher-order accuracy is expected to hold for linear (or approximately linear) statistics by analogy to the sample mean results. In any case, even if second-order accuracy is not achieved, the extrapolated distribution will always improve over each individual subsampling distribution, that is, an extrapolated distribution will always improve over a single distribution.

10.4.2 Case of Unknown Convergence Rate to the Asymptotic Approximation

A case of interest in some practical applications occurs when the order of the difference between the asymptotic and the true distribution is unknown. Indeed, the knowledge of the rate f_1 of the asymptotic approximation is implicit in the construction of both the interpolation and extrapolation. Consider for instance the simple case of estimating the mean. If

$E(X_i - EX_i)^3 \neq 0$, then (10.13) yields a second-order correct approximation and improves over the asymptotic normal distribution. But if $E(X_i - EX_i)^3 = 0$, then the asymptotic distribution is already second-order correct and (10.13) is less accurate; this follows from the fact that when the skewness is zero, the correct extrapolation (which will indeed improve upon the asymptotic distribution) should be built with $f_1(n) = n$ and not $n^{1/2}$. Other problematic situations occur when the common distribution of the $X_1, \ldots, X_n$ observations is concentrated on a lattice.

The above examples suggest that we either have to make a preliminary test on some parameter appearing in the Edgeworth expansion (depending on the statistic and the underlying distribution), or that we have to directly construct and employ an accurate estimator of f_1, which may then be used in forming the extrapolation. The first suggestion is not very satisfactory because it is highly problem-dependent and the Edgeworth expansion may be quite complicated.

In Chapter 8, it was shown that it is possible to estimate the convergence rate τ_n of the statistic $\hat{\theta}_n$, when τ_n is unknown, using several subsampling distributions. Employing a similar idea, let us now show how it is possible to estimate the accuracy rate f_1 by a simple regression. Under very general conditions on the statistic and under some conditions on b, the order of the error of the subsampling distribution approximation is $f_1(b)^{-1}$. Thus, if we study a collection of subsampling distributions, when b varies in its domain, we should be able to observe their convergence to the asymptotic distribution in connection to the rate f_1.

Let $\hat{\theta}_n$ and $\hat{\sigma}_n$ be as defined generally in Section 10.1; under regularity conditions (see, for instance, conditions A1–A5 of Bertail, 1997), it follows that $L^*_{n,b}(x)$ has the same Edgeworth expansion as $J^*_n(x, P)$, but in functions of b instead of n. For simplicity, we now focus on the case when $f_1(n)$ is of the form $n^{-\alpha}$, where α is unknown. If we choose b_1 and b_2 such that $b_1/b_2 \to 0$ as $n \to \infty$, then it is easy to see that

$$|L^*_{n,b_1}(x) - L^*_{n,b_2}(x)| = f_1(b_1)^{-1}(1 + o(1))|p(x, P)| + o_P(f_1(b)^{-1}). \quad (10.21)$$

Thus, uniformly in x,

$$\log |L^*_{n,b_1}(x) - L^*_{n,b_2}(x)| = -\alpha \log(b_1) + \log(|p(x, P)|) + o_P(1). \quad (10.22)$$

If now we compute several pairs of subsampling distributions with some different subsampling sizes $b_1^{(i)}, b_2^{(i)}$, for $i = 1, \ldots I$, we will then be able to estimate the index α by a simple linear regression of $|L^*_{n,b_1}(x) - L^*_{n,b_2}(x)|$ on $\log b_1$. Note that here for simplicity $b_2^{(i)} = b_2$ may be chosen to be the same for all i. If the $b_1^{(i)}, b_2^{(i)}$ are (different) powers of n, the resulting estimator $\hat{\alpha}$ can be shown to be such that

$$\hat{\alpha} = \alpha + o_P(\log(n)^{-1});$$

the proof parallels the results in Chapter 8 and is omitted. Finally, it is easy to see that if we now use $\widehat{f_1}(n) = n^{-\widehat{\alpha}}$ in place of $f_1(n)$, then all results in our previous sections still hold.

In the more general case, when the functional form of f_1 is unknown, we may consider (10.21), or rather, its logarithm, as a nonparametric regression for $f_1(\cdot)$. Under a monotonicity constraint, and assuming that $f_1(b) \to \infty$ as $b \to \infty$, consistent estimation of $f_1(\cdot)$ may still be possible, albeit more complicated and slower.

10.5 Strong Mixing Data

10.5.1 The Studentized Sample Mean

Let $\hat{\theta}_n = \bar{X}_n = n^{-1}\sum_{t=1}^{n} X_t$ be the sample mean, and let $\theta(P) = EX_1$ be the true mean. Also let $R(s) = E(X_t - \theta(P))(X_{t+|s|} - \theta(P))$, for $s = 0, \pm 1, \pm 2, \ldots$ be the autocovariance sequence. Both $\theta(P)$ and $R(\cdot)$ are generally unknown, and the objective is to obtain interval estimates for $\theta(P)$ based on the data in a nonparametric fashion. To achieve this goal, the sampling distribution of $\bar{X}_n = n^{-1}\sum_{t=1}^{n} X_t$ is required. Under regularity conditions—see, for example, our Theorem 3.4.1—it can be shown that $\bar{X}_n$ is asymptotically normal as in the i.i.d. case. Note that $\bar{X}_n$ is unbiased for $\theta(P)$ and that $\sigma_n^2 = Var(\sqrt{n}\bar{X}_n) = R(0) + 2\sum_{s=1}^{n}(1 - s/n)R(s) = \sigma_\infty^2 + O(1/n)$, provided the limit $\sigma_\infty^2 = \sum_{s=-\infty}^{\infty} R(s)$ exists.

As a matter of fact, we can say more about the normal approximation by means of an Edgeworth expansion. In this section, we assume that the strong mixing coefficients satisfy

$$\alpha_X(k) \leq d^{-1}e^{-dk} \tag{10.23}$$

for some $d > 0$, and that

$$E|X_1|^s < \infty \tag{10.24}$$

for some $s \geq 5$. Let $\hat{\sigma}_n^2$ be an estimator of σ_∞^2 based on $X_1, \ldots, X_n$, and accurate enough so that $\hat{\sigma}_n^2 = \sigma_\infty^2 + O_P(\sqrt{\log n/n})$ under conditions (10.23) and (10.24). For example, we can let $\hat{\sigma}_n^2 = \tilde{\sigma}_{0.5M,M,n}^2$ with $M = A\log n$, where the estimator $\tilde{\sigma}_{0.5M,M,n}^2$ is defined (by means of an extrapolated subsampling variance estimator) in Section 10.5.5 in what follows; see part (i) of Theorem 10.5.2.

Define the Edgeworth expansion

$$Q_n(x) = \Phi(x) + n^{-1/2}k_\infty^* \sigma_\infty^{-3/2} p(x),$$

where $p(x) = \frac{1}{6}\phi^{(2)}(x) - \frac{1}{2}\phi(x)$, and

$$k_\infty^* = \sum_{i,j} E[(X_0 - \theta(P))(X_i - \theta(P))(X_j - \theta(P))],$$

which is finite because of (10.23) and (10.24); here $\Phi(x)$, $\phi(x)$, and $\phi^{(k)}(x)$ denote the standard normal distribution, density, and k-th derivative of its density, respectively. Now, under some regularity conditions (e.g., under the Cramér-type conditions A3, A5, and A6 of Götze and Künsch, 1996), it follows that

$$\sup_x |J_n^*(x, P) - Q_n(x)| = O(\frac{M}{n^{1-2/s}}) + O(\beta_n) = O(\frac{\log n}{n^{1-2/s}}), \quad (10.25)$$

where $J_n^*(x, P) = P(\sqrt{n}\frac{\bar{X}_n - \theta(P)}{\hat{\sigma}_n} \leq x)$, and $\beta_n = E\hat{\sigma}_n^2 - \sigma_n^2$, is the bias of the variance estimator which is based on a block size $M = A \log n$; see Subsection 10.5.5 for more details. Using the choice $M = A \log n$ for a sufficiently large constant A implies that $\beta_n = O(n^{-1})$ and yields the rate in equation (10.25).

Consider now the subsampling distribution of the studentized sample mean, which is defined by

$$L_{n,b}^*(x) \equiv q^{-1} \sum_{i=1}^q 1\{\sqrt{b}\frac{\hat{\theta}_{n,b,i} - \hat{\theta}_n}{\hat{\sigma}_{n,b,i}} \leq x\}, \quad (10.26)$$

where $\hat{\theta}_{n,b,i} = b^{-1} \sum_{k=1}^{i+b-1} X_k$ and $\hat{\sigma}_{n,b,i}$ is the statistic $\hat{\sigma}_b$ computed on block $\{X_1, \ldots, X_{i+b-1}\}$. Similarly to the proof of Theorem 3.2.1, it can be shown that $Var(L_{n,b}^*(x)) = O(b/n)$, due to the geometric mixing rate (10.23).

Now,

$$q^{-1} \sum_{i=1}^q 1\{\sqrt{b}\frac{\hat{\theta}_{n,b,i} - \theta(P)}{\hat{\sigma}_{n,b,i}} \leq x\} = E(1\{\sqrt{b}\frac{\hat{\theta}_{n,b,1} - \theta(P)}{\hat{\sigma}_{n,b,1}} \leq x\}) + O_P(\sqrt{\frac{b}{n}})$$

$$= \Phi(x) + \frac{k_\infty^* p(x)}{b^{1/2}\sigma_\infty^{3/2}} + O_P(\frac{\log b}{b^{1-2/s}}) + O_P(\sqrt{\frac{b}{n}}).$$

The above, coupled with the fact that $\sqrt{b}(\theta(P) - \hat{\theta}_n)/\hat{\sigma}_{n,b,i} = O_P(\sqrt{\frac{b}{n}})$, yields that

$$L_{n,b}^*(x) = \Phi(x) + \frac{k_\infty^* p(x)}{b^{1/2}\sigma_\infty^{3/2}} + O_P(\sqrt{b/n}) + O_P(\log b/b^{1-2/s}). \quad (10.27)$$

It follows that, taking $b \propto \sqrt{n}$, we minimize the mean squared error (MSE) of $L_{n,b}^*(x)$, thus having $L_{n,b}^*(x) = \Phi(x) + O_P(n^{-1/4})$.

Note that the $O_P(n^{-1/4})$ rate is obviously inferior to the $O(n^{-1/2})$ rate of the Berry–Esseen bound, and to the $o_P(n^{-1/2})$ rate of the block-bootstrap, which is second-order accurate in this regular case as shown by Lahiri (1992) and Götze and Künsch (1996). Nevertheless, the loss in accuracy is counterbalanced by an increased generality and robustness in that $L_{n,b}^*(x)$ remains consistent even if the asymptotic normality (and associated Edgeworth expansion) does not hold, provided some (not necessarily normal) limiting distribution exists.

It is desirable, however, to improve upon the rather slow $O_P(n^{-1/4})$ rate. Interestingly, the interpolation of $L_{n,b}^*(x)$ with $\Phi(x)$, given by

$$L_{n,b}^{*,int}(x) \equiv (1 - \sqrt{\frac{b}{n}})\Phi(x) + \sqrt{\frac{b}{n}}L_{n,b}^*(x)$$

can be made second-order correct by minimizing

$$O_P(\frac{b^{1/2}\log b}{n^{1/2}b^{1-2/s}}) + O_P(\frac{b}{n}) + O_P(\sqrt{\frac{b}{n}}\frac{\log n}{n^{1-2/s}}),$$

that is, by choosing $b \propto n^{\frac{s}{3s-4}}(\log n)^{-1}$. With such a choice, we have a global accuracy of the interpolation $L_{n,b}^{*,int}(x)$ of order

$$O_P(\frac{\log n}{n^{(2s-4)/(3s-4)}}),$$

which is close to $O_P(n^{-2/3})$ when s is large. Note that this rate is still worse than the best rate of the block-bootstrap obtained by Götze and Künsch (1996), which is given by $O_P(n^{-\frac{3s-4}{4s}})$ and can become close to $O_P(n^{-3/4})$ when s is large. Nonetheless, note that for the interpolation to be correct, we need only $s \geq 5$, whereas at least $s \geq 24$ is needed for the block bootstrap to be second-order correct (see Götze and Künsch, 1996). Note, however, that the interpolation, being based on the asymptotic distribution, loses the robustness quality of the subsampling distribution. It is thus of interest to see how the extrapolation of subsampling distributions based on different block sizes performs in this setting.

So let $b_i = c_i^2 b$, where the c_is are some positive constants all different from each other for $i = 1, 2, \ldots, I$, and consider the subsampling distributions $L_{n,b_i}^*(x)$. We will generally define the extrapolated subsampling distribution $L_{(I)}^{*,ext}(x)$ by

$$L_{(I)}^{*,ext}(x) = \sum_{i=1}^{I} \lambda_i L_{n,b_i}^*(x), \tag{10.28}$$

where the λ_i's are some carefully chosen constants such that

$$\sum_{i=1}^{I} \lambda_i = 1. \tag{10.29}$$

Since we are conducting extrapolation as opposed to interpolation, the λ_i's will not all be positive. How to choose I and the λ_i's depends on our objective. For example, letting $I = 2$, imposing the additional condition

$$\sum_{i=1}^{I} \lambda_i/c_i = 0, \tag{10.30}$$

and solving for the λ_i's, we immediately see that the $1/\sqrt{b}$ term in (10.27) is removed, i.e.,

$$L_{(2)}^{*,ext}(x) \equiv \frac{-c_1}{c_2 - c_1} L_{n,b_1}^*(x) + \frac{c_2}{c_2 - c_1} L_{n,b_2}^*(x)$$

$$= \Phi(x) + O_P(\sqrt{b/n}) + O(\frac{\log b}{b^{1-2/s}}),$$

whereas $Var(L_{(2)}^{*,ext}(x))$ remains of order $O(b/n)$, and $L_{(2)}^{*,ext}(x)$ still retains the robustness property of being consistent, even if our asymptotic normality assumption breaks down and a different limiting distribution is in effect. Now, letting $b \propto (n \log^2 n)^{s/(3s-4)}$, minimizes the MSE of $L_{(2)}^{*,ext}(x)$, thus giving

$$L_{(2)}^{*,ext}(x) = \Phi(x) + O_P(n^{(2-s)/(3s-4)} (\log n)^{s/(3s-4)}).$$

The above rate of approximation becomes close to $n^{-1/3}$ for large s, thus improving upon the $n^{-1/4}$ rate of $L_{n,b}^*(x)$, but not achieving second-order correctness.

Letting $I = 3$, we may impose the condition

$$\sum_{i=1}^{I} \lambda_i c_i = 0 \qquad (10.31)$$

in addition to conditions (10.29) and (10.30). It is then easy to see that the actual MSE of the resulting $L_{(3)}^{*,ext}(x)$ will be smaller than that of $L_{(2)}^{*,ext}(x)$ although, unfortunately, the MSE is not reduced by an order of magnitude.

10.5.2 Estimation of a Two-Sided Distribution

For completeness, we now consider the accuracy of the two-sided subsampling distribution of the sample mean $\hat{\theta}_n = \bar{X}_n$ of strong mixing data. First note that equation (10.25) implies that

$$J_{n,|\cdot|}^*(x, P) = P\left\{n^{1/2} \hat{\sigma}_n^{-1} | \bar{X}_n - \theta(P)| \le x\right\} =$$

$$= J_n^*(x, P) - J_n^*(-x, P) = \Phi(x) - \Phi(-x) + O(\frac{\log n}{n^{1-2/s}}), \qquad (10.32)$$

where the even symmetry of the functions $p(x)$ and $\phi(x)$ was used. Similarly, equation (10.27) implies that

$$L_{n,b,|\cdot|}^*(x) = q^{-1} \sum_{i=1}^{q} 1\{b^{1/2} \hat{\sigma}_{n,b,i}^{-1} | \hat{\theta}_{n,b,i} - \hat{\theta}_n | \le x\} = L_{n,b}^*(x) - L_{n,b}^*(-x) =$$

$$= \Phi(x) - \Phi(-x) + O_P(\sqrt{b/n}) + O(\frac{\log b}{b^{1-2/s}}). \qquad (10.33)$$

Thus, $b \propto (n \log^2 n)^{s/(3s-4)}$ minimizes the MSE of $L^*_{n,b,|\cdot|}(x)$, yielding

$$L^*_{n,b,|\cdot|}(x) = \Phi(x) - \Phi(-x) + O_P(n^{(2-s)/(3s-4)}(\log n)^{s/(3s-4)}).$$

Consequently, the error in the approximation of $J^*_{n,|\cdot|}(x, P)$ by $L^*_{n,b,|\cdot|}(x)$ becomes close to $n^{-1/3}$ for large s, thus improving upon the corresponding $n^{-1/4}$ error rate of $L^*_{n,b}(x)$. It is noteworthy that the error of estimating $J^*_{n,|\cdot|}(x, P)$ by $L^*_{n,b,|\cdot|}(x)$ is of same magnitude as the error of estimating $J^*_n(x, P)$ by the extrapolation $L^{*,ext}_{(2)}(x)$.

However, the error of about $n^{-1/3}$ for the extrapolation $L^{*,ext}_{(2)}(x)$ should be compared to the error of $n^{-1/2}$ of the one-sided normal approximation, whereas the error of about $n^{-1/3}$ for the two-sided $L^*_{n,b,|\cdot|}(x)$ should be compared to the error of the two-sided normal approximation that is close to $O(n^{-1})$ for large s; see equation (10.32) that quantifies the error in approximating $J^*_{n,|\cdot|}(x, P)$ by $\Phi(x) - \Phi(-x)$. Going from $n^{-1/4}$ to $n^{-1/3}$ with $n^{-1/2}$ as target is certainly more important as compared to going from $n^{-1/4}$ to $n^{-1/3}$ with n^{-1} as target.

To summarize, $L^*_{n,b,|\cdot|}(x)$ cannot be considered higher-order accurate as it is still inferior to the normal approximation, given by $\Phi(x) - \Phi(-x)$ here. Nevertheless, the improved accuracy of the two-sided subsampling distribution over its one-sided analogue helps justify the improved finite-sample coverage levels associated with symmetric confidence intervals that are manifested in the simulations of Chapters 9 and 12.

10.5.3 The Unstudentized Sample Mean and the General Extrapolation Result

The discussion in the previous subsection has made apparent that perhaps second-order accuracy may be too much to hope for from extrapolated distributions in the case of strong mixing data. In particular, the $O_P(\sqrt{b/n})$ term in equation (10.27) is too big for second-order considerations in which the error has to be of order $o_P(1/\sqrt{n})$. Thus, we now turn to the simpler case of the unstudentized sample mean, and try to at least achieve close to full first-order accuracy (i.e., the $n^{-1/2}$ Berry–Esseen rate) via extrapolations.

Under conditions (2.3), (2.5) and (2.6) of Götze and Hipp (1983) as well as our conditions (10.23) and (10.24)—with $s > 3$ now—the following Edgeworth expansion holds uniformly in x:

$$P(\sqrt{n}(\bar{X}_n - \theta(P)) \leq x) = \Phi(x/\sigma_\infty) + \sum_{k=1}^{s-3} n^{-k/2} p_k(x/\sigma_\infty) + o(n^{-(s-3)/2}),$$

(10.34)

where the $p_k(x)$s are well-defined smooth functions.

Consider now the subsampling distribution of the sample mean, which is given by

$$L_{n,b}(x) = q^{-1} \sum_{i=1}^{q} 1\{\sqrt{b}(\hat{\theta}_{n,b,i} - \hat{\theta}_n) \leq x\},$$

where $\hat{\theta}_n = \bar{X}_n$, and $\hat{\theta}_{n,b,i} = b^{-1} \sum_{k=i}^{i+b-1} X_k$ as before. Due to the geometric mixing rate (10.23), it can again be shown that $Var(L_{n,b}(x)) = O(b/n)$. This combined with the fact that $\sqrt{b}(\hat{\theta}_n - \theta(P)) = O_P(\sqrt{\frac{b}{n}})$ yields

$$L_{n,b}(x) = E(1\{\sqrt{b}(\hat{\theta}_{n,b,1} - \theta(P)) \leq x\}) + O_P(\sqrt{\frac{b}{n}});$$

a rigorous proof relies on a Bernstein inequality as in Bertail (1997). We also have

$$P(\sqrt{b}(\hat{\theta}_{n,b,1} - \theta(P)) \leq x) = \Phi(x/\sigma_\infty) + \sum_{k=1}^{s-3} b^{-k/2} p_k(x) + o(b^{-(s-3)/2}).$$

The extrapolation is now easier to perform, and with better results, as compared to the studentized case. Again, we let $b_i = c_i^2 b$, where the c_is are some positive constants all different from each other for $i = 1, 2, \ldots, I$, and consider the subsampling distributions $L_{n,b_i}(x)$. We define the extrapolated subsampling distribution $L_{(I)}^{ext}(x)$ by

$$L_{(I)}^{ext}(x) = \sum_{i=1}^{I} \lambda_i L_{n,b_i}(x), \qquad (10.35)$$

where the λ_i's are some carefully chosen constants such that

$$\sum_{i=1}^{I} \lambda_i = 1. \qquad (10.36)$$

Now, let $I = s - 2$, and impose the conditions

$$\sum_{i=1}^{I} \lambda_i / c_i^k = 0, \qquad (10.37)$$

for $k = 1, \ldots, I - 1$. Solving for the λ_i's, it is easy to see that

$$L_{(I)}^{ext}(x) = \Phi(x/\sigma_\infty) + O_P(\sqrt{b/n}) + o(b^{-(s-3)/2}).$$

Thus letting $b = o(n^{1/(s-2)})$, we achieve the result

$$L_{(I)}^{ext}(x) - \Phi(x/\sigma_\infty) = o_P(n^{(1-s)/2(s-2)}),$$

which is just $o_P(1)$ if s is very close to 3, but can come very close to the Berry–Esseen rate $O_P(n^{-1/2})$ if s is large.

Note that $L_{(I)}^{ext}(x)$ retains the robustness property of being consistent even if the aforementioned Edgeworth does not hold provided some (not necessarily normal) limiting distribution exists. As a matter of fact, one can employ the aforementioned procedure without knowledge of s; for example, one can assume a large I, and construct $L^{ext}_{(I)}(x)$ based on some λ_is that satisfy (10.36) and (10.37) for $k = 1, \ldots, I-1$. Then, it is immediate that

$$L^{ext}_{(I)}(x) = \Phi(x/\sigma_\infty) + O_P(\sqrt{b/n}) + o_P(b^{-\min(I-1,s-3)/2}).$$

The sample mean example motivates the following general result involving an arbitrary parameter $\theta(P)$ and a general statistic $\hat{\theta}_n$, converging at an arbitrary rate τ_n to a continuous limiting distribution $J(x, P)$. In analogy to (10.1), we consider

$$J_n(x, P) \equiv \mathrm{Prob}_P\{\tau_n(\hat{\theta}_n - \theta(P)) \le x\} \to J(x, P). \tag{10.38}$$

Theorem 10.5.1. *Assume (10.38) with a continuous limiting distribution $J(x, P)$. Also assume (10.2), (10.3), and (10.23). Let $b_i = c_i^2 b$, where the c_is are some positive constants all different from each other for $i = 1, 2, \ldots, I$; here $I = 1 + R$ with R being a positive integer. Consider the subsampling distributions $L_{n,b_i}(x)$ as defined in (10.4) by letting $\hat{\sigma}_n \equiv 1$ (i.e., unstudentized), and construct the extrapolated subsampling distribution $L^{ext}_{(I)}(x)$ given by (10.35) with the λ_is being some constants such that (10.36) holds. Then,*

$$\sup_x |L^{ext}_{(I)}(x) - J(x, P)| = o_P(1).$$

If, in addition, we assume conditions $\alpha 3[I]$ and $\alpha 4[I]$ of Bertail (1997), and we also assume that $J(x, P)$ is continuously differentiable in x (with bounded derivative), and that $\hat{\theta}_n$ admits the following Edgeworth expansion on some function $f_1(n) = n^\varsigma$, for some $\varsigma > 0$, i.e., that

$$J_n(x, P) = J(x, P) + \sum_{k=1}^{R} f_1(n)^{-k} p_k(x) + O(f_1(n)^{-R-1}),$$

uniformly in x, where the $p_k(\cdot)$ are well-defined continuously differentiable functions, and if we impose on the λ_is the additional restrictions

$$\sum_{i=1}^{I} \lambda_i c_i^{-2k\varsigma} = 0 \tag{10.39}$$

for all $k = 1, 2, \ldots, R$, then,

$$\sup_x |L^{ext}_{(I)}(x) - J(x, P)| = O_P(\frac{\tau_b}{\tau_n} + \sqrt{b/n} + b^{-R\varsigma}).$$

The proof of the theorem is similar to what was discussed for the sample mean case and uses the assumed expansion of $J_n(x, P)$ in powers of $f_1(n)$, and results of Bertail (1997). Note in particular that conditions $\alpha 3[I]$, and $\alpha 4[I]$ of Bertail (1997) are easy to satisfy in our setting (see

Remark 6 of Bertail, 1997). For example, condition $\alpha 3[I]$ of Bertail (1997) is a mixing condition that is actually satisfied in our exponential mixing case provided we take $b = o(\sqrt{n/\log n})$. Similarly, condition $\alpha 4[I]$ of Bertail (1997) follows immediately by choosing b small enough to ensure $b^{I\zeta}\tau_b/\tau_n \to 0$.

For an easy interpretation of the conclusions of Theorem 10.5.1, consider the rather general case where $\tau_n = n^\xi$, for some $0 < \xi \leq 1/2$. Then, by letting $b \propto n^{1/(1+I\max(1,\zeta/\xi))}/\log n$, we immediately have that

$$\sup_x |L_{(I)}^{ext}(x) - J(x,P)| = O_P(f_1(n)^{-1} + (\frac{b}{n})^{\min(\xi,1/2)}).$$

Thus, if R is appropriately big, then the rate of estimation of $J(x,P)$ by $L_{(I)}^{ext}(x)$ can come very close to $O_P(f_1(n)^{-1} + \tau_n^{-1})$, even in general, non-$\sqrt{n}$ convergent cases. Obviously, extrapolated subsampling will definitely not be second-order correct in the unstudentized case; therefore it will be inferior to the studentized block-bootstrap when the block-bootstrap applies as well (see Künsch, 1989; Liu and Singh, 1992; Lahiri, 1992; Götze and Künsch, 1996; and Hall, Lahiri, and Jing, 1996).

Because of the i.i.d. randomization implicit in the block-bootstrap, the block-bootstrap distribution is 'more normal' (i.e., closer to the normal distribution) than the subsampling (or extrapolated subsampling) distribution. As noted in Politis and Romano (1994b), this fact can be explained by viewing the block-bootstrap distribution of the sample mean as the k-fold convolution of the subsampling distribution with itself (with $k = n/b$). Thus, it is not surprising that in regular cases with a normal limit distribution, e.g., the sample mean, linear statistics, or statistics approximable by linear statistics, the block-bootstrap may have an edge over subsampling. Nevertheless, the main advantage of subsampling is its generality and ability to handle nonregular difficult cases. The fact that the extrapolated subsampling distribution can be an accurate asymptotic approximation under very weak assumptions (e.g., weak moment assumptions, non-$\sqrt{n}$ convergent estimators, non-normal limit distributions, etc.) is rather remarkable.

10.5.4 Finite Population Correction in the Mixing Case

It is noteworthy that, in the strong mixing setting, there is evidence of a need for a finite population correction factor as in the i.i.d. case. Nevertheless, the extent of the influence of such a finite population correction factor on the accuracy of extrapolating distribution seems limited. To elaborate, recall that the finite population correction factor in the i.i.d. case was in particular due to the fact that

$$Var(b^{1/2}(\bar{X}_{n,b,i} - \bar{X}_n)) = (1 - \frac{b}{n})\sigma^2,$$

where $\bar{X}_{n,b,i}$ is the sample mean of subsample Y_i.

In the strong mixing case, it is quite interesting that a similar relation holds, thus indicating that a similar finite population correction factor (surprisingly, of the same form as in the i.i.d. case) may be appropriate. A straightforward calculation shows that we have

$$Var(b^{1/2}(\bar{X}_{n,b,i} - \bar{X}_n)) = (1 - \frac{b}{n})\sigma_\infty^2 + O(\frac{b}{n^2} + \frac{1}{b} + \frac{b^2}{n^3}). \quad (10.40)$$

Thus, it is obvious that it is generally advisable to take into account the finite population correction factor as it will reduce the error of the subsampling distribution in the mixing case as well. Note though that this finite population correction will be of real value here only if b is large with respect to $\sqrt{n}$, i.e., if $\sqrt{n} = o(b)$, in which case equation (10.40) reduces to

$$Var(b^{1/2}(\bar{X}_{n,b,i} - \bar{X}_n)) = (1 - \frac{b}{n})\sigma_\infty^2 + o(\frac{b}{n}). \quad (10.41)$$

The above is intuitive since, the larger b is, the more necessary the finite population correction should be. Nevertheless, as our previous results demonstrated, a choice of b that is very small (with respect to $\sqrt{n}$) is required in order to achieve higher-order properties of the extrapolated subsampling distributions. Therefore, the finite population correction, although recommendable in practice, will not render the second-order accuracy of the extrapolation of the finite-population-corrected subsampling distributions in the dependent case.

10.5.5 Bias-Corrected Variance Estimation for Strong Mixing Data

As before, let $\underline{X}_n = (X_1, \ldots, X_n)$ be an observed stretch of a stationary, strong mixing sequence of real-valued random variables $\{X_t, t \in \mathbb{Z}\}$ with mixing coefficients $\alpha_X(\cdot)$. For concreteness, we again consider the sample mean $\bar{X}_n = n^{-1}\sum_{i=1}^n X_i$ as an estimator of the true mean $\theta(P) = EX_1$; our results immediately extend to "smooth" statistics that are differentiable functions of the empirical first-marginal $\hat{F}(x) = n^{-1}\sum_{i=1}^n 1\{X_i \leq x\}$ through the use of influence functions. Again, let $R(k) = Cov(X_0, X_k)$ denote the autocovariance at lag k, and $f(w) = (2\pi)^{-1}\sum_{s=-\infty}^{\infty} R(s)\cos(ws)$ denote the spectral density function evaluated at point $w \in [-\pi, \pi]$ and assumed to exist. Note that a sufficient condition for existence of the spectral density function is simply the finiteness of the sum $\sum_{s=-\infty}^{\infty} |R(s)|$, which can in fact be guaranteed via an appropriate mixing and moment condition (see Lemma A.0.1 in Appendix A).

Similarly, under appropriate mixing and moment conditions (e.g., see Theorem 3.4.1), the CLT obtains, yielding $\sqrt{n}(\bar{X}_n - \theta(P))/\sigma_n \xrightarrow{\mathcal{L}} N(0,1)$, where $\sigma_n^2 = Var(\sqrt{n}\bar{X}_n) = R(0) + 2\sum_{s=1}^n (1-s/n)R(s)$. In order to actually use this Central Limit theorem to gauge the accuracy of $\bar{X}_n$ as an estimator of $\theta(P)$—e.g., by setting confidence intervals for $\theta(P)$—estimating the variance $\sigma_n^2 = R(0) + 2\sum_{s=1}^{n}(1 - s/n)R(s)$, or its asymptotic limit $\sigma_\infty^2 = \sum_{s=-\infty}^{\infty} R(s)$ becomes crucial. It is easy to see that, if the spectral density exists, then this limit also exists and is given by $\sigma_\infty^2 = 2\pi f(0)$, so that the problem of variance estimation is intimately related to the problem of nonparametric estimation of the spectral density function (evaluated at the origin).

Let $\hat{\sigma}_{b_M,n}^2$ be the subsampling estimator of σ_∞^2 as defined in equation (9.1) with $h_L = 1$. As mentioned in Section 9.2, under regularity conditions, $\hat{\sigma}_{b_M,n}^2$ is a consistent and asymptotically normal estimator. In particular, equations (9.2) and (9.3) give the asymptotic bias and variance of the subsampling estimator $\hat{\sigma}_{b_M,n}^2$. Note that, although the order of magnitude of $Var(\hat{\sigma}_{b_M,n}^2)$ is small, the order of magnitude of $Bias(\hat{\sigma}_{b_M,n}^2)$ is unnecessarily large.

For example, as mentioned in Section 9.2, a consideration of trade-off between bias and variance leads us to choose b_M of the order of $n^{1/3}$ so that we have: $\hat{\sigma}_{b_M,n}^2 = \sigma_\infty^2 + O_P(n^{-1/3})$. Realizing that the poor rate of convergence of $\hat{\sigma}_{b_M,n}^2$ is due to its bias, Politis and Romano (1995) proposed a bias-corrected version of the estimator $\hat{\sigma}_{b_M,n}^2$ that can also be viewed as an extrapolation. To perform a bias correction on the estimator $\hat{\sigma}_{b_M,n}^2$, we could use another round of subsampling to empirically estimate the bias of $\hat{\sigma}_{b_M,n}^2$ as described in Chapter 3. Note that the subsample replicate values of estimator $\hat{\sigma}_{b_M,n}^2$ (based on a subsample size of b, say) would be of type $\hat{\sigma}_{b_m,b}^2$, where b_m is of smaller order than b, as b_M is of smaller order than n; for example, if $b_M = n^{1/3}$, then $b_m = b^{1/3} = o(n^{1/3}) = o(b_M)$. In other words, the subsample replicate values of estimator $\hat{\sigma}_{b_M,n}^2$ that are used to perform the bias-correction are roughly of the same type as the original $\hat{\sigma}_{b_M,n}^2$ but with a block size of b_m instead of b_M, where $b_m < b_M$. Thus, the bias-corrected subsampling variance estimator can be generally formed by a linear combination of two subsampling variance estimators with different block sizes b_m and b_M, and is defined as

$$\tilde{\sigma}_{b_m,b_M,n}^2 \equiv (H+1)\hat{\sigma}_{b_M,n}^2 - H\hat{\sigma}_{b_m,n}^2 \qquad (10.42)$$

for some positive constant H. Note, however, that to achieve the bias correction, the choice of H should reflect the ratio b_m/b_M, or vice versa, the ratio b_m/b_M depends on H. For example, given a particular choice for H, b_m has to be chosen as $b_m = Hb_M/(1+H)$ to achieve the bias correction (or equivalently $H = b_m/(b_M - b_m)$). To see why, observe that for $b_M = o(\sqrt{n})$, equation (9.2) simplifies to

$$Bias(\hat{\sigma}_{b_M,n}^2) = E\hat{\sigma}_{b_M,n}^2 - \sigma_\infty^2 = -b_M^{-1} \sum_{k=-\infty}^{\infty} |k|R(k) + o(1/b_M). \qquad (10.43)$$

It is immediate that if (and only if) $b_m = Hb_M/(1+H)$, then the bias of $\tilde{\sigma}^2_{b_m,b_M,n}$ becomes of order $o(1/b_M)$ after cancellation of the $O(1/b_M)$ term involving $\sum_{k=-\infty}^{\infty} |k|R(k)$.

Note that equation (10.42) can be interpreted as an extrapolation (or a generalized jackknife) of the two subsampling variance estimators $\hat{\sigma}^2_{b_M,n}$ and $\hat{\sigma}^2_{b_m,n}$, and it has an improved asymptotic performance. To see this, consider the relation

$$E\hat{\sigma}^2_{b_M,n} \approx \sigma^2_{\infty} - b_M^{-1} \sum_{k=-\infty}^{\infty} |k|R(k)$$

as an approximate linear regression of $\hat{\sigma}^2_{b_M,n}$ on the "variable" b_M^{-1} with intercept (at $b_M^{-1} = 0$ that occurs when $b_M = \infty$) given by σ^2_{∞}. Taking two (or more) points of the type $\sigma^2_{b_M,n}$ can give us a linear regression line from which the intercept can be accurately extrapolated. The estimated intercept (from the line joining the two points $\hat{\sigma}^2_{b_M,n}$ and $\hat{\sigma}^2_{b_m,n}$) is exactly $\tilde{\sigma}^2_{b_m,b_M,n}$.

The generalized jackknife would imply that the bias of $\tilde{\sigma}^2_{b_m,b_M,n}$ becomes $o(1/b_M)$. Quite remarkably, an even further bias correction is achieved, which is quantified by the following theorem. The proof of the theorem is omitted; it closely parallels the results of Politis and Romano (1995) for the spectral density.

Theorem 10.5.2. *Assume that $E|X_t|^{2+\delta} < \infty$, for some $\delta > 0$.*

i. *Under the exponential strong mixing assumption (10.23), and letting $b_M = A \log n$, for some sufficiently large constant A, and*

$$b_m = b_M H/(H+1), \qquad (10.44)$$

where H is the positive constant H that figures in the definition of $\tilde{\sigma}^2_{b_m,b_M,n}$, we have that

$$\mathrm{Bias}(\tilde{\sigma}^2_{b_m,b_M,n}) = O(1/n). \qquad (10.45)$$

ii. *If $\alpha_X(k) = 0$ for all $|k|$ bigger than some $m > 0$, i.e., if the data are m-dependent, then letting b_m and b_M be two constants satisfying $b_M \geq b_m \geq m$ we have that*

$$\mathrm{Bias}(\tilde{\sigma}^2_{b_m,b_M,n}) = O(1/n). \qquad (10.46)$$

Note that equation (10.44) plays the same role as equation (10.30) in the problem of distribution estimation as it ensures a first-order bias reduction. Nevertheless, the extrapolation technique is much more successfull in the problem of variance estimation because it actually manages to do more, that is, reduce the bias by many orders of magnitude at once.

The effect of the extrapolation on the variance of the bias-corrected estimator $\tilde{\sigma}^2_{b_m,b_M,n}$ is the subject of the following theorem. It can be proved either using results of Politis and Romano (1995) on kernel spectral density estimation, or by simply using the fact that the large-sample correlation coefficient between $\hat{\sigma}^2_{b_m,n}$ and $\hat{\sigma}^2_{b_M,n}$ is approximately given by

$$(1 + \frac{b_M - b_m}{2b_M})\sqrt{b_m/b_M};$$

see, for example, Pedrosa and Schmeiser (1993).

Theorem 10.5.3. *Assume conditions strong enough to ensure that equation (9.3) holds true.*

Then,

$$Var(\tilde{\sigma}^2_{b_m,b_M,n}) = 2c\frac{b_M}{n}\frac{3H+1}{H+1}\sigma^4_\infty + o(b_M/n), \qquad (10.47)$$

where c was defined in equation (9.3).

As was to be expected, the bias-reduction comes at a price in terms of a variance increase, since $\frac{3H+1}{H+1}$ goes from 1 to 3 as H goes from 0 to ∞. Nevertheless, the order of magnitude of the MSE is decreased, because, while the variance is increased by a proportionality constant, the order of magnitude of the bias is made smaller. Note that there is a variety of sufficient conditions guaranteeing that equations (9.2) and (9.3) hold true (see Politis and Romano, 1995, and the references therein).

To elaborate on part (i) of Theorem 10.5.2, note that under the assumptions of Theorem 10.5.3 and the exponential strong mixing assumption (10.23), we have (for $H = 1$, say) that

$$\tilde{\sigma}^2_{0.5A \log n, A \log n, n} = \sigma^2_\infty + O_P(\sqrt{\log n/n}); \qquad (10.48)$$

thus $\tilde{\sigma}^2_{0.5A \log n, A \log n, n}$ may be used whenever an accurate variance estimator is needed. For example, it may be used for studentization in the context of subsampling distributions discussed in the previous section, or for studentization of moving blocks bootstrap distributions as in Götze and Künsch (1996).

Remark 10.5.1. Recall that estimating σ^2_∞ is tantamount to (nonparametric) estimation of the spectral density $f(0)$ at the origin. Although nonparametric functional estimation is notorious for its slow rates of convergence, we see here that $\tilde{\sigma}^2_{b_m,b_M,n}$ comes very close to the parametric $\sqrt{n}$ rate of convergence in the exponential mixing case, and does achieve the parametric rate in the m-dependent case (by taking b_m and b_M to be some big enough constants). This remarkable accuracy can be explained by the notion of higher-order (indeed, infinite-order) kernels for smoothing purposes (see Politis and Romano, 1995, 1996a, 1999, for more details).

Remark 10.5.2. Taking a linear combination of many (more than two) subsampling variance estimators is certainly possible, and can be achieved by the generalized jackknife/extrapolation principles discussed. However, its effect will be noticeable only in the asymptotic proportionality constants as the asymptotic MSE rate is made minimum by a linear combination of just two subsampling variance estimators.

Remark 10.5.3 (Choice of H). Note that the effect of the choice used for the constant H is limited to the proportionality constants in the asymptotic forms of $Bias(\tilde{\sigma}^2_{b_m, b_M, n})$ and $Var(\tilde{\sigma}^2_{b_m, b_M, n})$; the rates are not at all influenced by H (as long as $H > 0$). Nevertheless, since the bias-correction actually disappears if $H = 0$, we are well advised not to choose an H very close to 0 as in that case the bias correction effect will be small in practice (and in finite samples). Similarly, taking an H that is too large has an adverse effect in finite samples that can be explained by the analogy to the Dirichlet kernel (see Politis and Romano, 1995). A choice of H equal to or close to unity (e.g., $H \in [\frac{1}{2}, 2]$) has led to good performance in finite-sample simulation studies. Note that $H = 1$ corresponds to the simple choice $b_m = b_M/2$.

Remark 10.5.4 (Choice of block size b_M). We now revisit the problem of optimal, from the point of view of minimum mean squared error (MSE), choice of block size for variance estimation. We assume that the choice of H has already been fixed (e.g., $H = 1$), and only the choice of b_M is still open. Recall that in Section 9.2 we considered the issue of optimally choosing the block size b_M for the subsampling variance estimator $\hat{\sigma}^2_{b_M, n}$, and came to the conclusion that b_M should be taken proportional to $n^{1/3}$ but with an optimal proportionality constant that is difficult to estimate. It is a fortunate issue that the bias-corrected estimator $\tilde{\sigma}^2_{b_m, b_M, n}$ not only is more accurate than $\hat{\sigma}^2_{b_M, n}$ (the comparison being made with each estimator using its own MSE optimal block size), but comes with the added bonus of an easy way to estimate the optimal block size in practice. Note that part (ii) of Theorem 10.5.2 gives as an explicit formula, namely $b_M \geq b_m \geq m$, to minimize the $Bias(\tilde{\sigma}^2_{b_m, b_M, n})$ in the case of m-dependent data.

In addition, it is easy to see from equation (10.47) that to minimize the $Var(\tilde{\sigma}^2_{b_m, b_M, n})$ we should take b_m (and b_M) as small as possible. Since $b_m = b_M H/(H+1)$, where H has been previously decided upon, it follows that the recommendation stemming from part (ii) of Theorem 10.5.2 is to take $b_m = m$ (the smallest possible), and $b_M = b_m(H+1)/H$.

The practical problem now is that not all weakly dependent data are m-dependent, and even if they were, the value of m would be unknown. Nevertheless, weakly dependent data can at least approximately be thought of as m-dependent for some large enough m, since $\alpha_X(k) \to 0$ implies $\alpha_X(k) \approx 0$ for all k bigger than some m. The issue now is simply to estimate this approximate threshold m for a particular data set at hand. To do this one may look at the correlogram, that is, a plot of the sample autocovariances $\hat{R}(k)$, as advocated in Politis and Romano (1995). Thus, a reasonable estimate of m is the smallest integer after which the correlogram plot seems to settle down to the value zero (except for noise fluctuations). Consequently, for some previously chosen (and fixed) value of H in the interval $[\frac{1}{2}, 2]$, the estimator $\tilde{\sigma}^2_{\hat{m},\hat{m}(H+1)/H,n}$ will be near-optimal, where $\hat{m}$ is our best estimate of m from the correlogram plot, i.e., m is the smallest integer such that $\hat{R}(k) \approx 0$ for all $k \geq \hat{m}$.

For an illustration of this proposal, let $H = 1$, and consider Figure 10.1 that shows a plot of the sample autocovariance $\hat{R}(k) = Cov(X_0, X_k)$ calculated from the same MA(2) data set of size $n = 100$ that led to Figure 9.2. The straight lines correspond to the usual $\pm 1.96 \hat{R}(0)/\sqrt{n}$ bands that are customarily used as a simple test (at level 0.05) of the hypothesis that the underlying sequence $\{X_t\}$ is i.i.d. It is apparent that the i.i.d. hypothesis is rejected for this data set. We can then proceed to build 95% bands under the assumption that the data follow a linear MA(q) model—in i.i.d. residuals—using Bartlett's formula in Brockwell and Davis (1991, p. 221); for example, one can try increasing values of q with the objective of having about 95% of the $\hat{R}(k)$s (for $k > q$) fall within the MA(q) 95% band. Note, however, that Bartlett's formula can be misleading if the residuals in the true model are not i.i.d.; if uncorrelated (but not i.i.d.) residuals are a possibility, then the 95% band can still be constructed using subsampling or resampling (see, e.g., Romano and Thombs, 1996).

The dotted lines in the figure represent the 95% band under the hypothesis of a linear MA(2) model with i.i.d. residuals. Since about 95% of the $\hat{R}(k)$s (for $k > 2$) indeed fall within the dotted band, the assumption of MA(2) is a plausible one, notwithstanding that some points, such as $\hat{R}(7)$, fall outside the band; thus, we may estimate m by $\hat{m} = 3$. As a consequence, the estimate $\tilde{\sigma}^2_{3,6,n} = 10.28$ would be employed, which is a reasonably accurate estimate of the true asymptotic variance that is equal to 9.

Nevertheless, any choice for $\hat{m}$ (and b_m) in the neighborhood of 3 gives good results as Figure 10.2 shows. In particular, the optimal choice of b_m for this particular data set is $b_m = 4$, and it leads to the estimate $\tilde{\sigma}^2_{4,8,n} = 9.53$. Figure 10.2 shows the bias-corrected estimator $\tilde{\sigma}^2_{b_m,2b_m,n}$ as a function of b_m for the particular data set at hand; this figure should be compared to Figure 9.1 that shows the estimator $\hat{\sigma}^2_{b_M,n}$ as a function of b_M. Notably, both $\tilde{\sigma}^2_{b_m,2b_m,n}$ and $\hat{\sigma}^2_{b_M,n}$ can be inaccurate with a poor choice of the block size; the distinguishing features are that (a) there is a simple and practical way of picking a good block size for $\tilde{\sigma}^2_{b_m,2b_m,n}$, and that (b) with a poor choice of b_m, $\tilde{\sigma}^2_{b_m,2b_m,n}$ can even become negative. The negativity is a problem but could also be considered to be a blessing in disguise, as it alerts us on the poor block size choice; this aspect is further elaborated upon in the next remark.

Remark 10.5.5 (Negativity issues). As demonstrated in Figure 10.2, $\tilde{\sigma}^2_{b_m,b_M,n}$ is not almost surely nonnegative. Note that this is not a problem with $\tilde{\sigma}^2_{b_m,b_M,n}$ in particular, but rather of all higher-order accurate variance

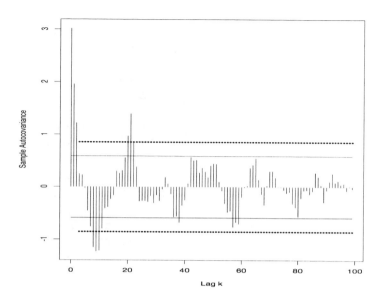

FIGURE 10.1. The sample autocovariance $\hat{R}(k)$ vs. the lag k, for $k = 0, 1, \ldots, 99$, calculated from the same MA(2) data set of size $n = 100$ that led to Figure 9.2; the straight lines correspond to approximate 95% (pointwise) confidence intervals under the i.i.d. assumption for the underlying sequence $\{X_t\}$ sequence, while the dotted lines correspond to approximate 95% (pointwise) confidence intervals under the MA(2) assumption for $\{X_t\}$ via Bartlett's formula.

(or spectral) estimators; see Politis and Romano (1995). Although this problem disappears asymptotically as $n \to \infty$, in finite samples it might pose a real concern. In the spectral estimation setup, however, there exists an easy fix by taking the positive part, i.e., if the estimator turns up to be nonpositive, use zero as your estimator instead. Unfortunately, this fix does not work if we want to studentize using $\tilde{\sigma}^2_{b_m,b_M,n}$. So, since zero is not an acceptable estimator, there are two practical ways out:

(a) Try out different choices for b_M (or for A, if we take $b_M = A \log n$), and use the corresponding $\tilde{\sigma}^2_{b_m,b_M,n}$ if it turns out to be positive, or

(b) use a fraction of $\hat{\sigma}^2_{b_M,n}$ (e.g., $l\hat{\sigma}^2_{b_M,n}$ for some $l \in (0,1]$) as our variance estimator, i.e., 'shrink' the original uncorrected estimator $\hat{\sigma}^2_{b_M,n}$ toward zero.

Regarding (b), note that $\tilde{\sigma}^2_{b_m,b_M,n}$ can be interpreted as the difference $\hat{\sigma}^2_{b_M,n} - \widehat{Bias}(\hat{\sigma}^2_{b_M,n})$; a negative $\tilde{\sigma}^2_{b_m,b_M,n}$ indicates that our estimate of the bias of $\hat{\sigma}^2_{b_M,n}$ is positive and large (actually too large). Nevertheless, we might take the hint that $\hat{\sigma}^2_{b_M,n}$ has a positive bias and attempt to reduce it by taking $l\hat{\sigma}^2_{b_M,n}$ as our estimator.

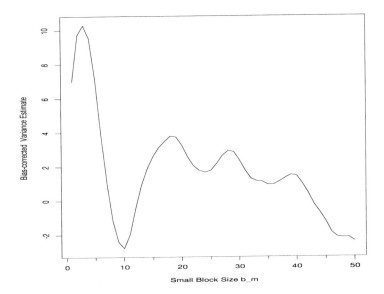

FIGURE 10.2. The bias-corrected variance estimator $\tilde{\sigma}^2_{b_m, 2b_m, n}$ as a function of b_m, calculated from the same MA(2) data set of size $n = 100$ that led to Figure 9.2.

The additional difficult question of choosing l actually prompts us to favor method (a) of solving the problem of a negative estimate. To further support the suggestion that a negative $\tilde{\sigma}^2_{b_m, b_M, n}$ can occur as a result of a poor choice of b_M, consider the following small simulation example: 190 independent time series stretches of $\{X_t\}$, each of length $n = 1000$, were generated from the normal autoregressive-moving average (ARMA) model that satisfies the difference equation

$$X_t - 1.325X_{t-1} + 1.338X_{t-2} - 0.662X_{t-3} + 0.240X_{t-4} =$$

$$= Z_t - 0.2Z_{t-1} + 0.04Z_{t-2},$$

for all t, where the sequence $\{Z_t\}$ is i.i.d. N(0,1). Estimators $\tilde{\sigma}^2_{b_m, b_M, n}$ were computed from each of the 190 stretches and for different combinations of b_m and b_M. This ARMA model example was discussed in detail in Politis and Romano (1995), where it was shown that the optimal (from a MSE point of view) choices of b_m and b_M are 5 and 10, respectively. Note that there are no occurrences (out of the 190 trials) of the event $\{\tilde{\sigma}^2_{b_m, b_M, n} \leq 0\}$ for the "optimal" choices $b_m = 5$ and $b_M = 10$, and this nonnegativity holds for a variety of other reasonable combinations of b_m and b_M as the entries of Table 10.1 illustrate. Incidents of negativity start occuring if $b_m = 20$ or larger. Nevertheless, a choice of $b_m \geq 20$ (and $b_M = 30$, 40 or 50) is

TABLE 10.1. Entries are the empirically estimated probabilities of the event $\{\tilde{\sigma}^2_{b_m,b_M,n} \leq 0\}$ for different combinations of b_m and b_M.

	$b_M=5$	$b_M=10$	$b_M=20$	$b_M=30$	$b_M=40$	$b_M=50$
$b_m=5$		0.000	0.000	0.000	0.000	0.000
$b_m=10$			0.000	0.000	0.000	0.000
$b_m=20$				0.042	0.005	0.016
$b_m=30$					0.200	0.080
$b_m=40$						0.221

blatantly suboptimal in this case, as the simulation results of Politis and Romano (1995) demonstrate.

10.6 Moderate Deviations in Subsampling Distribution Estimation

In this section, we assume that $\underline{X}_n = (X_1, \ldots, X_n)$ is an observed stretch of a stationary, m-dependent sequence of real random variables $\{X_t, t \in \mathbb{Z}\}$ with finite second moments. Recall that the m-dependence condition means that the set of random variables $\{X_t, t \leq k\}$ is independent of the set $\{X_t, t \geq s\}$ as long as $s - k > m$.

It is assumed that the true mean $\theta(P) = EX_1$ is unknown and estimated by the sample mean $\bar{X}_n = n^{-1} \sum_{i=1}^{n} X_i$ of the observed stretch $\underline{X}_n = (X_1, \ldots, X_n)$. For notational convenience, we assume in this section that

$$\sum_{k=-m}^{m} Cov(X_0, X_k) = 1,$$

so that the following Central Limit theorem (CLT) holds:

$$J_n(x, P) \equiv P\{\sqrt{n}(\bar{X}_n - \theta(P)) \leq x\} \longrightarrow \Phi(x), \quad \text{as} \quad n \to \infty, \quad (10.49)$$

where $\Phi(\cdot)$ is the distribution function of a standard normal random variable. As a matter of fact, the CLT for stationary m-dependent sequences given by equation (10.49) holds true under the sole assumption that the sequence $\{X_t, t \in \mathbb{Z}\}$ has finite second moments (see, for example, Liu and Singh, 1992).

As before, let b be an integer depending on n such that $1 < b < n$. Also let $\bar{X}_{n,b,i}$ be the sample mean of the block $(X_i, X_{i+1}, \ldots, X_{i+b-1})$, i.e., $\bar{X}_{n,b,i} = b^{-1} \sum_{k=i}^{i+b-1} X_k$. The subsampling distribution of the (unstudentized) sample mean is defined as before by

$$L_{n,b}(x) \equiv q^{-1} \sum_{i=1}^{q} 1\{\sqrt{b}(\bar{X}_{n,b,i} - \bar{X}_n) \leq x\}, \quad (10.50)$$

10.6 Moderate Deviations in Subsampling Distribution Estimation

where $q = n - b + 1$.

As follows from Theorem 3.2.1, the CLT (10.49) together with the design requirements that $b \to \infty$ as $n \to \infty$ with $b/n \to 0$ are sufficient to ensure that the subsampling distribution is consistent, i.e., that

$$\sup_x |L_{n,b}(x) - \Phi(x)| = o_P(1) \tag{10.51}$$

and

$$\sup_x |L_{n,b}(x) - J_n(x, P)| = o_P(1). \tag{10.52}$$

The question of moderate deviations in a convergence such as (10.51) or (10.52) has to do with investigating the rate of convergence as x itself is allowed to increase with n. In particular, the moderate deviation rate of convergence in equation (10.52) is especially interesting in statistical applications, where the objective typically is to approximate the quantiles of the (generally unknown) sampling distribution $J_n(\cdot, P)$ by the corresponding quantiles of $L_{n,b}(\cdot)$. In the remainder of this chapter, we will focus, without loss of generality, on the right tail of the unknown sampling distribution by considering large values of x. In an analogous manner, the rate of convergence in the left tail can be examined.

To study moderate deviations, we will consider the estimator $L_{n,b}(x)$ as x increases with n at an appropriate rate; thus, we consider a nondecreasing sequence $x_n > 0$. Our exposition follows the report by Bertail, Gamst, and Politis, (1998), and our first result concerns the unbiased version of the subsampling distribution given by

$$\underline{L}_{n,b}(x_n) \equiv q^{-1} \sum_{i=1}^{q} 1\{\sqrt{b}(\bar{X}_{n,b,i} - \theta(P)) \le x_n\}.$$

Lemma 10.6.1. *Assume that $\{X_t, t \in \mathbb{Z}\}$ is a stationary, m-dependent sequence of real random variables with finite second moments. Let $n \to \infty$ and let $x_n > 0$ be any nondecreasing sequence; then*

$$\underline{L}_{n,b}(x_n) = J_b(x_n, P) + \sqrt{1 - J_b(x_n, P)}\, O_P\left(\sqrt{\frac{b}{n}}\right). \tag{10.53}$$

Proof. Consider the quantity

$$\underline{L}_{n,b}^{(j)}(x_n) \equiv \frac{b+m}{n} \sum_{i=0}^{\lfloor n/(b+m) \rfloor - 1} 1\{\sqrt{b}(\bar{X}_{n,b,i(b+m)+j} - \theta(P)) \le x_n\},$$

for $j = 1, \ldots, (b+m)$, where $\lfloor \cdot \rfloor$ denotes the integer part. Note that due to m-dependence, $\underline{L}_{n,b}^{(j)}(x_n)$ is an average of $\lfloor n/(b+m) \rfloor$ independent and identically distributed Bernoulli random variables with mean $J_b(x_n, P)$. It follows that

$$Var(\underline{L}_{n,b}^{(j)}(x_n)) = \frac{b+m}{n} J_b(x_n, P)(1 - J_b(x_n, P))$$

$$= O\left(\frac{b+m}{n}(1 - J_n(x_n, P))\right).$$

Now, note that

$$\underline{L}_{n,b}(x_n) = \frac{1}{b+m} \sum_{j=1}^{b+m} \underline{L}_{n,b}^{(j)}(x_n).$$

But,

$$\left| Cov\left(\underline{L}_{n,b}^{(j)}(x_n), \underline{L}_{n,b}^{(k)}(x_n)\right) \right| \leq \sqrt{Var\left(\underline{L}_{n,b}^{(j)}(x_n)\right) Var\left(\underline{L}_{n,b}^{(k)}(x_n)\right)}$$
$$= Var\left(\underline{L}_{n,b}^{(j)}(x_n)\right)$$
$$= O\left(\frac{b+m}{n}(1 - J_b(x_n, P))\right),$$

where we used the Cauchy–Schwarz inequality together with the fact that

$$Var\left(\underline{L}_{n,b}^{(j)}(x_n)\right) = Var\left(\underline{L}_{n,b}^{(k)}(x_n)\right).$$

Now, since

$$Var\left(\underline{L}_{n,b}(x_n)\right) = (b+m)^{-2} \sum_{j=1}^{b+m} \sum_{k=1}^{b+m} Cov\left(\underline{L}_{n,b}^{(j)}(x_n), \underline{L}_{n,b}^{(k)}(x_n)\right),$$

it follows that $Var\left(\underline{L}_{n,b}(x_n)\right) = O\left(\frac{b+m}{n}(1 - J_b(x_n, P))\right)$ as well. Since

$$E\underline{L}_{n,b}(x_n) = J_b(x_n, P),$$

the lemma is proved. ∎

Lemma 10.6.1 holds true even in the case of fixed (nonincreasing) b. However, in order to estimate the asymptotic distribution via subsampling, b must tend to infinity as well. In addition, in order to obtain a moderate deviations consistency result, we must impose some limitation on the growth of x_n; this growth limitation may be considered a "moderate deviation."

Note also that the quantity $\underline{L}_{n,b}(x_n)$ is not a proper statistic, as it depends on the unknown parameter $\theta(P)$. By substituting $\bar{X}_n$ in place of $\theta(P)$ in $\underline{L}_{n,b}(x_n)$, we obtain the familiar subsampling distribution estimator $L_{n,b}(x_n)$. Interestingly, the nonexact centering in $L_{n,b}(x_n)$ does not present a problem, as the following theorem shows.

Theorem 10.6.1. *Assume that $\{X_t, t \in \mathbb{Z}\}$ is a stationary, m-dependent sequence of real random variables such that $E|X_t|^3 < \infty$. Let $b \to \infty$ as $n \to \infty$ with $b = o(n)$, and let $x_n > 0$ be any nondecreasing sequence such that $x_n = O(b^{1/6})$.*

10.6 Moderate Deviations in Subsampling Distribution Estimation

Then,

$$L_{n,b}(x_n) = \underline{L}_{n,b}(x_n) + O_P\left(\sqrt{\frac{b}{n}}(x_n + \sqrt{\ln \frac{n}{b}})\phi(x_n)\right). \quad (10.54)$$

Proof. We have $\underline{L}_{n,b}(x_n) - L_{n,b}(x_n) = q^{-1}\sum_{i=1}^{q} Y_i$, where
$$Z_i = 1\{\sqrt{b}(\bar{X}_{n,b,i} - \theta(P)) \leq x_n\} - 1\{\sqrt{b}(\bar{X}_{n,b,i} - \bar{X}_n) \leq x_n\}.$$
Therefore, for any $\eta > 0$ (perhaps depending on n) we have

$$|L_{n,b}(x_n) - \underline{L}_{n,b}(x_n)| = q^{-1}\left|\sum_{i=1}^{q} Z_i\right|$$

$$\leq q^{-1}\left|\sum_{i=1}^{q} Z_i 1\left\{|\bar{X}_n - \theta(P)| > \frac{\eta}{\sqrt{n}}\right\}\right|$$

$$+ q^{-1}\left|\sum_{i=1}^{q} Z_i 1\left\{|\bar{X}_n - \theta(P)| \leq \frac{\eta}{\sqrt{n}}\right\}\right|$$

$$\leq 1\left\{|\bar{X}_n - \theta(P)| > \frac{\eta}{\sqrt{n}}\right\}$$

$$+ q^{-1}\left|\sum_{i=1}^{q} 1\left\{x_n < \sqrt{b}(\bar{X}_{n,b,i} - \theta(P))\right.\right.$$
$$\left.\left. \leq x_n + \eta\sqrt{\frac{b}{n}}\right\}\right|$$

$$+ q^{-1}\left|\sum_{i=1}^{q} 1\left\{x_n - \eta\sqrt{\frac{b}{n}}\right.\right.$$
$$\left.\left. < \sqrt{b}(\bar{X}_{n,b,i} - \theta(P)) \leq x_n\right\}\right|,$$

where the last inequality comes from splitting the event $\{Z_i \neq 0\}$ on the value of $(\bar{X}_n - \theta(P))$. Taking expectations makes it clear that

$$E|L_{n,b}(x_n) - \underline{L}_{n,b}(x_n)| \leq P\left\{|\bar{X}_n - \theta(P)| > \frac{\eta}{\sqrt{n}}\right\}$$

$$+ q^{-1}\sum_{i=1}^{q} P\left\{x_n < \sqrt{b}(\bar{X}_{n,b,i} - \theta(P))\right.$$
$$\left. \leq x_n + \eta\sqrt{\frac{b}{n}}\right\}$$

$$+ q^{-1}\sum_{i=1}^{q} P\left\{x_n - \eta\sqrt{\frac{b}{n}}\right.$$
$$\left. < \sqrt{b}(\bar{X}_{n,b,i} - \theta(P)) \leq x_n\right\}$$

$$= P_I + P_{II} + P_{III}.$$

Now, from results of Heinrich (1982, 1984), we have for $x_n = o(b^{1/2})$ that

$$\frac{1 - P\{\sqrt{b}(\bar{X}_b - \theta(P)) \leq x_n\}}{1 - \Phi(x_n)} = e_b(x_n), \qquad (10.55)$$

where

$$e_b(x_n) = \exp\left(\frac{x_n^3 k_3}{6\sqrt{b}} + o\left(\frac{x_n^3}{\sqrt{b}}\right)\right)\left(1 + O\left(\frac{1 + x_n}{\sqrt{b}}\right)\right),$$

and $k_3 = E(X_1 - \theta(P))^3/(Var(X_1))^{3/2}$ is the skewness of X_1. Thus we have, uniformly in i,

$$\frac{1 - P\{\sqrt{b}(\bar{X}_{n,b,i} - \theta(P)) \leq x_n\}}{1 - \Phi(x_n)} = e_b(x_n),$$

and we can use (10.55) to bound P_{II} and P_{III}. For notational convenience, define

$$z_n = x_n + \eta\sqrt{\frac{b}{n}}.$$

Clearly,

$$\begin{aligned}P_{II} &= (1 - \Phi(x_n))e_b(x_n) - (1 - \Phi(z_n))e_b(z_n) \\ &\leq (\Phi(z_n) - \Phi(x_n))e_b(x_n)(1 + o(1)) \\ &= O\left(\sqrt{\frac{b}{n}}\eta\phi(x_n)\right),\end{aligned} \qquad (10.56)$$

where the last equality comes from a Taylor expansion about x_n, and the monotonicity of $\phi(\cdot)$. Note that we have used the facts

$$e_b(x_n) \leq e_b(z_n)(1 + o(1)), \quad \text{and} \quad e_b(x_n) = O(1),$$

since $x_n = O(b^{1/6})$ by assumption. The P_{III} term is handled similarly.

Using (10.55) again, with n replacing b, we get a bound for the P_I-term as well:

$$\begin{aligned}P_I &\leq (1 - \Phi(\eta))e_n(\eta) \\ &= O\left(\frac{\phi(\eta)}{\eta}e_n(\eta)\right).\end{aligned} \qquad (10.57)$$

Now let

$$\eta = x_n + \sqrt{\ln\frac{n}{b}},$$

and note that, under the theorem's assumptions, $e_n(\eta) = O(1)$. An easy calculation gives

$$P_I = O\left(\sqrt{\frac{b}{n}}\frac{\phi(x_n)}{\eta}\right) = O(P_{II}),$$

10.6 Moderate Deviations in Subsampling Distribution Estimation

where
$$P_{II} = O\left(\sqrt{\frac{b}{n}}(x_n + \sqrt{\ln\frac{n}{b}})\phi(x_n)\right).$$

The above bounds, in conjunction with Markov's inequality, imply that
$$L_{n,b}(x_n) = \underline{L}_{n,b}(x_n) + O_P\left(\sqrt{\frac{b}{n}}(x_n + \sqrt{\ln\frac{n}{b}})\phi(x_n)\right)$$

and this proves the theorem. ∎

The following theorem summarizes our findings regarding moderate deviations in the subsampling distribution $L_{n,b}(x_n)$.

Theorem 10.6.2. *Under the conditions of Theorem 10.6.1, we have*
$$L_{n,b}(x_n) = J_n(x_n, P) + O\left(\frac{\phi(x_n)}{x_n}\left(e^{x_n^3 k_3/(6\sqrt{b})} - 1\right)\right)$$

$$+ O_P\left(\sqrt{\frac{b}{n}\frac{\phi(x_n)}{x_n}}\right) + O_P\left(\sqrt{\frac{b}{n}}(x_n + \sqrt{\ln\frac{n}{b}})\phi(x_n)\right), \quad (10.58)$$

where k_3 is the skewness of X_1.

Proof. By Lemma 10.6.1, we have that
$$\underline{L}_{n,b}(x_n) - J_b(x_n, P) = \sqrt{1 - J_b(x_n, P)}\, O_P\left(\sqrt{\frac{b}{n}}\right)$$
$$= \sqrt{(1 - \Phi(x_n))O(1)}\, O_P\left(\sqrt{\frac{b}{n}}\right)$$
$$= O_P\left(\sqrt{\frac{b}{n}\frac{1}{\sqrt{x_n}}}\exp\left(-\frac{x_n^2}{4}\right)\right),$$

where the second equality comes from the results of Heinrich (1982, 1984) and the fact that $x_n = O(b^{1/6})$, and the third equality uses the fact that $1 - \Phi(x_n) = O(\phi(x_n)/x_n)$. Now, with $x_n = O(b^{1/6})$ and $b = o(n)$, the results of Heinrich (1982, 1984) also imply that
$$\frac{1 - J_n(x_n, P)}{1 - \Phi(x_n)} = \frac{1 - P\{\sqrt{n}(\bar{X}_n - \theta(P)) \le x_n\}}{1 - \Phi(x_n)} = e_n(x_n)$$

$$= \exp\left(\frac{x_n^3 \gamma}{6\sqrt{n}} + o\left(\frac{x_n^3}{\sqrt{n}}\right)\right)\left(1 + O\left(\frac{1 + x_n}{\sqrt{n}}\right)\right) = O(1). \quad (10.59)$$

Finally, combining equation (10.59) with the results of Theorem 10.6.1 and Lemma 10.6.1, gives

$$L_{n,b}(x_n) - J_n(x_n, P) = L_{n,b}(x_n) - \underline{L}_{n,b}(x_n)$$
$$+ \underline{L}_{n,b}(x_n) - J_b(x_n, P)$$
$$+ J_b(x_n, P) - J_n(x_n, P)$$
$$= O_P\left(\sqrt{\frac{b}{n}}(x_n + \sqrt{\ln\frac{n}{b}})\exp\left(-\frac{x_n^2}{2}\right)\right)$$
$$+ O_P\left(\sqrt{\frac{b}{n}}\frac{1}{\sqrt{x_n}}\exp\left(-\frac{x_n^2}{4}\right)\right)$$
$$+ (1 - \Phi(x_n))(e_b(x_n) - e_n(x_n));$$

note that $e_b(x_n) - e_n(x_n) = O(e_b(x_n) - 1)$, and the proof is completed using the definition of $e_b(x_n)$. ∎

Finally, note that our previous results are valid as stated even if x_n is a constant, in which case the following easy corollary obtains.

Corollary 10.6.1. *Under the same conditions as Theorem 10.6.1, but in addition assuming that $x_n = x$ (a constant), we have*

$$L_{n,b}(x) = J_n(x, P) + O\left(\sqrt{\frac{1}{b}}\right) + O_P\left(\sqrt{\frac{b}{n}\ln\frac{n}{b}}\right) \qquad (10.60)$$

uniformly in x.

It is possible that a separate argument, specific to the case $x_n = $ constant, may improve the O_P term above by removing the logarithmic term. Note that the $O\left(\sqrt{1/b}\right)$ term can not be improved as it is nothing other than the Berry-Esseen rate in the Central Limit Theorem, i.e., the rate in the approximation of $\Phi(x)$ (or of $J_n(x, P)$) by $J_b(x, P)$.

Nevertheless, if we assume that x_n diverges to infinity (even at a slow, logarithmic rate), i.e., a *bona fide* moderate deviation setup occurs, then Theorem 10.6.2 has the following important specialization exemplifying the moderate deviation property of the subsampling distribution $L_{n,b}(x_n)$. Corollary 10.6.2 may be compared with an analogous result of Jing (1997) on the moderate deviation property of the block-bootstrap distribution estimator.

Corollary 10.6.2. *Under the same conditions as Theorem 10.6.1, but in addition assuming that $x_n \geq \sqrt{2\ln(n/b)}$, we have*

$$L_{n,b}(x_n) = J_n(x_n, P) + O_P\left(\sqrt{\frac{b\,\phi(x_n)}{n\,x_n}}\right). \qquad (10.61)$$

To give an example, equation (10.61) holds true if we choose $x_n = n^\beta$, for some $\beta \in (0, 1/6)$, with the concurrent choice $b = n^\varsigma$, for $\varsigma \in [6\beta, 1)$.

Remark 10.6.1. Note that the assumption of m-dependence of the sequence $\{X_t\}$ used in connection with moderate deviations results could be relaxed to a weaker mixing condition if/when stronger versions of probabilistic results such as Heinrich's (1982, 1984) are developed.

Remark 10.6.2. Finally, note that if

$$\sum_{k=-m}^{m} Cov(X_0, X_k) = \sigma_m^2 \neq 1,$$

then all results of this section hold true by substituting $\Phi(x_n/\sigma_m)$ instead of $\Phi(x_n)$, and $\phi(x_n/\sigma_m)/\sigma_m$ instead of $\phi(x_n)$ throughout.

10.7 Conclusions

This chapter's results are summarized epigrammatically as follows:

i. The one-sided subsampling distribution estimator of the studentized sample mean of i.i.d. data and/or mixing data cannot be made higher-order accurate by careful choice of b; the two-sided subsampling distribution is more accurate than the one-sided but cannot outperform the normal approximation.
ii. Extrapolation of subsampling distributions for the sample mean of i.i.d. data achieves second-order accuracy if (and only if) the finite population correction is taken into account.
iii. Extrapolation of subsampling distributions for general statistics from i.i.d. data certainly improves over each individual subsampling distribution estimator, and may achieve higher-order correctness by again taking into account the finite population correction, that is, using τ_r instead of τ_b in the subsampling estimator.
iv. Knowing a priori the rate of convergence of the sampling distribution of a general statistic to its asymptotic limit is not necessary in order for extrapolation to be performed; the rate of convergence can be estimated via a preliminary round of subsampling, and the estimated rate can be effectively used in the extrapolation.
v. Interpolation of subsampling distributions for the sample mean of strong mixing observations achieves second-order accuracy; this finding extends the i.i.d. result of Booth and Hall (1993). Another such extension is given by the notion of "robust interpolation" (in the i.i.d. case) that achieves higher-order accuracy under the usual Edgeworth conditions, but at the same time maintains asymptotic consistency even if the asymptotic normality assumption breaks down with a limiting stable law being in effect instead.

vi. Extrapolation of subsampling distributions for the sample mean and general statistics calculated from strong mixing data is not second-order correct, but is definitely an improvement over a single subsampling distribution estimator.

vii. Extrapolation of subsampling variance estimators for the sample mean of strong mixing data actually succeeds in achieving the smallest possible asymptotic mean squared error (MSE). In addition, it provides a practical solution to the difficult problem of choosing the block size(s) for variance estimation; specifically, the estimator $\tilde{\sigma}^2_{\hat{m},2\hat{m},n}$ is near-optimal from a practical point of view.

viii. Extrapolation of many (more than two) subsampling distributions for general statistics calculated from strong mixing data can achieve close to "full first-order" accuracy, while at the same time retaining the robustness property of each individual subsampling distribution estimator of being asymptotically consistent even if the assumed Edgeworth-type conditions turn out not to be valid.

ix. Extrapolation of many (more than two) subsampling variance estimators for the sample mean of strong mixing data is possible but its effect is noticeable only in the asymptotic proportionality constants as the asymptotic MSE rate is already minimized by an extrapolation of just two subsampling variance estimators.

x. The subsampling distribution estimator of the sample mean of m-dependent data exemplifies a moderate deviation property; as a consequence, the subsampling distribution estimator is expected to be quite accurate in the tails, which is a very desirable property for the construction of accurate confidence intervals.

11
Subsampling the Mean with Heavy Tails

11.1 Introduction

It has been two decades since Efron (1979) introduced the bootstrap procedure for estimating sampling distributions of statistics based on independent and identically distributed (i.i.d.) observations. While the bootstrap has enjoyed tremendous success and has led to something like a revolution in the field of statistics, it is known to fail for a number of counterexamples. One well-known example is the case of the mean when the observations are heavy-tailed. If the observations are i.i.d. according to a distribution in the domain of attraction of a stable law with index $\alpha < 2$ (see Feller, 1971), then the sample mean, appropriately normalized, converges to a stable law. However, Athreya (1987) showed that the bootstrap version of the normalized mean has a limiting random distribution, implying inconsistency of the bootstrap. An alternative proof of Athreya's result was presented by Knight (1989). Kinateder (1992) gave an invariance principle for symmetric heavy-tailed observations. It has been realized that taking a smaller bootstrap sample size can result in consistency of the bootstrap, but knowledge of the tail index of the limiting law is needed (see Section 1.3 and Athreya, 1985, and Arcones, 1990; also see Wu, Carlstein, and Cambanis, 1993, and Arcones and Giné, 1989).

In this chapter, we describe how the subsampling method can be used to make asymptotically correct inference for the mean when the observations are heavy-tailed. The presentation closely follows Romano and Wolf (1998a).

In Section 11.2, a brief summary of results pertaining to stable distributions is presented. Section 11.3 extends previous theory for constructing confidence intervals for general parameters. In particular, a theorem applying to studentized statistics when the estimator of scale, appropriately standardized, converges to a nondegenerate distribution is presented. This result is needed in Section 11.4, where two concrete approaches for making inference for the mean in the heavy-tailed case are introduced. The first one appeals to the limiting stable distribution when the sample mean is standardized accordingly. This involves knowledge or estimation of the tail index of the underlying distribution. The second approach uses the idea of self-normalizing sums (e.g., Logan et al., 1973), avoiding the explicit estimation of the tail index. Section 11.5 briefly discusses the choice of the block size. Finite sample behavior is examined via a simulation study in Section 11.6. All tables appear at the end of the chapter.

11.2 Stable Distributions

The aim of this section is to give a brief overview of stable distributions. We will focus on results that are necessary for the theory of subsequent sections and the simulation study at the end of the chapter. For a more detailed discussion of stable distributions, see Zolotarev (1986) and Samorodnitsky and Taqqu (1994).

Definition 11.2.1. A random variable X has a *stable* distribution if for any positive numbers A and B, there is a positive number C and a real number D such that

$$\mathcal{L}(AX_1 + BX_2) = \mathcal{L}(CX + D). \tag{11.1}$$

Here, X_1 and X_2 are independent copies of X and $\mathcal{L}(\cdot)$ denotes the distribution of a random variable.

Note that when talking about stable distributions, we implicitly exclude trivial point masses. In case (11.1) holds with $D = 0$, the random variable X is called *strictly stable*. If X is stable and X and $-X$ have the same distribution, X is called *symmetric stable*. Clearly, a symmetric stable random variable is also strictly stable.

Fact 11.2.1. *If X is a stable random variable, there exists a real number $\alpha \in (0, 2]$ such that the constant C in (1.1) satisfies*

$$C^\alpha = A^\alpha + B^\alpha. \tag{11.2}$$

A proof can be found in Feller (1971, Section VI.1). The number α is called the *index of stability* or *characteristic exponent*. A stable random variable X with index α is also called α-*stable*. For example, one can easily see that a Gaussian random variable is stable with index $\alpha = 2$.

It is important to note that only stable random variables can serve as nondegenerate limiting distributions of sums of i.i.d. random variables.

Fact 11.2.2. *A random variable X is stable if and only if it has a domain of attraction; that is, there exists a sequence of i.i.d. random variables $Z_1, Z_2, Z_3, \ldots$ and sequences of positive numbers $\{d_n\}$ and real numbers $\{a_n\}$ such that:*

$$\frac{Z_1 + Z_2 + \cdots + Z_n}{d_n} + a_n \xrightarrow{\mathcal{L}} X. \tag{11.3}$$

Here, $\xrightarrow{\mathcal{L}}$ denotes convergence in distribution. A proof can be found in Feller (1971, Chapter VI), for example. In general, if X is α-stable, then $d_n = n^{1/\alpha} L(n)$, where $L(\cdot)$ is a slowly varying function at infinity. Such a function satisfies $\lim_{x \to \infty} L(tx)/L(x) = 1$ for all $t > 0$. In the standard case, $d_n = n^{1/\alpha}$ simply, and the Z_i are said to belong to the *normal* domain of attraction of X. For example, if the Z_i have finite variance, then $d_n = n^{1/2}$ and they belong to the normal domain of attraction of a Gaussian random variable. An example of a nonstandard situation is when the Z_i have density $f(x) = 1/|x|^3$, for $|x| \geq 1$. In that case it is known (e.g., Romano and Siegel, 1986, Example 5.47) that $(Z_1 + \cdots + Z_n)/[n \log(n)]^{1/2} \xrightarrow{\mathcal{L}} N(0, 1)$. Hence, $d_n = n^{1/2} \log(n)^{1/2}$. Note that $\log(\cdot)^{1/2}$ varies slowly at infinity. For a broader discussion of domains of attraction of stable distributions, see Feller (1971, Section XVII.5).

While it is known that the densities of α-stable random variables exist and are continuous, they can only be written down in closed form in some special cases (see below). However, formulas for their characteristic functions are available.

Fact 11.2.3. *If a random variable X is stable, then there exist parameters $0 < \alpha \leq 2$, $\sigma \geq 0$, $-1 \leq \beta \leq 1$, and μ such that its characteristic function is given by*

$$E[(\exp i\theta X)] = \begin{cases} \exp(-\sigma^\alpha |\theta|^\alpha (1 - i\beta(\text{sign } \theta) \tan \frac{\pi \alpha}{2}) + i\mu\theta) & \text{if } \alpha \neq 1, \\ \exp(-\sigma |\theta| (1 + i\beta \frac{2}{\pi}(\text{sign } \theta) \ln |\theta|) + i\mu\theta) & \text{if } \alpha = 1. \end{cases} \tag{11.4}$$

The parameter α is the index of stability and the parameters σ, β, and μ are unique (β is irrelevant if $\alpha = 2$).

It follows from (11.4) that $\alpha = 1$ represents a special case. Due to the logarithmic factor $\ln |\theta|$, it is a more difficult case as well. In discussions of stable distributions, $\alpha = 1$ often has to be treated separately. For applications in later sections of the chapter, we will be interested in the case $1 < \alpha \leq 2$ only, as for $\alpha \leq 1$ the mean is infinite (see Fact 11.2.6).

Since (11.4) is characterized by four parameters, the corresponding stable distribution is usually denoted by $S_\alpha(\sigma, \beta, \mu)$. Here, μ is the location parameter, σ the scale parameter, and β the skewness parameter. These names are due to the following properties (Samorodnitsky and Taqqu, 1994).

i. If $X \sim S_\alpha(\sigma, \beta, \mu)$ and a is a constant, then $X + a \sim S_\alpha(\sigma, \beta, \mu + a)$.
ii. If $X \sim S_\alpha(\sigma, \beta, \mu)$ and a is a nonzero constant, then for $\alpha \neq 1$, $aX \sim S_\alpha(|a|\sigma, \beta, a\mu)$.
iii. $X \sim S_\alpha(\sigma, \beta, \mu)$ is symmetric if and only if $\beta = 0$ and $\mu = 0$. It is symmetric about μ if and only if $\beta = 0$.
iv. $X \sim S_\alpha(\sigma, \beta, \mu)$ is strictly stable if and only if $\mu = 0$, as long as $\alpha \neq 1$.

The special scenarios where the densities of stable distributions are known in closed form are $\alpha = 2$, $\alpha = 1$, and $\alpha = 0.5$. For $\alpha = 2$, $S_2(\sigma, 0, \mu) = N(\mu, 2\sigma^2)$. Note the additional factor of 2 in the variance of the Gaussian distribution. As said before, when $\alpha = 2$ the value of β does not matter, so it is usually set equal to zero. For $\alpha = 1$, $S_1(\sigma, 0, \mu)$ is the Cauchy distribution with density $\sigma/[\pi((x-\mu)^2) + \sigma^2]$. Finally, for $\alpha = 0.5$, $S_{0.5}(\sigma, 1, \mu)$ is the Lévy distribution with density $(\sigma/2\pi)^{1/2} 1/(x-\mu)^{3/2} \exp[-\sigma/2(x-\mu)]$, for $x > 0$.

Stable random variables have a characteristic tail behavior. In the Gaussian case $\alpha = 2$ and with $\mu = 0$ (Feller, 1957),

$$P(X < -\lambda) = P(X > \lambda) \sim \frac{1}{2\sqrt{\pi}\lambda} \exp(-\lambda^2/(4\sigma^2)) \quad \text{as } \lambda \to \infty.$$

However, for $\alpha < 2$ the tail properties behave like a power of $-\alpha$ (Samorodnitsky and Taqqu, 1994).

Fact 11.2.4. *Let $X \sim S_\alpha(\sigma, \beta, \mu)$ with $0 < \alpha < 2$. Then,*

$$\lim_{\lambda \to \infty} \lambda^\alpha P(X > \lambda) = C_\alpha \frac{1+\beta}{2} \sigma^\alpha,$$
$$\lim_{\lambda \to \infty} \lambda^\alpha P(X < -\lambda) = C_\alpha \frac{1-\beta}{2} \sigma^\alpha, \quad (11.5)$$

where C_α is a constant only depending on α.

It is not surprising that the domain of attraction of a stable law is characterized by the tail behavior as well.

Fact 11.2.5. *Let X be a random variable with distribution function F. Then, the distribution of X belongs to the domain of attraction of a stable law with tail index $0 < \alpha < 2$ if and only if the following two conditions are satisfied:*

i. *The tail sum varies regularly with exponent $-\alpha$, that is,*

$$P(|X| > x) = 1 - F(x) + F(-x) = x^{-\alpha} \tilde{L}(x),$$

where $\tilde{L}(\cdot)$ is a slowly varying function at infinity.

ii. *The tails are balanced; that is, as $x \to \infty$*

$$\frac{1 - F(x)}{1 - F(x) + F(-x)} \to p, \quad \frac{F(-x)}{1 - F(x) + F(-x)} \to q,$$

where $0 \leq p \leq 1$ and $p + q = 1$.

Proof. See Feller (1971, Section XVII.5, Corollary 2). ∎

Since $E|X|^p = \int_0^\infty P(|X|^p > \lambda)\,d\lambda$, the following is an immediate consequence of Fact 11.2.5.

Fact 11.2.6. *Let X have distribution in the domain of attraction of a stable law with tail index $0 < \alpha < 2$. Then,*

$$E|X|^p < \infty \quad \text{for any} \quad 0 < p < \alpha,$$
$$E|X|^p = \infty \quad \text{for any} \quad p \geq \alpha.$$

This implies that for $\alpha < 2$ an α-stable random variable does not have an α-th moment, but all smaller moments exist. In particular, apart from Gaussian distributions, stable random variables have infinite variance. The rest of the chapter will focus on distributions in the domain of attraction of a stable law with tail index $\alpha > 1$, where the mean is finite.

11.3 Extension of Previous Theory

The basic subsampling methodology in the context of i.i.d. observations was discussed in detail in Chapter 2. However, it turns out that we need an extension of the theory presented there, giving a more general result for studentized roots.

We often subsample the root $\tau_n(\hat{\theta}_n - \theta(P))$. But for some applications the use of a studentized root $\tau_n(\hat{\theta}_n - \theta(P))/\hat{\sigma}_n$, where $\hat{\sigma}_n$ is some nonnegative estimate of scale, is preferable. Define $J_n^*(P)$ to be the sampling distribution of $\tau_n(\hat{\theta}_n - \theta(P))/\hat{\sigma}_n$ based on a sample of size n from P. Also, define the corresponding cumulative distribution function

$$J_n^*(x, P) = Prob_P\{\tau_n(\hat{\theta}_n - \theta(P))/\hat{\sigma}_n \leq x\}. \tag{11.6}$$

Subsampling for scaled or studentized statistics has previously been considered in the case where $\hat{\sigma}_n$ converges to a positive constant in probability (see Subsection 2.5.1). The generalization required must allow for $\hat{\sigma}_n$, appropriately standardized, to converge to a nondegenerate distribution.

The essential assumption needed to construct asymptotically valid confidence regions for $\theta(P)$ now becomes slightly more involved than for the nonstudentized case.

Assumption 11.3.1. *$J_n^*(P)$ converges weakly to a limit law $J^*(P)$. In addition, $a_n(\hat{\theta}_n - \theta(P))$ converges weakly to V, and $d_n\hat{\sigma}_n$ converges weakly to W, for positive sequences $\{a_n\}$ and $\{d_n\}$ satisfying $\tau_n = a_n/d_n$. Here, V and W are two random variables, where W does not have positive mass at zero.*

The subsampling method for the studentized case is described in Subsection 2.5.1. In particular, the subsampling approximation to (11.6) is defined by

$$L_{n,b}^*(x) = N_n^{-1} \sum_{i=1}^{N_n} 1\{\tau_b(\hat{\theta}_{n,b,i} - \hat{\theta}_n)/\hat{\sigma}_{n,b,i} \leq x\}. \quad (11.7)$$

Subsection 2.5.1 discusses the case where $\hat{\sigma}_n$ converges in probability to a positive constant. This encompasses the t-statistic in the context of i.i.d. observations with a finite second moment. In that case, the estimate of scale $\hat{\sigma}_n$ is equal to the sample standard deviation, which converges in probability to the true standard deviation. This no longer holds for heavy-tailed observations and it turns out we need the extension, handling the case of $d_n \hat{\sigma}_n$ converging to a nondegenerate distribution.

Theorem 11.3.1. *Assume Assumption 11.3.1. Also, assume $a_b/a_n \to 0$, $\tau_b/\tau_n \to 0$, $b/n \to 0$, and $b \to \infty$ as $n \to \infty$.*

i. *If x is a continuity point of $J^*(\cdot, P)$, then $L_{n,b}^*(x) \to J^*(x, P)$ in probability.*
ii. *If $J^*(\cdot, P)$ is continuous, then $\sup_x |L_{n,b}^*(x) - J^*(x, P)| \to 0$ in probability.*
iii. *For $\alpha \in (0,1)$, let*

$$c_{n,b}^*(1-\alpha) = \inf\{x : L_n^*(x) \geq 1-\alpha\}.$$

Correspondingly, define

$$c^*(1-\alpha, P) = \inf\{x : J^*(x, P) \geq 1-\alpha\}.$$

If $J^*(\cdot, P)$ is continuous at $c^*(1-\alpha, P)$, then

$$Prob_P\{\tau_n(\hat{\theta}_n - \theta(P))/\hat{\sigma}_n \leq c_{n,b}^*(1-\alpha)\} \to 1-\alpha \text{ as } n \to \infty. \quad (11.8)$$

Thus, the asymptotic coverage probability under P of the interval $I_1 = [\hat{\theta}_n - \hat{\sigma}_n \tau_n^{-1} c_{n,b}^*(1-\alpha), \infty)$ is the nominal level $1-\alpha$.

Proof. To prove (i), note that

$$L_{n,b}^*(x) = N_n^{-1} \sum_{i=1}^{N_n} 1\{\tau_b(\hat{\theta}_{n,b,i} - \hat{\theta}_n)/\hat{\sigma}_{n,b,i} \leq x\}$$

$$= N_n^{-1} \sum_{i=1}^{N_n} 1\{\tau_b(\hat{\theta}_{n,b,i} - \theta(P))/\hat{\sigma}_{n,b,i} \leq x + \tau_b(\hat{\theta}_n - \theta(P))/\hat{\sigma}_{n,b,i}\}.$$

(11.9)

We want to show that the terms $\tau_b(\hat{\theta}_n - \theta(P))/\hat{\sigma}_{n,b,i}$ are negligible in the last equation. To this end, for $t > 0$, let

$$R_n(t) = N_n^{-1} \sum_{i=1}^{N_n} 1\{\tau_b(\hat{\theta}_n - \theta(P))/\hat{\sigma}_{n,b,i} \leq t\}$$

$$= N_n^{-1} \sum_{i=1}^{N_n} 1\{d_b \hat{\sigma}_{n,b,i} \geq d_b \tau_b(\hat{\theta}_n - \theta(P))/t\}$$

$$= N_n^{-1} \sum_{i=1}^{N_n} 1\{d_b \hat{\sigma}_{n,b,i} \geq a_b(\hat{\theta}_n - \theta(P))/t\}.$$

Here, we are assuming without loss of generality that both the sequences a_n and b_n are nonnegative. By Assumption 11.3.1 and $a_b/a_n \to 0$, we have that for any $\delta > 0$, $a_b(\hat{\theta}_n - \theta(P)) \leq \delta$ with probability tending to one. Therefore, with probability tending to one,

$$R_n(t) \geq N_n^{-1} \sum_{i=1}^{N_n} 1\{d_b \hat{\sigma}_{n,b,i} \geq \delta/t\}.$$

We need to consider the case $t > 0$ only, as the scale estimates $\hat{\sigma}_{n,b,i}$ are nonnegative. Due to the usual subsampling argument of Theorem 2.2.1, $N_n^{-1} \sum_{i=1}^{N_n} 1\{d_b \hat{\sigma}_{n,b,i} \geq \delta/t\}$ converges in probability to $P(W \geq \delta/t)$, as long as δ/t is a continuity point of W. Hence, we can make sure that $R_n(t)$ is arbitrarily close to one by choosing δ small enough; remember we assume that W does not have positive mass at zero. In other words, for any $t > 0$, we have $R_n(t) \to 1$ in probability. Let us now rewrite (11.9) in the following way:

$$L_{n,b}^*(x) = N_n^{-1} \sum_{i=1}^{N_n} 1\{\tau_b(\hat{\theta}_{n,b,i} - \theta(P))/\hat{\sigma}_{n,b,i} \leq x + \tau_b(\hat{\theta}_n - \theta(P))/\hat{\sigma}_{n,b,i}\}$$

$$\leq N_n^{-1} \sum_{i=1}^{N_n} 1\{\tau_b(\hat{\theta}_{n,b,i} - \theta(P))/\hat{\sigma}_{n,b,i} \leq x + t\} + (1 - R_n(t)),$$

for any positive number t. The last inequality follows because the i-th term in (11.9) is less than or equal to

$$1\{\tau_b(\hat{\theta}_{n,b,i} - \theta(P))/\hat{\sigma}_{n,b,i} \leq x + t\} + 1\{\tau_b(\hat{\theta}_n - \theta(P))/\hat{\sigma}_{n,b,i} > t\}; \quad (11.10)$$

then, sum over all i. We have seen that $(1 - R_n(t)) \to 0$ in probability and hence by a standard subsampling argument again we get, for any $\epsilon > 0$, $L_{n,b}^*(x) \leq J^*(x + t, P) + \epsilon$ with probability tending to one, provided that $x + t$ is a continuity point of $J^*(\cdot, P)$. Letting t tend to zero shows that $L_{n,b}^*(x) \leq J^*(x, P) + \epsilon$ with probability tending to one. A similar argument leads to $L_{n,b}^*(x) \geq J^*(x, P) - \epsilon$ with probability tending to one. Since ϵ is arbitrary, this implies $L_{n,b}^*(x) \to J^*(x, P)$ in probability, and thus we have proved (i).

The proofs of (ii) and (iii) given (i) are very similar to the proofs of (ii) and (iii) given (i) in Theorem 2.2.1, and thus are omitted. ∎

Remark 11.3.1. Because $\binom{n}{b}$ may be large, $L_{n,b}$ may be difficult to compute. Instead, a stochastic approximation can be employed (see Section 2.4).

11.4 Subsampling Inference for the Mean with Heavy Tails

Suppose the X_i are i.i.d. univariate random variables in the domain of attraction of a stable law with index $1 < \alpha < 2$. It follows that the underlying distribution P possesses a finite mean but that its variance is infinite (see Section 11.2). The goal is to find a confidence interval for $\theta(P) = E(X_i)$. Our choice for the estimator is the sample mean $\hat{\theta}_n = \bar{X}_n = n^{-1} \sum_{i=1}^{n} X_i$. The subsampling methodology requires a normalization resulting in a nondegenerate limiting distribution. In this section, we will discuss two possible approaches, one that relies on a stable limiting law and another one which uses self-normalizing sums.

11.4.1 Appealing to a Limiting Stable Law

In case the underlying distribution belongs to the normal domain of attraction of a stable law, we can make use of the following result.

Proposition 11.4.1. *Assume $X_1, X_2, \ldots$ is a sequence of random variables in the normal domain of attraction of a stable law with index of stability $1 < \alpha \leq 2$. Denote the common mean by $\theta(P)$.*

Then, $n^{-1/\alpha}(X_1 + \cdots + X_n - n\theta(P)) = n^{1-1/\alpha}(\bar{X}_n - \theta(P))$ converges weakly to an α-stable distribution with mean zero.

Proof. The proof follows immediately from the CLT when $\alpha = 2$. For the case of $1 < \alpha < 2$, it is a consequence of Theorem 3 in Section XVII.5 of Feller (1971). ∎

It might be possible to use Proposition 11.4.1 to make some semistandard nonparametric inference for $\theta(P)$ by constructing a consistent estimator of the limiting distribution of the properly standardized sample mean. Note, however, that this distribution depends on three unknown parameters, namely α, σ, and β (see Fact 11.2.3). Consistently identifying the limiting distribution, unfortunately, solves only part of the problem. To find a confidence interval for $\theta(P)$, the quantiles of a stable law have to be obtained. Since the distribution function of most stable laws is not available in closed from, one has to resort to tables, which are necessarily limited to a finite number of parameter combinations, or one has to appeal to cumbersome simulation techniques.

11.4 Subsampling Inference for the Mean with Heavy Tails

On the other hand, the subsampling technique only hinges on the index of stability α, since the normalizing constants are given by $\tau_n = n^{1-1/\alpha}$ and $\tau_b = b^{1-1/\alpha}$, respectively. Let $\hat{\alpha}_n = \hat{\alpha}_n(X_1, \ldots, X_n)$ denote an estimator of α based on the segment $X_1, \ldots, X_n$. This notation includes the (rare) case where α is known, since $\hat{\alpha}_n \equiv \alpha$ is a valid estimator. The idea is to use the estimated index instead of the (unknown) true index in the subsampling method. Then, the subsampling approximation to $J_n(x, P)$ is defined by

$$L^{\hat{\alpha}}_{n,b}(x) = N_n^{-1} \sum_{i=1}^{N_n} 1\{b^{1-1/\hat{\alpha}_n}(\bar{X}_{n,b,i} - \bar{X}_n) \leq x\}, \quad (11.11)$$

where $\bar{X}_{n,b,i}$ is the sample mean of the i-th subsample of size b

The following Proposition shows that the approximation (11.11) is still asymptotically valid, as long as $\hat{\alpha}_n$ converges to α in probability at rate $\log n$.

Proposition 11.4.2. *Assume $\{X_i\}$ is a sequence of i.i.d. random variables in the normal domain of an α-stable law with index $1 < \alpha \leq 2$. Denote the common mean by $\theta(P)$. Also assume that $b \to \infty$ and $b/n \to 0$ as $n \to \infty$. Finally, assume that $\hat{\alpha}_n = \alpha + o_P((\log n)^{-1})$*

Then, the conclusions of Theorem 2.2.1 and Corollary 2.4.1 still hold if $L_{n,b}(x)$ is replaced by $L^{\hat{\alpha}}_{n,b}(X)$, and if τ_n and τ_b are replaced by $n^{1-1/\hat{\alpha}_n}$ and $b^{1-1/\hat{\alpha}_n}$, respectively.

Proof. The proof is a consequence of Theorem 8.3.1.

Therefore, applying the subsampling method only requires a $\log n$ consistent estimator for the tail index α. Several such estimators are known, among them the Pickands (1975), Hill (1975), and de Haan and Resnick (1980) estimators. Tail index estimators typically are based upon a number q of extreme order statistics. Asymptotic consistency of the estimators requires that $q \to \infty$ but $q/n \to 0$ as $n \to \infty$.

Perhaps the most widely used of these estimators is the Hill estimator, defined in the following way. Let $X_{(i)}$ denote the i-th largest value of the sample $X_1, \ldots, X_n$, so that $X_{(1)} \geq X_{(2)} \geq \ldots \geq X_{(n)}$. Then, the Hill estimator based on the q upper order statistics is given by

$$H_{q,n} \equiv \frac{1}{q} \sum_{i=1}^{q} \log\left(\frac{Y_{(i)}}{Y_{(i+1)}}\right).$$

If $q = q(n)$ satisfies $q \to \infty$ and $q/n \to 0$ as $n \to \infty$, $H_{q,n}$ converges in probability to α^{-1}. Consequently, $H_{q,n}^{-1}$ is a consistent estimator of α. Under additional assumptions, such as second-order regular variation,

$$q^{1/2}(H_{n,q} - \alpha^{-1}) \stackrel{\mathcal{L}}{\Longrightarrow} N(0, \alpha^{-2})$$

(see Hall, 1982).

A practical problem with the Hill estimator, but also the other two aforementioned estimators, lies in choosing the number of order statistics q for a finite sample size n. There do not appear to exist any reliable guidelines that work well in generality. These estimators are based on the fact that in the tails the distribution function behaves like a power of $-\alpha$ (see Fact 11.2.5). Indeed, they work best if the "behaves like" can be strengthened to a "behaves exactly like" as in the case of a Pareto distribution, which has distribution function $F(x) = 1 - kx^{-\alpha}$, for $x > c > 0$. For i.i.d. observations from a Pareto distribution, the Hill estimator is known to be very insensitive, meaning that its dependence on the number of order statistics q is very small (see Mittnik et al., 1996, and Resnick, 1997). Note that for $\alpha < 2$, the Pareto distribution belongs to the domain of attraction of a stable distribution with index α. However, if the observations are i.i.d. from a stable distribution, then the Hill estimator becomes extremely sensitive to the choice of q and is therefore impractical for reasonable sample sizes (again see Mittnik, Paolella, and Rachev, 1996, and Resnick, 1997). The same holds true for the Pickands and de Haan and Resnick estimators. Mittnik, Paolella, and Rachev (1996) devise a new estimator that is custom-tailored to stable distributions and show that it works well in that scenario. However, it is reasonable to expect that it will not perform as well in general, such as for observations having a Pareto distribution.

At this point, an alternative tail index estimator, based on the subsampling technique, is proposed. As noticed before, when the underlying distribution is in the normal domain of attraction of a stable law, the proper normalizing constant is $n^{1-1/\alpha}$, so that the rate of convergence is $\beta \equiv 1 - 1/\alpha$. Chapter 8 discusses consistent subsampling estimators for β. These estimators depend on a number I of subsampling distributions with different block sizes, where $I \geq 2$, and on a number J of estimated quantiles, where $J \geq 1$. Assume that Assumption 2.2.1 is satisfied for $\tau_n = n^\beta$ for some positive β. The goal is to consistently estimate β. For this purpose, we employ the estimator $\beta_{I,J}$ of Theorem 8.2.2.

For the application of the mean in the heavy-tailed context, it was seen that $\beta = 1 - 1/\alpha$, as long as the underlying distribution is in the normal domain of attraction of a stable law with tail index α. Hence, an obvious estimator of α is given by

$$\hat{\alpha}_{I,J} = 1/(1 - \beta_{I,J}). \tag{11.12}$$

It immediately follows that, under the conditions of Theorem 8.2.2, we have $\hat{\alpha}_{I,J} = \alpha + o_P((\log n)^{-1})$. Therefore, Proposition 11.4.2 allows for an application of the subsampling method in conjunction with the tail index estimator $\hat{\alpha}_{I,J}$ to construct asymptotically valid confidence intervals for the mean.

The application of the estimators $\beta_{I,J}$ and $\hat{\alpha}_{I,J}$ requires choices of I, J, the block sizes b_i, and the quantiles t_{2j} and t_{2j-1}. These choices are discussed in general in Chapter 8. We will be specific about our choices in the context of our simulation studies in Section 11.6.

Remark 11.4.1. As an alternative to subsampling, the bootstrap with resample size $m < n$ may be considered. However, as in the case of subsampling, the proper standardization of the bootstrap distribution depends on the underlying tail index. In the case where the standardization is known, Athreya (1985) showed that the bootstrap distribution converges to the right limit in probability, given that $m \to \infty$ and $m/n \to 0$. Arcones and Giné (1989) strengthened this result to almost sure convergence under the condition of $m(\log \log n)/n \to 0$. Neither paper discusses the validity of the bootstrap approach in conjunction with an estimated rate. Also, no suggestion of how to pick the resample size m in practice is made.

Wu, Carlstein, and Cambanis (1993) introduced an averaged bootstrap that overcomes the randomness in the limiting law of the bootstrap with resample size $m = n$. They showed that the averaged bootstrap converges to the correct limit in the case of (exactly) stable data, provided that appropriate sample-based adjustments for scale and skewness are made. However, this approach may not extend to distributions in the domain of attraction of stable laws only.

11.4.2 Using Self-Normalizing Sums

It is well known that if the observations are i.i.d. from a distribution with finite second moment, then the t-statistic

$$T_n = n^{1/2} \frac{\bar{X}_n - \theta(P)}{\hat{\sigma}_n} \tag{11.13}$$

has a limiting standard normal distribution. Here, $\hat{\sigma}_n$ is the square root of the usual unbiased estimate of the variance,

$$\hat{\sigma}_n^2 = \frac{1}{n-1} \sum_{i=1}^{n} (X_i - \bar{X}_n)^2. \tag{11.14}$$

The fact that T_n has, under fairly general conditions, a nondegenerate limiting distribution, even if the underlying distribution has an infinite second moment, makes it a *self-normalizing sum*. The limiting behavior of T_n for heavy-tailed distributions has, among others, been studied by Hotelling (1961), Efron (1969), and Logan et al. (1973). In the paper of Logan et al. (1973), exact densities of the limiting distribution of T_n are derived for the case of the underlying distribution belonging to the domain of attraction of a stable law. It is seen that the density not only depends in a complicated way on α, but it also depends on the tail balance parameters p and q from Fact 11.2.5. Again, this greatly diminishes the appeal of any inference based on explicit estimation of the limiting distribution. On the other hand, the following proposition allows for an easy application of the subsampling method.

Proposition 11.4.3. *Assume $\{X_i\}$ is a sequence of i.i.d. random variables in the domain of attraction of an α-stable law with $1 < \alpha \leq 2$. Denote the common mean by $\theta(P)$. Define $\hat{\theta}_n = \bar{X}_n$, the usual sample mean, and $\hat{\sigma}_n$ the usual sample standard deviation. Also, let $\tau_n = n^{1/2}$ and $\tau_b = b^{1/2}$. Assume that $b \to \infty$ and $b/n \to 0$ as $n \to \infty$.*

Then, the conclusions of Theorem 11.3.1 hold.

Proof. We have to show that the conditions of Assumption 11.3.1 are met. To this end, define

$$U_n = n^{1/2} \frac{\bar{X}_n - \theta(P)}{(n^{-1}\sum_{i=1}^n (X_i - \theta(P))^2)^{1/2}}$$

$$= \frac{V_n}{W_n},$$

where

$$V_n = \frac{X_1 + \cdots + X_n - n\theta(P)}{n^{1/\alpha} L(n)}$$

and

$$W_n = \left(\frac{(X_1 - \theta(P))^2 + \cdots + (X_n - \theta(P))^2}{n^{2/\alpha} L^2(n)} \right)^{1/2}.$$

Here, $L(\cdot)$ is the slowly varying function ensuring that V_n converges to a stable law G (see Section 11.2).

First consider the case where the X_i have a stable distribution. It follows that $L(\cdot) \equiv 1$. Logan et al. (1973) show that in this case (V_n, W_n) has a nondegenerate joint limiting distribution, where the limiting distribution of W_n does not have positive mass at zero. Indeed, the limiting distribution of W_n^2 is a positive stable law with index $\alpha/2$.

It turns out that, in the general case, where the X_i are in the domain of attraction of G, the joint limiting distribution of (V_n, W_n) is identical to that of the stable case (again, see Logan et al., 1973).

By simple algebra, finally

$$T_n = U_n \left(\frac{n-1}{n - W_n^2} \right)^{1/2},$$

where the second term converges to one in probability. Hence the conditions of Assumption 11.3.1 are satisfied. ∎

The power of Proposition 11.4.3 lies in the fact that we always can subsample the t-statistic T_n, regardless of the tail index α of the underlying distribution. Therefore, it is not necessary to know or to estimate α. In addition, this approach is not restricted to distributions in the *normal* domain of attraction of stable laws and therefore it is more general than the method of Subsection 11.4.1.

Remark 11.4.2. Arcones and Giné (1991) showed that bootstrapping the t-statistic with resampling size $m < n$ also gives asymptotically correct results, provided that $m \to \infty$ and $m/n \to 0$ as $n \to \infty$.

11.5 Choice of the Block Size

Since Efron's bootstrap is not consistent for the mean in the heavy-tailed case, the calibration method of Subsection 9.3.1 cannot be used. Therefore, we propose the minimum volatility method of Subsection 9.3.2 to choose the block size b.

The algorithm is now illustrated with the help of two simulated data sets. The goal is to find a confidence interval for the mean of a possibly heavy-tailed distribution, given a sample of i.i.d. observations. (See Section 11.4 for the theory on this application.) First, we generated a data set of size $n = 100$ i.i.d. from a symmetric stable distribution with mean zero and tail index $\alpha = 1.5$. The range of b was chosen as $b_{small} = 4$ and $b_{big} = 40$. We computed symmetric subsampling intervals according to the approaches of Subsection 11.4.1, taking α to be known, and of Subsection 11.4.2, avoiding the explicit estimation of α. Since a stochastic approximation of the kind (2.12) was employed with $N = 1000$, the endpoints of the intervals were smoothed according to Step 2 with $m = 2$. The minimization of the volatility in Step 3 was done using $k = 2$. The results are shown at the top of Figure 11.1. The left plot corresponds to the approach appealing to a limiting stable law, while the right plot corresponds to the self-normalizing approach. The block sizes b chosen by the algorithm are highlighted by a star. The resulting final confidence intervals are included in the plots.

This exercise was repeated for another data set of size $n = 500$ i.i.d. from a symmetric stable distribution with mean zero and tail index $\alpha = 1.7$. The range of b was there chosen as $b_{small} = 4$ and $b_{big} = 100$.

The plots show that, for the two simulated data sets, the self-normalizing approach is somewhat less sensitive to the choice of the block size, that is, the confidence interval endpoints change more slowly as b changes. This behavior is typical and was observed for many other simulations as well.

Remark 11.5.1. Arcones and Giné (1991) considered bootstraping the t-statistic with a smaller bootstrap size $m < n$, which corresponds to the block size b of the subsampling method. They suggested choosing $m = n/(\log \log n)^{1+\delta}$ for some small $\delta > 0$. For example, they used $m = 35$ with $n = 50$, and $m = 65$ with $n = 100$. This seems to correspond to $\delta \approx 0.02$.

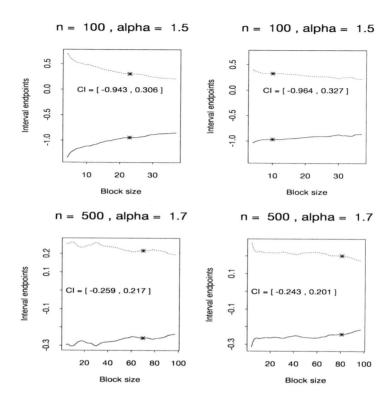

FIGURE 11.1. Illustration of the minimizing confidence interval volatility algorithm for two data sets. The plots on the left correspond to the approach appealing to a stable limit, while the plots on the right correspond to the self-normalizing approach. The block sizes selected by the algorithm are highlighted by a star. The final confidence intervals appear within the plots.

11.6 Simulation Study

The purpose of this section is to shed some light on the small sample performance of the subsampling method by means of a simulation study. In particular, it is of interest to compare the two approaches of Subsections 11.4.1 and 11.4.2. Performance is judged by coverage probabilities of nominal 95% two-sided confidence intervals. We include both equal-tailed and symmetric subsampling intervals in the study. The normalizing constant for the approach of Subsection 11.4.1 depends on the tail index α.

The application of the estimator $\hat{\alpha}_{I,J}$ requires choices of I, J, the block sizes b_i, and the quantiles t_{2j} and t_{2j-1}. We chose $I = 5$, $J = 10$, the t_{2j-1} equally spaced between 0.01 and 0.25 and $t_{2j} = 1 - t_{2j-1}$, for $j = 1\ldots J$. Finally, the block size b_i was chosen as $n^{0.5\gamma_i}$, rounded to the nearest integer. Here, $\gamma_i = 1 + (\log i/I)/(\log 100)$, for $i = 1\ldots I$. For example, with

$I = 5$ and a sample size of $n = 100$, this yields block sizes of 4, 6, 7, 8, and 9.

Two data-generating mechanisms are considered. First, the stable distribution with mean zero, varying tail index parameter α, and varying skewness parameter β. Second, the Pareto distribution with distribution function $P(X \leq x) = 1 - x^{-\alpha}$, for $x > 1$, and varying tail index parameter α. Note that the mean of this distribution is given by $\alpha/(\alpha-1)$. Pareto observations can be easily generated by applying the inverse of the distribution function to Uniform(0,1) observations. For the generation of stable observations, we first used the function $rstab()$ of the statistical package *Splus*. However, we experienced some problems, since, for skewed distributions ($\beta \neq 0$), $rstab()$ does not seem to produce variables with the specified mean. In the end, we used the program *Stable* 2.11 provided by John Nolan at http://www.cas.american.edu/~jpnolan/stable.html.

The simulations use sample size of $n = 100$. The model parameters for block size selection Algorithm 9.3.3 were $b_{small} = 4$, $b_{big} = 30$, $m = 2$, and $k = 2$. Estimated coverage probabilities of nominal 95% confidence intervals are based on 1000 repetitions for each scenario. The results are presented in Table 11.1. SL stands for the approach of Subsection 11.4.1, appealing to a stable limit, while SN stands for the self-normalizing approach of Subsection 11.4.2. The subscripts ET and SYM denote equal-tailed and symmetric intervals, respectively (see Section 11.3). CLT stands for the Central Limit theorem approach, falsely assuming a finite variance.

The results for symmetric, stable observations are overall quite satisfactory, although the difference between equal-tailed and symmetric SN intervals is noteworthy. For skewed, stable observations, coverage decreases with α and this is even more true for Pareto observations. The overall best choice appear to be symmetric SN intervals and while their performance is far from perfect, they present a significant improvement over the CLT intervals. However, it appears that in the context of heavy-tailed observations far bigger sample sizes are needed to achieve overall satisfactory performance as compared to the finite variance case.

11.7 Conclusions

In this chapter, it was demonstrated that the subsampling method can be used to construct asymptotically correct confidence intervals for the mean when the observations are i.i.d. from a distribution with infinite variance. Two different approaches were proposed. The first one is based on the fact that the sample mean, properly standardized, has a limiting stable law, provided that the underlying distribution belongs to the normal domain of attraction of a stable law. This approach has the practical disadvantage that the tail index of the underlying distribution has to be estimated. The second approach consists of subsampling the usual t-statistic, which turns out to be a self-normalizing sum. It is more general, in the sense that it is not restricted to distributions in the normal domain of attraction of a stable law. Moreover, it avoids the problem of having to estimate the tail index. A theorem was stated that shows the validity of this approach, extending the theory of Chapter 2.

By means of a simulation study, the small sample performance was examined. As to be expected, the results vary with the underlying distribution. The second approach to construct symmetric confidence intervals, which was based on subsampling the t-statistic, yielded the best overall results.

11.8 Tables

TABLE 11.1. Estimated coverage probabilities of various nominal 95% subsampling confidence intervals. The sample size is $n = 100$ always. The estimates are based on 1000 replications for each scenario.

Stable observations, $\beta = 0$					
α	SL_{ET}	SL_{SYM}	SN_{ET}	SN_{SYM}	CLT
1.9	0.93	0.92	0.93	0.94	0.94
1.7	0.94	0.95	0.87	0.95	0.94
1.5	0.95	0.94	0.79	0.96	0.92
1.3	0.97	0.96	0.73	0.96	0.98
1.1	0.98	0.98	0.66	0.97	0.98
Stable observations, $\beta = 0.5$					
α	SL_{ET}	SL_{SYM}	SN_{ET}	SN_{SYM}	CLT
1.9	0.93	0.92	0.93	0.94	0.95
1.7	0.94	0.94	0.87	0.95	0.94
1.5	0.94	0.93	0.81	0.94	0.92
1.3	0.89	0.90	0.75	0.89	0.80
1.1	0.52	0.60	0.53	0.59	0.42
Pareto observations					
α	SL_{ET}	SL_{SYM}	SN_{ET}	SN_{SYM}	CLT
1.9	0.82	0.86	0.90	0.92	0.80
1.7	0.82	0.86	0.89	0.91	0.75
1.5	0.74	0.83	0.87	0.88	0.68
1.3	0.61	0.72	0.83	0.82	0.52
1.1	0.41	0.27	0.64	0.61	0.24

12
Subsampling the Autoregressive Parameter

12.1 Introduction

This chapter is concerned with making inference for ρ in the simple AR(1) model

$$X_t = \mu + \rho X_{t-1} + \epsilon_t, \tag{12.1}$$

where $\{\epsilon_t\}$ is a strictly stationary white noise innovation sequence and $\rho \in (-1, 1]$. It is well known that if $|\rho| < 1$, the sequence $\{X_t\}$ is strictly stationary with mean $\mu/(1 - \rho)$. However, for $\rho = 1$ it is a random walk with drift. The parameter ρ can be consistently estimated using ordinary least squares (OLS). Unfortunately, inference for ρ is nontrivial, since the limiting distribution of the estimator depends on the underlying parameters ρ and μ and on the distribution of $\{\epsilon_t\}$. If $|\rho| < 1$, the OLS estimator converges to a normal distribution at rate $n^{1/2}$. For $\rho = 1$ and $\mu = 0$, it converges to a nonstandard distribution at rate n. Finally, for $\rho = 1$ and $\mu \neq 0$, it converges to a normal distribution at rate $n^{3/2}$. To construct an asymptotic confidence interval for ρ, it therefore seems necessary to know which the correct model is. The discontinuity of the limiting distribution causes the bootstrap (based on resampling estimated residuals) to fail (see Basawa et al., 1991).

Note that the situation can be improved somewhat, but not remedied, by basing the inference on the OLS t-statistic. It will converge to a standard normal distribution in case $|\rho| < 1$ and the ϵ_t are i.i.d. or in case $\rho = 1$ and $\mu \neq 0$. On the other hand, if $|\rho| < 1$ and the ϵ_t are uncorrelated only, the limiting distribution is normal with mean zero but unknown variance. Finally, if $\rho = 1$ and $\mu = 0$, the limiting distribution is nonstandard (see Section 12.3 for details).

For this reason, much of the econometrics literature—under the heading of "unit root tests"—has been concerned with simply testing the null hypothesis of $\rho = 1$. One still needs to know whether $\mu = 0$ to derive the sampling distribution of the test statistic under the null. However, it turns that this problem can be avoided by including a time trend in the estimation process, that is, by using OLS to estimate ρ from the following model

$$X_t = \mu + \delta t + \rho X_{t-1} + \epsilon_t. \qquad (12.2)$$

In this case it turns out that the estimator converges to a nonstandard, but fully known distribution at rate n no matter what the value of μ. The foundations of the unit root test literature were laid by Dickey and Fuller (1979) and Phillips and Perron (1988). A nice overview can be found in Chapter 17 of Hamilton (1994).

While testing for $\rho = 1$ is a worthwhile endeavor, it often leaves something to be desired. As Stock (1991) points out, "reporting only unit root tests and point estimates of the largest root is unsatisfying as a description of the data: this fails to convey information about the sampling uncertainty or, more precisely, the range of models (i.e., values of ρ) that are consistent with the observed data."

He proposes a method for finding confidence intervals for ρ based on the related model

$$X_t = \alpha + \delta t + V_t, \quad V_t = \rho V_{t-1} + \epsilon_t. \qquad (12.3)$$

The method relies upon so-called local-to-unity asymptotics, which assume that ρ shrinks to one as the sample size tends to infinity. More specifically,

$$\rho = 1 + \frac{c}{n}, \qquad (12.4)$$

for some constant c. Typically, c is thought to be less than or equal to zero, although the theory also works for positive c. It can be shown that under this model the OLS estimator converges at rate n to a nonstandard distribution that depends on c only (the distribution is a functional of an Ornstein-Uhlenbeck process). Therefore, one can test $H_0 : c = c_0$ for any arbitrary value of c_0 and a confidence interval for c can be obtained as the collection of c_0 values not rejected by the corresponding test. Given the sample size n, one can convert the confidence interval for c to one for ρ using relation (12.4). Alternatively, confidence intervals can be based on the OLS t-statistic in a similar fashion. (See Stock, 1991, for details.)

From simulation studies in Stock (1991), it appears that this method works well if $c = 0$ but gets worse the further c is away from zero. This would imply that the method works well if $\rho = 1$ or if the sample size is small and ρ is very close to one. For any $\rho < 1$, coverage probability of confidence intervals will gradually deteriorate as the sample size increases. Of course, this may be considered a philosophical rather than a practical problem, since in reality one is always faced with a fixed sample size.

It is the aim of this chapter to provide a new way of constructing confidence intervals for ρ. A method is proposed that gives asymptotically correct results for all three cases of model (12.1) *without* knowing anything about the underlying parameters. Moreover, ρ is considered to be fixed rather than a function of the sample size. The exposition of the chapter closely follows Romano and Wolf (1998b).

Section 12.2 provides some extensions of previous theory needed for our application. In Section 12.3, it is described how the extended theory can be applied to construct asymptotically correct confidence intervals for ρ. The issue of the choice of the block size is briefly discussed in Section 12.4. Finally, a simlation study sheds some light on the small sample behavior of the subsampling method in Section 12.5. All tables appear at the end of the chapter.

12.2 Extension of Previous Theory

12.2.1 The Basic Method

The subsampling method for stationary observations was introduced in Chapter 3. Those results could be used to make inference for the autoregressive parameter ρ in the case $|\rho| < 1$. Clearly, a more general result is needed that also works for the random walk case $\rho = 1$. Rather than presenting a custom-tailored answer pertaining to ρ, we shall derive a more general theorem for general, univariate parameters θ.

Suppose $\{\ldots, X_{-1}, X_0, X_1, \ldots\}$ is a sequence of random variables taking values in an arbitrary sample space S, and defined on a common probability space. Denote the joint probability law governing the infinite sequence by P. The goal is to construct a confidence interval for some real-valued parameter $\theta(P)$, on the basis of observing $\{X_1, \ldots, X_n\}$. We assume the existence of a consistent estimator $\hat{\theta}_n = \hat{\theta}_n(X_1, \ldots, X_n)$.

For time series data, the gist of the subsampling method is to recompute the statistic of interest on smaller blocks of the observed sequence $X_1, \ldots, X_n$. Define $\hat{\theta}_{n,b,t} = \hat{\theta}_b(X_t, \ldots, X_{t+b-1})$, the estimator of $\theta(P)$ based on the subsample $\{X_t, \ldots, X_{t+b-1}\}$. Note that $\hat{\theta}_n = \hat{\theta}_{n,n,1}$. Let $J_{b,t}(P)$ be the sampling distribution of $\tau_b(\hat{\theta}_{n,b,t} - \theta(P))$, where τ_b is an appropriate normalizing constant. Also, define the corresponding cumulative distribution function:

$$J_{b,t}(x, P) = Prob_P\{\tau_b(\hat{\theta}_{b,t} - \theta(P)) \leq x\}. \tag{12.5}$$

12.2 Extension of Previous Theory

As in previous chapters, let $J_n(P) = J_{n,1}(P)$. We will state a theorem in greater generality than needed for our application by allowing for some heteroskedasticity in the underlying process; for example, this would allow us to extend the main results of this chapter to the case where the innovation sequence of the AR(1) model exhibits nonconstant variance. A major assumption that is needed to construct asymptotically valid confidence intervals for $\theta(P)$ is the following.

Assumption 12.2.1. *There exists a limiting law $J(P)$ such that*

i. $J_n(P)$ *converges weakly to* $J(P)$ *as* $n \to \infty$,
ii. *for every continuity point x of $J(P)$ and for any sequences n, b with $n, b \to \infty$ and $b/n \to 0$, we have $\frac{1}{n-b+1} \sum_{t=1}^{n-b+1} J_{b,t}(x, P) \to J(x, P)$.*

Remark 12.2.1. Note that condition (ii) follows trivially from condition (i) if the process $\{X_t\}$ is strictly stationary or, which is weaker, if the subsample statistics $\tau_b(\hat\theta_{n,b,t} - \theta(P))$ are strictly stationary.

The subsampling approximation to $J_n(x, P)$ we study is defined by

$$L_{n,b}(x) = \frac{1}{n-b+1} \sum_{t=1}^{n-b+1} 1\{\tau_b(\hat\theta_{n,b,t} - \hat\theta_n) \leq x\}. \tag{12.6}$$

Since $J_n(x, P)$ is approximated by $L_{n,b}(x)$, both should weakly converge to the same limit, namely, $J(x, P)$. To ensure that $L_{n,b}(x)$ converges to $J(x, P)$ in probability, it is necessary that the information in the $n - b + 1$ subsample statistics $\tau_b(\hat\theta_{n,b,t} - \hat\theta_n)$ tend to infinity with the sample size n. In previous chapters, this followed from a weak dependence condition on the underlying sequence $\{X_t\}$, that is, an α-mixing condition (see Definition A.0.1 in the Appendices). Rather than requiring the underlying sequence to be weakly dependent, we now make this requirement for the subsample statistics only. To this end, let $Z_{n,b,t} = \tau_b(\hat\theta_{n,b,t} - \theta(P))$, and denote the α-mixing coefficients corresponding to the $\{Z_{n,b,t}, t = 1, \ldots, n - b + 1\}$ sequence by $\alpha_{n,b}(\cdot)$. The following theorem shows that the approach imposing a mixing condition on the subsample statistics rather than the underlying sequence is sufficient.

Theorem 12.2.1. *Assume Assumption 12.2.1 and that $\tau_b/\tau_n \to 0$, $b/n \to 0$, $b \to \infty$, and $n^{-1} \sum_{h=1}^{n} \alpha_{n,b}(h) \to 0$ as $n \to \infty$.*

i. *If x is a continuity point of $J(\cdot, P)$, then $L_{n,b}(x) \to J(x, P)$ in probability.*
ii. *If $J(\cdot, P)$ is continuous, then $\sup_x |L_{n,b}(x) - J(x, P)| \to 0$ in probability.*
iii. *For $\alpha \in (0, 1)$, let*

$$c_{n,b}(1 - \alpha) = \inf\{x : L_{n,b}(x) \geq 1 - \alpha\}.$$

Correspondingly, define

$$c(1-\alpha, P) = \inf\{x : J(x, P) \geq 1-\alpha\}.$$

If $J(\cdot, P)$ is continuous at $c(1-\alpha, P)$, then

$$Prob_P\{\tau_n(\hat{\theta}_n - \theta(P)) \leq c_{n,b}(1-\alpha)\} \to 1-\alpha \text{ as } n \to \infty.$$

Thus, the asymptotic coverage probability under P of the interval $I_1 = [\hat{\theta}_n - \tau_n^{-1}c_{n,b}(1-\alpha), \infty)$ is the nominal level $1-\alpha$.

Proof. The proof is almost identical to the proof of Theorem 4.2.1. To prove (i), follow the proof from there until (and including) the part where it is shown that it is sufficient to restrict our attention to $U_n(x)$.

Since $E(U_n(x)) = \frac{1}{q}\sum_{a=t}^{q} J_{b,t}(x)$, the proof of (i) reduces by Assumption 12.2.1 to showing that $Var(U_n(x))$ tends to zero. Define

$$I_{b,t} = 1\{\tau_b(\hat{\theta}_{n,b,t} - \theta(P)) \leq x\}, \quad t = 1, \ldots, q,$$

$$s_{q,h} = \frac{1}{q}\sum_{t=1}^{q-h} Cov(I_{b,t}, I_{b,t+h}).$$

According to Lemma A.0.2 of Appendix A,

$$|Cov(I_{b,t}, I_{b,t+h})| \leq 4\alpha_{n,b}(h) \qquad (12.7)$$

and therefore,

$$Var(U_n(x)) = \frac{1}{q}\left(s_{q,0} + 2\sum_{h=1}^{q-1} s_{q,h}\right)$$

$$\leq \frac{1}{q}\left(1 + 2\sum_{h=1}^{q-1} \alpha_{n,b}(h)\right) \to 0. \qquad (12.8)$$

This completes the proof of (i).

The proofs of (ii) and (iii) are identical to the corresponding proofs for Theorem 4.2.1. ∎

12.2.2 Subsampling Studentized Statistics

The application of Theorem 12.2.1 requires knowledge of the rate of convergence τ_n. In standard cases, this is simply $n^{1/2}$. However, nonstandard cases exist and of prime interest here is the unit root case where the rate of convergence can be given by n or even $n^{3/2}$ (see Section 12.3). Therefore, subsampling inference for the autoregressive root seems to require knowledge about the underlying model. But this is exactly the problem that we are trying to solve! One way around this dilemma would be to use the rate estimation approach of Chapter 8. Another way, and the one pursued here, is to base inference on a studentized statistic instead. This requires another

12.2 Extension of Previous Theory

extension of the theory so far. The previous theorem was stated nevertheless, since it is interesting in its own right, but also since it provides the basis for the proof of the theorem handling the studentized case.

The focus is now on a studentized root $\tau_n(\hat{\theta}_n - \theta(P))/\hat{\sigma}_n$, where $\hat{\sigma}_n$ is some nonnegative estimate of scale. Define $J^*_{b,t}(P)$ to be the sampling distribution of $\tau_b(\hat{\theta}_{n,b,t} - \theta(P))/\hat{\sigma}_{n,b,t}$, the studentized statistic based on the subsample $\{X_t, \ldots, X_{t+b-1}\}$. Also, define the corresponding cumulative distribution function

$$J^*_{b,t}(x, P) = Prob_P\{\tau_b(\hat{\theta}_{n,b,t} - \theta(P))/\hat{\sigma}_{n,b,t} \leq x\}.$$

As usual, let $J^*_n(P) = J^*_{n,1}(P)$.

Subsampling for scaled or studentized statistics in the context of i.i.d. data has previously been considered in Section 2.5 in the case where $\hat{\sigma}_n$ converges to a positive constant in probability, and in Section 11.3 in the case where $\hat{\sigma}_n$, appropriately standardized, converges weakly. For the purposes of this chapter, a result covering dependent data is needed.

The subsampling method is modified to the studentized case in the obvious way. Analogously to (12.6), define

$$L^*_{n,b}(x) = \frac{1}{n-b+1} \sum_{t=1}^{n-b+1} 1\{\tau_b(\hat{\theta}_{n,b,t} - \hat{\theta}_n)/\hat{\sigma}_{n,b,t} \leq x\}. \qquad (12.9)$$

$L^*_{n,b}(x)$ then represents the subsampling approximation to $J^*_n(x, P)$.

The essential assumption needed to construct asymptotically valid confidence regions for $\theta(P)$ now becomes more involved than for the non-studentized case.

Assumption 12.2.2.

i. $J^*_n(P)$ converges weakly to a nondegenerate limit law $J^*(P)$. In addition, $a_n(\hat{\theta}_n - \theta(P))$ converges weakly to V, and $d_n\hat{\sigma}_n$ converges weakly to W, for positive sequences $\{a_n\}$ and $\{d_n\}$ satisfying $\tau_n = a_n/d_n$. Here, V and W are two random variables, with distributions $V(P)$ and $W(P)$, where $W(P)$ does not have positive mass at zero.
ii. For every continuity point x of $J^*(P)$ and for any sequences n, b with $n, b \to \infty$ and $b/n \to 0$, $\frac{1}{n-b+1} \sum_{t=1}^{n-b+1} J^*_{b,t}(x, P) \to J^*(x, P)$.
iii. For every continuity point x of $W(P)$ and for any sequences n, b with $n, b \to \infty$ and $b/n \to 0$, $\frac{1}{n-b+1} \sum_{t=1}^{n-b+1} W_{b,t}(x, P) \to W(x, P)$. Here, $W_{b,t}(x, P) = Prob_P\{d_b\hat{\sigma}_{n,b,t} \leq x\}$.

Remark 12.2.2. Note that conditions (ii) and (iii) follow trivially from condition (i) if the process $\{X_t\}$ is strictly stationary or, which is weaker, if the subsample statistics $\tau_b(\hat{\theta}_{n,b,t} - \theta(P))/\hat{\sigma}_{n,b,t}$ and $\hat{\sigma}_{n,b,t}$ are strictly stationary, meaning that their distributions are independent of t.

Let $Z_{n,b,t} = (\tau_b(\hat{\theta}_{n,b,t} - \theta(P))/\hat{\sigma}_{n,b,t}, \hat{\sigma}_{n,b,t})'$, and denote the α-mixing coefficients corresponding to the $\{Z_{n,b,t}, t = 1,\ldots,n-b+1\}$ sequence by $\alpha_{n,b}(\cdot)$.

Theorem 12.2.2. *Assume Assumption 12.2.2 and that $a_b/a_n \to 0$, $\tau_b/\tau_n \to 0$, $b/n \to 0$, $b \to \infty$, and $n^{-1}\sum_{h=1}^{n} \alpha_{n,b}^*(h) \to 0$ as $n \to \infty$.*

  i. *If x is a continuity point of $J^*(\cdot, P)$, then $L_{n,b}^*(x) \to J^*(x, P)$ in probability.*
  ii. *If $J^*(\cdot, P)$ is continuous, then $\sup_x |L_{n,b}^* - J^*(x, P)| \to 0$ in probability.*
  iii. *For $\alpha \in (0, 1)$, let*

$$c_{n,b}^*(1-\alpha) = \inf\{x : L_{n,b}^*(x) \geq 1-\alpha\}.$$

Correspondingly, define

$$c^*(1-\alpha, P) = \inf\{x : J^*(x, P) \geq 1-\alpha\}.$$

If $J^*(\cdot, P)$ is continuous at $c^*(1-\alpha, P)$, then

$$\mathrm{Prob}_P\{\tau_n(\hat{\theta}_n - \theta(P))/\hat{\sigma}_n \leq c_{n,b}^*(1-\alpha)\} \to 1-\alpha \text{ as } n \to \infty.$$

Thus, the asymptotic coverage probability under P of the interval $I_1^* = [\hat{\theta}_n - \hat{\sigma}_n \tau_n^{-1} c_{n,b}^*(1-\alpha), \infty)$ is the nominal level $1-\alpha$.

Proof. Again, let $q = n - b + 1$. To prove (i), note that

$$L_{n,b}^*(x) = \frac{1}{q}\sum_{t=1}^{q} 1\{\tau_b(\hat{\theta}_{n,b,t} - \hat{\theta}_n)/\hat{\sigma}_{n,b,t} \leq x\}$$

$$= \frac{1}{q}\sum_{t=1}^{q} 1\{\tau_b(\hat{\theta}_{n,b,t} - \theta(P))/\hat{\sigma}_{n,b,t} \leq x + \tau_b(\hat{\theta}_n - \theta(P))/\hat{\sigma}_{n,b,t}\}.$$

(12.10)

The goal is to show that the terms $\tau_b(\hat{\theta}_n - \theta(P))/\hat{\sigma}_{n,b,t}$ are negligible in the last equation. To this end, for $u > 0$, let

$$R_n(u) = \frac{1}{q}\sum_{t=1}^{q} 1\{\tau_b(\hat{\theta}_n - \theta(P))/\hat{\sigma}_{n,b,t} \leq u\}$$

$$= \frac{1}{q}\sum_{t=1}^{q} 1\{d_b \hat{\sigma}_{n,b,t} \geq a_b(\hat{\theta}_n - \theta(P))/u\}.$$

Here, without loss of generality it is assumed that both the sequences a_n and b_n are nonnegative. By Assumption 12.2.2 and the assumption $a_b/a_n \to 0$, it follows that, for any $\delta > 0$, $a_b(\hat{\theta}_n - \theta(P)) \leq \delta$ with probability tending

12.2 Extension of Previous Theory

to one. Therefore, with probability tending to one,

$$R_n(u) \geq \frac{1}{q} \sum_{t=1}^{q} 1\{d_b \hat{\sigma}_{n,b,t} \geq \delta/u\}.$$

We need to consider the case $u > 0$ only, as the scale estimates $\hat{\sigma}_{n,b,t}$ are nonnegative. According to the usual subsampling argument of Theorem 12.2.1, $\frac{1}{q} \sum_{t=1}^{q} 1\{d_b \hat{\sigma}_{n,b,t} \geq \delta/u\}$ converges in probability to $1 - W(\delta/u, P)$, as long as δ/u is a continuity point of $W(P)$; note that here it is not required that $d_b/d_n \to 0$, since the subsample statistics are of the form $d_b \hat{\sigma}_{n,b,t}$ rather than $d_b(\hat{\sigma}_{n,b,t} - \hat{\sigma}_n)$. Hence, it can be ensured that $R_n(u)$ is arbitrarily close to one by choosing δ small enough; recall that $W(P)$ does not have positive mass at zero. In other words, for any $u > 0$, we have that $R_n(u) \to 1$ in probability. Now rewrite (12.10) in the following way:

$$L_{n,b}^*(x) = \frac{1}{q} \sum_{t=1}^{q} 1\{\tau_b(\hat{\theta}_{n,b,t} - \theta(P))/\hat{\sigma}_{n,b,t} \leq x + \tau_b(\hat{\theta}_n - \theta(P))/\hat{\sigma}_{n,b,t}\}$$

$$\leq \frac{1}{q} \sum_{t=1}^{q} 1\{\tau_b(\hat{\theta}_{n,b,t} - \theta(P))/\hat{\sigma}_{n,b,t} \leq x + u\} + (1 - R_n(u)),$$

for any positive number u. The last inequality follows because the t-th term in (12.10) is less than or equal to

$$1\{\tau_b(\hat{\theta}_{n,b,t} - \theta(P))/\hat{\sigma}_{n,b,t} \leq x + u\} + 1\{\tau_b(\hat{\theta}_n - \theta(P))/\hat{\sigma}_{n,b,t} > u\}; \quad (12.11)$$

then, sum over all t. It has been seen that $(1 - R_n(u)) \to 0$ in probability and hence by a standard subsampling argument again we get, for any $\epsilon > 0$, $L_{n,b}^*(x) \leq J^*(x + u, P) + \epsilon$ with probability tending to one, provided that $x + u$ is a continuity point of $J^*(\cdot, P)$. Letting u tend to zero shows that $L_{n,b}^*(x) \leq J^*(x, P) + \epsilon$ with probability tending to one. A similar argument leads to $L_{n,b}^*(x) \geq J^*(x, P) - \epsilon$ with probability tending to one. Since ϵ is arbitrary, this implies $L_{n,b}^* \to J^*(x, P)$ in probability, and thus we have proved (i).

The proofs of (ii) and (iii) given (i) are very similar to the proofs of (ii) and (iii) given (i) in Theorem 12.2.1 and thus are omitted. ∎

As mentioned in previous chapters, one can use equal-tailed or symmetric confidence intervals when two-sided intervals are desired. To construct two-sided symmetric subsampling intervals, one estimates the two-sided sampling distribution function. Let $J_{n,|\cdot|}^*(P)$ be the sampling distribution of $\tau_n|\hat{\theta}_n - \theta(P)|/\hat{\sigma}_n$. Define

$$L_{n,b,|\cdot|}^*(x) = \frac{1}{n-b+1} \sum_{t=1}^{n-b+1} 1\{\tau_b \left|\hat{\theta}_{n,b,t} - \hat{\theta}_n\right|/\hat{\sigma}_{n,b,t} \leq x\}. \quad (12.12)$$

$L_{n,b,|\cdot|}^*(x)$ then represents the subsampling approximation to $J_{n,|\cdot|}^*(x)$.

Corollary 12.2.1. *Assume Assumption 12.2.2, $a_b/a_n \to 0$, $\tau_b/\tau_n \to 0$, $b/n \to 0$, and $b \to \infty$ as $n \to \infty$. Also assume that $n^{-1}\sum_{h=1}^{n} \alpha_{n,b}(h) \to 0$. Denote by $J^*_{|\cdot|}(P)$ the limiting distribution of $J^*_{n,|\cdot|}(P)$.*

i. *If x is a continuity point of $J^*_{|\cdot|}(\cdot, P)$, then $L^*_{n,b,|\cdot|}(x) \to J^*_{|\cdot|}(x, P)$ in probability.*
ii. *If $J^*_{|\cdot|}(\cdot, P)$ is continuous, then $\sup_x |L^*_{n,b,|\cdot|} - J^*_{|\cdot|}(x, P)| \to 0$ in probability.*
iii. *For $\alpha \in (0, 1)$, let*

$$c^*_{n,b,|\cdot|}(1-\alpha) = \inf\{x : L^*_{n,b,|\cdot|}(x) \geq 1 - \alpha\}.$$

Correspondingly, define

$$c^*_{|\cdot|}(1-\alpha, P) = \inf\{x : J^*_{|\cdot|}(x, P) \geq 1 - \alpha\}.$$

If $J^*_{|\cdot|}(\cdot, P)$ is continuous at $c^*_{|\cdot|}(1 - \alpha, P)$, then

$$Prob_P\{\tau_n \left|\hat{\theta}_n - \theta(P)\right|/\hat{\sigma}_n \leq c^*_{n,b,|\cdot|}(1-\alpha)\} \to 1 - \alpha \text{ as } n \to \infty.$$

Thus, the asymptotic coverage probability under P of the interval $I^*_{SYM} = [\hat{\theta}_n - \hat{\sigma}_n \tau_n^{-1} c^*_{n,b,|\cdot|}(1-\alpha), \hat{\theta}_n + \hat{\sigma}_n \tau_n^{-1} c^*_{n,b,|\cdot|}(1-\alpha)]$ is the nominal level $1 - \alpha$.

Proof. The proof follows immediately from Theorem 12.2.2 and the continuous mapping theorem.

12.3 Subsampling Inference for the Autoregressive Root

The goal of this section is to show that the subsampling method for studentized statistics as outlined in the previous section can be used to make inference for the autoregressive root. We observe a sample $X_1, \ldots, X_n$ from the AR(1) model

$$X_t = \mu + \rho X_{t-1} + \epsilon_t, \tag{12.13}$$

where $\{\epsilon_t\}$ is a strictly stationary white noise innovation sequence with mean zero and variance σ_ϵ^2. Hence, we assume that the ϵ_t are uncorrelated, but possibly dependent. Note that the assumption of strict stationarity is made primarily for convenience of the proofs but could be relaxed considerably; see Remark 12.3.6. Denote the α-mixing coefficients corresponding to $\{\epsilon_t\}$ by $\alpha_\epsilon(\cdot)$. The estimator for ρ is obtained by OLS regression of X_t on X_{t-1}, including an intercept in the regression. Let $\bar{X}_{i,j} = (j - i + 1)^{-1}\sum_{t=i}^{j} X_t$, the sample mean of the block of observations

12.3 Subsampling Inference for the Autoregressive Root

$\{X_i, X_{i+1}, \ldots, X_j\}$. Then,

$$\hat{\rho}_n = \frac{\sum_{t=1}^{n-1}(X_{t+1} - \bar{X}_{2,n})(X_t - \bar{X}_{1,n-1})}{\sum_{t=1}^{n-1}(X_t - \bar{X}_{1,n-1})^2}, \quad \hat{\mu}_n = \bar{X}_{2,n} - \hat{\rho}_n \bar{X}_{1,n-1}. \tag{12.14}$$

The corresponding studentized statistic is based on the OLS estimate of the standard deviation of $(\hat{\rho}_n)$,

$$\widehat{SD}(\hat{\rho}_n) = \frac{s_n}{[\sum_{t=1}^{n-1}(X_t - \bar{X}_{1,n-1})^2]^{1/2}}, \tag{12.15}$$

where

$$s_n^2 = \frac{1}{n-3} \sum_{t=1}^{n-1}(X_{t+1} - \bar{X}_{2,n} + \hat{\rho}_n \bar{X}_{1,n-1} - \hat{\rho}_n X_t)^2.$$

To fit this application in the framework of the theory of Subsection 12.2.2, introduce

$$\hat{\sigma}_n = n^{1/2} \widehat{SD}(\hat{\rho}_n), \tag{12.16}$$

so the studentized statistic and the corresponding subsample statistics are of the form

$$n^{1/2} \frac{(\hat{\rho}_n - \rho)}{\hat{\sigma}_n}, \quad b^{1/2} \frac{(\hat{\rho}_{n,b,t} - \hat{\rho}_n)}{\hat{\sigma}_{n,b,t}}. \tag{12.17}$$

Here, $\hat{\rho}_{n,b,t}$ and $\hat{\sigma}_{n,b,t}$ are the statistics $\hat{\rho}_b$ and $\hat{\sigma}_b$ computed on the subsample $\{X_t, \ldots, X_{t+b-1}\}$.

The following proposition shows that the ideas of Subsection 12.2.2 can be used to construct asymptotic confidence intervals for ρ.

Proposition 12.3.1. *Assume $\{X_t\}$ is a sequence of random variables according to model (12.13). Assume that for some $\delta > 0$, $E|\epsilon_t|^{4+\delta} < \infty$, $\sum_{h=1}^{\infty} \alpha_\epsilon(h)^{\delta/(2+\delta)} < \infty$, and, if $|\rho| < 1$, $\sum_{h=1}^{\infty} \alpha_X(h)^{\delta/(2+\delta)} < \infty$. Take $\hat{\rho}_n$ and $\hat{\sigma}_n$ as defined in (12.14) and (12.16), respectively. Also, let $\tau_n = n^{1/2}$ and assume that $b \to \infty$ and $b/n \to 0$ as $n \to \infty$.*

Then, the conclusions of Theorem 12.2.2 and Corollary 12.2.1 hold, provided we replace $\theta(P)$ by ρ, $\hat{\theta}_n$ by $\hat{\rho}_n$, and $\hat{\theta}_{n,b,t}$ by $\hat{\rho}_{n,b,t}$.

Proof. We need to show that Assumption 12.2.2 holds, that the subsample statistics satisfy the mixing condition $n^{-1} \sum_{h=1}^{n} \alpha_{n,b}(h) \to 0$, and that $a_b/a_n \to 0$.

We start with Assumption 12.2.2 and $a_b/a_n \to 0$. First consider the stationary case $|\rho| < 1$. It is well known (e.g., Chapter 8 of Hamilton, 1994) that if the ϵ_t are i.i.d., then

$$n^{1/2}(\hat{\rho}_n - \rho) \xrightarrow{\mathcal{L}} N(0, 1 - \rho^2),$$

$$\hat{\sigma}_n^2 \xrightarrow{P} 1 - \rho^2,$$

and
$$n^{1/2}\frac{(\hat{\rho}_n - \rho)}{\hat{\sigma}_n} \overset{\mathcal{L}}{\Longrightarrow} N(0,1).$$

Here, $\overset{\mathcal{L}}{\Longrightarrow}$ denotes convergence in distribution and $\overset{P}{\longrightarrow}$ denotes convergence in probability. Actually, these results are frequently used to make asymptotic inference for ρ in the stationary case. However, that inference can be arbitrarily misleading if the ϵ_t are uncorrelated but dependent, as was pointed out by Romano and Thombs (1996). Under the conditions of Proposition 12.3.1, they proved

$$n^{1/2}(\hat{\rho}_n - \rho) \overset{\mathcal{L}}{\Longrightarrow} N(0, \xi^2). \tag{12.18}$$

Here, the limiting variance is given by

$$\xi^2 = \lim_n nVar(\hat{\rho}_n)$$
$$= Var^{-2}(X_1)[c_{2,2} - 2\rho c_{1,2} + \rho^2 c_{1,1}], \tag{12.19}$$

where $c_{i,j}$ is defined as follows

$$c_{i+1,j+1} = \sum_{d=-\infty}^{\infty} \text{Cov}(X_0 X_i, X_d X_{d+j}). \tag{12.20}$$

If the ϵ_t are i.i.d., then $\xi^2 = 1 - \rho^2$. But, as is clear from equations (12.19) and (12.20), in the dependent case this can change to any positive value, as a function of the underlying dependence structure; see Romano and Thombs (1996) for some explicit examples. On the other hand, the relation

$$\hat{\sigma}_n^2 \overset{P}{\longrightarrow} 1 - \rho^2 \tag{12.21}$$

remains unchanged, so that for the studentized statistic we get

$$n^{1/2}\frac{(\hat{\rho}_n - \rho)}{\hat{\sigma}_n} \overset{\mathcal{L}}{\Longrightarrow} N(0, \frac{\xi^2}{1-\rho^2}). \tag{12.22}$$

Since the limiting variance can be very different from one, basing inference for ρ on limiting standard normality of the OLS t-statistic can be completely misleading if the ϵ_t are uncorrelated only. Subsampling, on the other hand, will give correct answers. According to stationarity, it follows that the distributions of $b^{1/2}(\hat{\rho}_{n,b,t} - \rho)/\hat{\sigma}_{n,b,t}$ and $\hat{\sigma}_{n,b,t}$ are independent of t. And by (12.22) and (12.21) it follows that their respective limiting distributions are $N(0, \frac{\xi^2}{1-\rho^2})$ and point mass at $1 - \rho^2$. Therefore, all three conditions of Assumption 12.2.2 are met. Since $\tau_n = n^{1/2}$ and $d_n \equiv 1$, $a_n = n^{1/2}$. Thus, $\tau_b/\tau_n \to 0$ and $a_b/a_n \to 0$ trivially follow from $b/n \to 0$.

Next, consider the random walk case $\rho = 1$. Surprisingly at first, the asymptotics in this case strongly depend upon whether $\mu = 0$. We discuss the instance $\mu = 0$ first. In this case, it is known (e.g., Chapter 17 of

Hamilton, 1994) that

$$n(\hat{\rho}_n - 1) \overset{\mathcal{L}}{\Longrightarrow} \frac{\frac{1}{2}\{[W(1)]^2 - 1\} - W(1)\int_0^1 W(r)dr}{\int_0^1 [W(r)]^2 dr - \left[\int_0^1 W(r)dr\right]^2}, \quad (12.23)$$

$$n^{1/2}\hat{\sigma}_n \overset{\mathcal{L}}{\Longrightarrow} \frac{1}{\left\{\int_0^1 [W(r)]^2 dr - \left[\int_0^1 W(r)dr\right]^2\right\}^{1/2}}, \quad (12.24)$$

and

$$n^{1/2}\frac{(\hat{\rho}_n - 1)}{\hat{\sigma}_n} \overset{\mathcal{L}}{\Longrightarrow} \frac{\frac{1}{2}\{[W(1)]^2 - 1\} - W(1)\int_0^1 W(r)dr}{\left\{\int_0^1 [W(r)]^2 dr - \left[\int_0^1 W(r)dr\right]^2\right\}^{1/2}}, \quad (12.25)$$

where $W(\cdot)$ denotes a standard Brownian motion. These results follow from a functional Central Limit theorem for the $\{X_t\}$ process and were first proved by Dickey and Fuller (1979) for i.i.d. innovations ϵ_t. Phillips and Perron (1988) extended them to uncorrelated observations. It is easy to see that the distributions of $b^{1/2}(\hat{\rho}_{n,b,t} - \rho)/\hat{\sigma}_{n,b,t}$ and $b^{1/2}\hat{\sigma}_{n,b,t}$ are independent of t. Their respective limiting distributions are given by (12.25) and (12.24); note that the distribution defined by (12.24) does not have positive mass at zero. Therefore, all three conditions of Assumption 12.2.2 are met. Furthermore, we have $\tau_n = n^{1/2}$, $d_n = n^{1/2}$, and $a_n = n$. Thus, $\tau_b/\tau_n \to 0$ and $a_b/a_n \to 0$ trivially are implied by $b/n \to 0$.

The situation is very different for the case $\mu \neq 0$. The asymptotic results are

$$n^{3/2}(\hat{\rho}_n - 1) \overset{\mathcal{L}}{\Longrightarrow} N(0, \sigma_\epsilon^2 \frac{\mu^2}{12}), \quad (12.26)$$

$$n\hat{\sigma}_n \overset{P}{\longrightarrow} \sigma_\epsilon \frac{\mu}{\sqrt{12}}, \quad (12.27)$$

and

$$n^{1/2}\frac{(\hat{\rho}_n - 1)}{\hat{\sigma}_n} \overset{\mathcal{L}}{\Longrightarrow} N(0,1). \quad (12.28)$$

The proof follows from Section 2 of West (1988). The reason for the difference from the previous case is that for $\mu \neq 0$ the regressor X_{t-1} is asymptotically dominated by the time trend $\mu(t-1)$. In large samples, it is as if the explanatory variable X_{t-1} were replaced by the time trend $\mu(t-1)$. Superconsistency of $\hat{\rho}_n$ at rate $n^{3/2}$ and asymptotic standard normality of the studentized statistic ensue. Note that, in contrast to the stationary case $|\rho| < 1$, the limiting variance of the t-statistic is always one and is not affected by the dependence structure of the ϵ_t. It is easy to see that the distributions of $b^{1/2}(\hat{\rho}_{n,b,t} - \rho)/\hat{\sigma}_{n,b,t}$ and $b\hat{\sigma}_{n,b,t}$ are independent of t,

282   12. Subsampling the Autoregressive Parameter

with their respective limiting distributions being $N(0,1)$ and point mass at $\sigma_\epsilon \mu / \sqrt{12}$. Therefore, all three conditions of Assumption 12.2.2 are met. Furthermore, $\tau_n = n^{1/2}$, $d_n = n$, and $a_n = n^{3/2}$. Thus, $\tau_b/\tau_n \to 0$ and $a_b/a_n \to 0$ are trivially are implied by $b/n \to 0$.

Finally, we show that the subsample statistics satisfy the necessary mixing condition. Note that, for the stationary case $|\rho| < 1$, this condition immediately follows from the mixing condition of the $\{X_t\}$ process. However, for the case $\rho = 1$, this is no longer true. For example, if the innovation sequence $\{\epsilon_t\}$ is i.i.d. with unit variance, then $Cov(X_t, X_{t+s}) = t$, which does not vanish as $s \to \infty$. For this reason, the weak dependence structure of the subsampling statistics needs to be verified from first principles in the case $\rho = 1$. To this end, we claim that in this case $\hat{\rho}_{n,b,t}$ and $\hat{\sigma}_{n,b,t}$ are functions of μ and $\epsilon_t, \ldots, \epsilon_{t+b-1}$ only. From this claim and with μ being a constant, it then follows that $\alpha_{n,b}(b+h) \leq \alpha_\epsilon(h)$ and therefore that

$$\frac{1}{n}\sum_{h=1}^{n} \alpha_{n,b}(h) \leq \frac{b}{n} + \frac{1}{n}\sum_{h=1}^{n} \alpha_\epsilon(h).$$

Both terms on the right-hand side converge to zero by the assumptions. To demonstrate the claim, note that for $i \leq k \leq j$,

$$X_k - \bar{X}_{i,j} = X_k - \frac{1}{j-i+1}\sum_{t=i}^{j} X_t$$

$$= X_i + (k-i)\mu + \sum_{l=i+1}^{k} \epsilon_l$$

$$- \frac{1}{j-i+1}\sum_{t=i}^{j}(X_i + (t-i)\mu + \sum_{l=i+1}^{t} \epsilon_t)$$

$$= (k - i + \frac{j-i}{2})\mu + \sum_{l=i+1}^{k} \epsilon_l - \frac{1}{j-i+1}\sum_{t=i}^{j}(\sum_{l=i+1}^{t} \epsilon_t).$$
(12.29)

Hence, for $i \leq k \leq j$, $X_k - \bar{X}_{i,j}$ is a function of μ and $\epsilon_i, \ldots, \epsilon_j$ only. From this observation and equation (12.14), it follows immediately that $\hat{\rho}_{n,b,t}$ is a function of μ and $\epsilon_t, \ldots, \epsilon_{t+b-1}$ only. To see that the same is true for $\hat{\sigma}_{n,b,t}$, conclude from equations (12.14) to (12.16) that it is left to show that $s_{n,b,t}^2$ is a function of μ and $\epsilon_t, \ldots, \epsilon_{t+b-1}$ only. But, this is implied by

$$s_{n,b,t}^2 = \frac{1}{b-3}\sum_{a=t}^{t+b-2}(X_{a+1} - \bar{X}_{2,t+b-1} + \hat{\rho}_{n,b,t}\bar{X}_{t,t+b-2} - \hat{\rho}_{n,b,t}X_a)^2$$

$$= \frac{1}{b-3}\sum_{a=t}^{t+b-2}(X_{a+1} - \bar{X}_{2,t+b-1} + \hat{\rho}_{n,b,t}(X_a - \bar{X}_{t,t+b-2}))^2,$$

12.3 Subsampling Inference for the Autoregressive Root 283

relationship (12.29), and $\hat{\rho}_{n,b,t}$ being a function of μ and $\epsilon_t, \ldots, \epsilon_{t+b-1}$ only. ∎

Remark 12.3.1. Basing the inference on the OLS t-statistic allows us to get around the problem of unknown convergence rate. This is because the t-statistic has a proper limiting distribution in each of the three cases. Therefore, it is a *self-normalizing sum*, a concept previously utilized in Subsection 11.4.2.

Remark 12.3.2. An alternative approach would be to base the inference on the unstudentized statistic $\tau_n(\hat{\rho}_n - \rho)$. As has been demonstrated, the proper normalizing constant τ_n will be $n^{1/2}$, n, or $n^{3/2}$, depending on the underlying parameters ρ and μ. Therefore, the proper rate is unknown; but, as outlined in Chapter 8, it can be consistently estimated and the estimated rate may be employed in constructing subsampling confidence intervals without affecting their asymptotic validity. However, we will not pursue this approach here.

Remark 12.3.3. Equations (12.22), (12.25), and (12.28) imply that basing asymptotic inference for ρ on the limiting distribution of the studentized statistic requires knowledge of the underlying parameters ρ and μ, which defeats the purpose. Moreover, in the case $|\rho| < 1$, one also needs to consistently estimate the dependence structure of the $\{\epsilon_t\}$ process. This would be very difficult in practice, as can be seen from (12.19) and (12.20). On the other hand, it has been shown that subsampling works in all three cases and therefore allows to make asymptotically correct inference no matter what the underlying parameters and dependence structure of the $\{\epsilon_t\}$ process.

Remark 12.3.4. If focus were only on the case of $|\rho| < 1$ and i.i.d. innovations, asymptotic inference for ρ could also be based on the standard normality of the OLS t-statistic. However, it is known that finite sample properties of this approach are not good when ρ is close to one. The intuition here is that for ρ close to one, the distribution of the t-statistic in a finite sample should not be all that different from its distribution for $\rho = 1$, a case which the normal method cannot handle. The subsampling approach, on the other hand, also works for $\rho = 1$ and it is therefore expected to enjoy better finite sample properties. See Section 12.5 for some simulation studies that confirm this conjecture.

Remark 12.3.5. Note that, unlike the bootstrap, the subsampling method can handle discontinuities of the limiting distribution of (standardized or studentized) estimators as a function of the underlying model parameters. The intuition here is that the subsampling approximation of the sampling distribution of an estimator is based on subsample statistics computed from smaller blocks of the observed data. The subsample statistics are therefore always generated from the true model. The bootstrap, on the other hand, bases its approximation on pseudo statistics computed from bootstrap data according to a bootstrap distribution estimated from the observed time series. The bootstrap data come from a model close to the truth, but not exactly the truth. This can cause the bootstrap to fail. A case in point is inference for the AR(1) parameter. The residual-based bootstrap works if $|\rho| < 1$, but it does not work if $\rho = 1$ (Basawa et al., 1991).

Remark 12.3.6. The assumption of strict stationary of the innovation process $\{\epsilon_t\}$ can be relaxed to the extent that conditions (ii) and (iii) of Assumption 12.2.2 are still satisfied. For instance, this would allow for seasonal effects reflected in nonconstant variances and/or changing distributions of the ϵ_t.

Remark 12.3.7. The mixing condition on the X_t sequence in the case of $|\rho| < 1$ can actually be relaxed. Indeed, it is known that an AR(1) sequence with mixing innovations may not be α-mixing at all (for example, see Andrews, 1984). However, in those cases the $\{X_t\}$ sequence will still be a mixingale, a concept introduced by McLeish (1975a,b). Our proof would have to be modified in two places. First, we need to show that subsampling in general still works if the underlying sequence is a mixingale. However, this only hinges on a bound like (12.8) that in turn is implied by a covariance inequality of the type (12.7). Such bounds are also valid for mixingales (see Chapter 4 of Gallant and White, 1988). Second, we need to show that the convergences in law (12.18) and (12.22) are still valid. But again, this is implied by $\{X_t\}$ being a mixingale (again, see Chapter 4 of Gallant and White, 1988).

Remark 12.3.8. In case one is willing to assume i.i.d. rather than uncorrelated innovations, the assumption of a finite $2 + \delta$ moment of ϵ_t in Proposition 12.3.1 is sufficient. Note also that in this case the mixing condition on the $\{X_t\}$ process follows from the innovations being strong mixing, provided that the distribution of ϵ_t is continuous (e.g., see Subsection 2.4.1 of Doukhan, 1994).

12.4 Choice of the Block Size

Since the residual-based bootstrap is not consistent for the autoregressive parameter, the calibration method of Subsection 9.3.1 cannot be employed. Therefore, we use the minimum volatility method of Subsection 9.3.2 to choose the block size b.

The algorithm is illustrated with the help of two simulated data sets. First, we generated a time series of size $n = 200$ according to model (12.13), using $\rho = 0.95$, $\mu = 0$, and i.i.d. standard normal innovations. The range of b was chosen as $b_{small} = 10$ and $b_{big} = 36$. The minimization of the volatility in Step 2 was done using $k = 2$. The results are shown at the

12.4 Choice of the Block Size   285

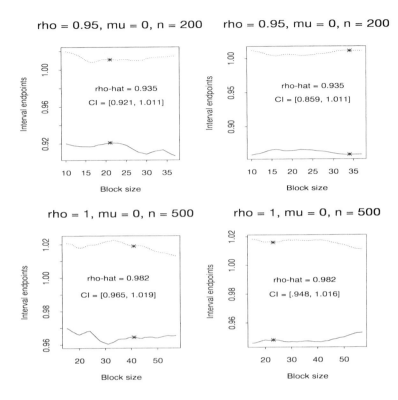

FIGURE 12.1. Illustration of the minimizing confidence interval volatility algorithm for two data sets. The plots on the left correspond to equal-tailed confidence intervals, while the plots on the right correspond to symmetric confidence intervals. The block sizes selected by the algorithm are highlighted by a star. The final confidence intervals are listed within the plots, together with the point estimates.

top of Figure 12.1. The left plot corresponds to equal-tailed confidence intervals while the right plot corresponds to symmetric confidence intervals. The block sizes b chosen by the algorithm are highlighted by a star. The resulting final confidence intervals are included in the plots together with the point estimate $\hat{\rho}_n$.

This exercise was repeated for another data set of size $n = 500$, using $\rho = 1$, $\mu = 0$, and i.i.d. standard normal innovations. The range of b chosen was $b_{small} = 15$ and $b_{big} = 58$. The results are shown at the bottom of Figure 12.1.

The plots show that symmetric intervals are somewhat more stable, that is, the endpoints change less as b is varied. This behavior is typical and was observed for other simulations as well.

12.5 Simulation Study

The purpose of this section is to shed some light on the small sample performance of the subsampling method by means of a simulation study. Model (12.13) serves as the data-generating process. The constant μ is either equal to 0 or 0.1; the value 0.1 was chosen to make the time trend not overly dominating when $\rho = 1$. The AR(1) parameter ρ is one of the following: 1, 0.99, 0.95, 0.9, or 0.5. As innovation process $\{\epsilon_t\}$, we consider i.i.d. standard normal and $\epsilon_t = Z_t Z_{t-1}$, where the Z_t are i.i.d. standard normal. The latter process is uncorrelated but 1-dependent. The sample sizes used are $n = 200$ and $n = 500$.

Performance of confidence intervals is measured by empirical coverage probability of nominal two-sided 95% intervals. We compute equal-tailed as well as symmetric subsampling intervals, denoted by Sub_{ET} and Sub_{SYM}, respectively. Estimated coverage is based on 1000 replications for each scenario.

The subsampling method is compared with the CLT and Stock's (1991) method. The CLT uses asymptotic standard normality of the t-statistic for $\hat{\rho}_n$. This is the correct limiting distribution if $|\rho| < 1$ and the ϵ_t are i.i.d. or if $\rho = 1$ and $\mu \neq 1$. The inference of Stock (1991) is limited to ρ values close to one, where "close" depends on the sample size. For $n = 200$, it works as low as $\rho = 0.9$, while for $n = 500$, it works as low as $\rho = 0.95$. Inference is based on the model (12.3) by rewriting it in the following way

$$X_t = \tilde{\alpha} + \tilde{\delta} t + \rho X_{t-1} + \epsilon_t. \tag{12.30}$$

Stock's test statistic is the OLS t-statistic for $\rho = 1$ in (12.30). This statistic is denoted by $t_{\hat{\rho}}$ here, since Stock's notation $\hat{\tau}^\tau$ can be confused with the normalizing sequence τ_n of the subsampling method. Table A.1 in Stock (1991) allows one to check, up to some mild interpolation, whether a certain ρ value is contained in the two-sided 95% confidence interval, given the outcome of $t_{\hat{\rho}}$ and the sample size n.

The simulation results are presented in Tables 12.1 and 12.2 and show the following. As expected, the CLT does not work very well in cases where it should not, that is, if $\rho = 1$ or very close to one and $\mu = 0$, or if $\rho < 1$ and the innovations are dependent. However, it also does not work very well in the case $\rho = 1$ and $\mu = 0.1$, where it should, at least asymptotically. It does work well when ρ is not very close to one and the innovations are i.i.d.

The next thing to note is that equal-tailed subsampling intervals tend to undercover and are not competitive. On the other hand, symmetric subsampling intervals work rather well (see Subsection 10.5.2 for a related theoretical explanation). Stock's intervals also work well when they apply, although they seem to be somewhat less robust against dependent innovations than subsampling.

As a competitor to symmetric subsampling intervals that can be used universally, one might think of the following approach: use Stock's intervals when they apply—that is, if ρ is close to one relative to the sample size—and use the CLT intervals otherwise. The tables show that this approach is comparable to symmetric subsample intervals if the innovations are i.i.d. However, it fails when the innovations are correlated, since the CLT intervals undercover significantly (which is not surprising, since they are known to asymptotically fail from our results in Section 12.3).

12.6 Conclusions

In this chapter, a new method was proposed for constructing asymptotic confidence intervals for ρ in the basic AR(1) model

$$X_t = \mu + \rho X_{t-1} + \epsilon_t,$$

where $\{\epsilon_t\}$ is strictly stationary white noise and $\rho \in (-1, 1]$. This is a difficult problem, since the limiting distribution of the OLS estimator depends in a discontinuous way not only on ρ itself but also on μ and the dependence structure of the innovations ϵ_t. For example, this discontinuity causes the residual-based bootstrap to fail.

The method proposed in this chapter is the subsampling method for studentized statistics, applied to the OLS estimator $\hat{\rho}_n$. Unlike bootstrapping, subsampling can handle the discontinuities of the limiting distribution of the t-statistic.

An extension of previous theory was provided which is needed for the case $\rho = 1$ when the $\{X_t\}$ sequence is no longer weakly dependent. Another extension was needed for studentized statistics computed from dependent data.

Some simulation studies showed that the finite sample properties of subsampling intervals are encouraging, at least when symmetric intervals are used. When ρ is equal or close to one, their coverage is about as good as the coverage of Stock's (1991) intervals. When ρ is far away from one, subsample intervals cover about as well as CLT intervals for i.i.d. innovations and much better than CLT intervals for uncorrelated but dependent innovations. The last observation is due to the often neglected fact that CLT intervals are inconsistent when innovations are uncorrelated but dependent.

Finally, it should be pointed out that subsampling is a very general and powerful technique and is not restricted to models as simple as the AR(1) model discussed in this chapter. For instance, if the goal is to find confidence intervals for the largest autoregressive root of a time series, one can base inferences on the usual augmented Dickey–Fuller (1979) regression, which controls for short-term dynamics by including higher-order autoregressive terms in the regression. Subsampling still applies, but the proofs are more cumbersome due to additional notation. Moreover, there appear to be many other difficult econometric problems where subsampling could be used beneficially, especially when the limiting distribution of the estimator depends on the underlying model parameters in a complicated and maybe even discontinuous way. One of many examples is inference in models with integrated or nearly integrated regressors as discussed in Elliot and Stock (1994), Cavanagh, Elliot, and Stock (1995), and Elliot (1998). The main condition for subsampling to work is that the underlying sequence or, more generally, the subsample statistics are weakly dependent in a sufficient way.

12.7 Tables

TABLE 12.1. Estimated coverage probabilities of various nominal 95% confidence intervals when the innovations are i.i.d. standard normal. CLT denotes intervals which approximate the sampling distribution of the t-statistic by a standard normal distribution. Sub_{ET} and Sub_{SYM} denote equal-tailed and symmetric subsampling intervals, respectively. Stock denotes the intervals according to Stock (1991). The estimates are based on 1000 replications for each scenario.

$\mu = 0, n = 200, \epsilon_t = Z_t$				
ρ	CLT	Sub_{ET}	Sub_{SYM}	Stock
1	0.70	0.92	0.92	0.96
0.99	0.82	0.91	0.93	0.95
0.95	0.89	0.85	0.94	0.94
0.90	0.93	0.84	0.95	0.95
0.50	0.95	0.87	0.93	NA

$\mu = 0, n = 500, \epsilon_t = Z_t$				
ρ	CLT	Sub_{ET}	Sub_{SYM}	Stock
1	0.71	0.95	0.95	0.95
0.99	0.87	0.92	0.96	0.95
0.95	0.94	0.84	0.96	0.96
0.90	0.95	0.85	0.96	NA
0.50	0.94	0.90	0.94	NA

$\mu = 0.1, n = 200, \epsilon_t = Z_t$				
ρ	CLT	Sub_{ET}	Sub_{SYM}	Stock
1	0.79	0.92	0.94	0.96
0.99	0.81	0.91	0.93	0.95
0.95	0.88	0.85	0.95	0.95
0.90	0.93	0.84	0.95	0.94
0.50	0.95	0.87	0.93	NA

$\mu = 0.1, n = 500, \epsilon_t = Z_t$				
ρ	CLT	Sub_{ET}	Sub_{SYM}	Stock
1	0.88	0.91	0.97	0.95
0.99	0.87	0.91	0.96	0.94
0.95	0.93	0.86	0.97	0.95
0.90	0.94	0.85	0.96	NA
0.50	0.94	0.90	0.94	NA

TABLE 12.2. Estimated coverage probabilities of various nominal 95% confidence intervals when the innovations are $\epsilon_t = Z_t Z_{t-1}$, where the Z_t are i.i.d. standard normal. CLT denotes intervals which approximate the sampling distribution of the t-statistic by a standard normal distribution. Sub_{ET} and Sub_{SYM} denote equal-tailed and symmetric subsampling intervals, respectively. Stock denotes the intervals according to Stock (1991). The estimates are based on 1000 replications for each scenario.

	$\mu = 0, n = 200, \epsilon_t = Z_t Z_{t-1}$			
ρ	CLT	Sub_{ET}	Sub_{SYM}	Stock
1	0.70	0.96	0.95	0.93
0.99	0.82	0.96	0.96	0.94
0.95	0.88	0.88	0.96	0.93
0.90	0.88	0.84	0.94	0.94
0.50	0.81	0.77	0.90	NA

	$\mu = 0, n = 500, \epsilon_t = Z_t Z_{t-1}$			
ρ	CLT	Sub_{ET}	Sub_{SYM}	Stock
1	0.69	0.96	0.95	0.94
0.99	0.87	0.92	0.97	0.93
0.95	0.91	0.84	0.96	0.93
0.90	0.92	0.85	0.97	NA
0.50	0.79	0.82	0.91	NA

	$\mu = 0.1, n = 200, \epsilon_t = Z_t Z_{t-1}$			
ρ	CLT	Sub_{ET}	Sub_{SYM}	Stock
1	0.80	0.94	0.96	0.94
0.99	0.81	0.95	0.96	0.93
0.95	0.90	0.88	0.97	0.92
0.90	0.87	0.84	0.95	0.93
0.50	0.82	0.79	0.91	NA

	$\mu = 0.1, n = 500, \epsilon_t = Z_t Z_{t-1}$			
ρ	CLT	Sub_{ET}	Sub_{SYM}	Stock
1	0.86	0.93	0.97	0.94
0.99	0.88	0.93	0.97	0.94
0.95	0.91	0.85	0.97	0.93
0.90	0.90	0.82	0.96	NA
0.50	0.78	0.82	0.91	NA

13
Subsampling Stock Returns

13.1 Introduction

There has been considerable debate in the recent finance literature over whether stock returns are predictable. A number of studies appear to provide empirical support for the use of the current dividend-price ratio, or dividend yield, as a measure of expected stock returns (see, for example, Rozeff, 1984; Campbell and Shiller, 1988b; Fama and French, 1988; Hodrick, 1992; and Nelson and Kim, 1993). The problem with such studies is that stock return regressions face several kinds of statistical problems, among them strong dependency structures and biases in the estimation of regression coefficients. These problems tend to make findings against the no predictability hypothesis appear more significant than they really are.

Having recognized this, Goetzmann and Jorion (1993) argue that previous findings might be spurious and are largely due to the poor small sample performance of commonly used inference methods. They employ a bootstrap approach and conclude that there is no strong evidence indicating that dividend yields can be used to forecast stock returns. One should note, however, that their special approach is not shown to be backed up by theoretical properties. Also, it requires a lot of custom-tailoring to the specific situation at hand. For other scenarios, a different tailoring would be needed.

We intend to help in resolving some of the disagreement by applying the subsampling method, given its model-free nature and robustness against dependence and heteroskedasticity (see Chapter 4). Much of the material presented is extracted from Wolf (2000).

Section 13.2 provides some background that is needed to formulate and appreciate the problem. Also, some previous approaches are considered. Section 13.3 discusses how the subsampling method can be applied to the problem. Small-sample performance is examined via two simulation studies in Section 13.4. Results of applying subsampling to three real data sets are presented in Section 13.5. Finally, some additional results, based on a reorganization of long-horizon regressions and a joint test for multiple return horizons, are provided in Section 13.6. All tables appear at the end of the chapter.

13.2 Background and Definitions

We start out defining the stock return problem and looking at some of the previous studies in more detail. Most of the empirical studies use monthly data. Define the one-period real total return as

$$R_{t+1} = (P_{t+1} + d_{t+1})/P_t, \tag{13.1}$$

where P_t is the end-of-month real stock price and d_t is the real dividends paid during month t. The total return can be decomposed into capital and income return:

$$R_{t+1} = R_{t+1}^C + R_{t+1}^I \equiv P_{t+1}/P_t + d_{t+1}/P_t. \tag{13.2}$$

In computing the dividend yield, the approach of Hodrick (1992) is followed. Since dividend payments are highly seasonal, a monthly annualized dividend series D_t is computed from compounding 12 monthly dividends at the one-month Treasury bill rate r_t:

$$D_t = d_t + (1 + r_t)d_{t-1} + (1 + r_t)(1 + r_{t-1})d_{t-2} + \cdots$$
$$\cdots + (1 + r_t)(1 + r_{t-1})\cdots(1 + r_{t-10})d_{t-11}.$$

Then, annual dividend yield is defined as D_t/P_t. The historic random walk model specifies that the returns R_t are i.i.d. (independent and identically distributed) according to some distribution. The distribution is often assumed to be lognormal, implying that log returns are normal. One implication of this particular model, but also other models implying constant expected returns, is that future returns are unpredictable. In particular, the linear regression model

$$\ln(R_{t+k,t}) = \alpha_k + \beta_k(D_t/P_t) + \epsilon_{t+k,k} \tag{13.3}$$

would have a true β_k coefficient of zero. Here,

$$\ln(R_{t+k,t}) = \ln(R_{t+1}) + \cdots + \ln(R_{t+k})$$

is the continuously compounded k-period return. All of the aforementioned studies are concerned with testing the the null hypothesis $H_0: \beta_k = 0$. Typically, a number of return horizons k are considered, since for some theoretical reasons (e.g., present value model) predictability might be suspected to increase with the return horizon. Most studies are able to reject the null hypothesis at conventional significant levels for all horizons considered, suggesting that future returns can be partially forecasted using present dividend yields. The empirical evidence typically increases with the return horizon.

It is clear that under the null hypothesis the stochastic behavior of the error variables $\epsilon_{t+k,k}$ in (13.3) is completely determined by the stochastic behavior of the R_t process. Even under the random walk model, which is stronger than the null hypothesis of $\beta_k = 0$, the $\epsilon_{t+k,k}$ are uncorrelated only for $k = 1$. For $k > 1$, the errors will always exhibit serial correlation due to the resulting overlap. For example, under the random walk model, they follow a MA$(k-1)$ process. In case the log returns are correlated, or under the alternative hypothesis, the $\epsilon_{t+k,k}$ can be arbitrarily serially correlated for all values of k. The estimation of β_k can be easily done by ordinary least squares, but testing the null hypothesis is nontrivial for a number of reasons.

First, in the case of correlated residuals, as in the case of long-horizon regressions, the usual ordinary least squares (OLS) standard errors are not valid. Second, the independent variable in the regression (13.3) is predetermined but not exogenous. That is to say, D_t/P_t is uncorrelated with the current error term $\epsilon_{t+k,k}$, but it is not generally uncorrelated with past error terms $\epsilon_{t+k-j,k}$, $j > 1$. This is because

$$\epsilon_{t+k-j,k} = \ln(R_{t+k-j,t}) - \alpha_k - \beta_k(D_{t-j}/P_{t-j})$$

and the dividend yield series D_t/P_t is highly persistent at monthly intervals. It is well known that regressions with predetermined independent variables can lead to biased, although consistent, estimates. A standard reference is Stambaugh (1986). In the case of stock return horizons, the OLS estimator $\hat{\beta}_k$ is typically upward biased. Third, there is evidence that the finite sampling distribution of $\hat{\beta}_k$ is skewed to the right (see, for example, Goetzmann and Jorion, 1993).

In the remainder of this chapter, we will discuss various inference methods for β_k according to two criteria: asymptotic consistency and small sample properties.

13.2.1 The GMM Approach

A common approach for making inference for β_k in the context of dependent and possibly heteroskedastic observations is to correct the standard errors of regression coefficient estimates for serial correlation according to the generalized method of moments (GMM) by Hansen and Hodrick (1980) and Hansen (1982). The so corrected standard errors are then utilized in computing the t-statistic corresponding to H_0: $\beta_k = 0$. Most papers that follow this idea base the correction on the additional hypothesis that log returns are uncorrelated, in which case the residuals of a k-horizon regression follow a simple MA($k-1$) process. The GMM method fares well in terms of asymptotic consistency. It has been shown to work under weak and very general conditions (see above references).

Small sample properties, on the other hand, might pose a problem. Since GMM uses asymptotic normality, centered at the true β_k, it accounts neither for the finite sample bias of $\hat{\beta}_k$ nor for its skewness to the right. In addition, there is evidence that the GMM corrections of standard errors for serial correlation are often insufficient in finite samples (for example, see Richardson and Stock, 1989), Ferson and Foerster, 1994, and our Subsection 9.5.2). It is therefore expected that the GMM approximation to the true sampling distribution of $\hat{\beta}_k$ is centered at too small a value and has a right tail that is too short. The consequence is that observed (positive) values of $\hat{\beta}_k$ will be judged as overly significant and hence tests for β_k will be biased toward false rejection of H_0.

Two examples of studies employing the GMM method can be found in Fama and French (1988) and in Chapter 7 of Campbell, Lo, and MacKinley (1997). Both studies reject the null hypothesis $\beta_k = 0$ at conventional significance levels, at least for return horizons of one year and beyond.

13.2.2 The VAR Approach

An alternative approach is to estimate the finite sampling distribution of $\hat{\beta}_k$ under the null hypothesis, and to use this estimated distribution to attach a P-value to the observed value of $\hat{\beta}_k$. To this end, some data generating mechanism that imposes the constraints of the null hypothesis has to be specified.

Campbell and Shiller (1988a), Hodrick (1992), Nelson and Kim (1993), and Goetzmann and Jorion (1995), among others, consider a first-order vector autoregression (VAR) in at least the following two variables: log return and dividend yield. Sometimes, additional variables are included. For example, Campbell and Shiller (1988a) include a term corresponding to earnings price ratio. Hodrick (1992) includes the one-month Treasury-bill return relative to its previous 12-month moving average which is denoted rb_t. To describe his model, let

$$Z_t \equiv [\ln(R_t) - E(\ln(R_t)), D_t/P_t - E(D_t/P_t), rb_t - E(rb_t)]^T.$$

Then a first-order VAR, or VAR(1), is given by

$$Z_{t+1} = AZ_t + u_{t+1}, \tag{13.4}$$

where A is a 3×3 matrix and u_t is a three-dimensional white noise innovation sequence.

Hodrick (1992) fits this model to the observed data and then sets the first row of the estimated VAR(1) matrix equal to zero, and the constant term corresponding to log returns equal to the unconditional mean implied by the original VAR. Of course, specifying the VAR parameters is not sufficient, as an innovation sequence u_t has to be fed to the VAR model. Since there is strong empirical evidence for return data to exhibit conditional heteroskedasticity, Hodrick fits a generalized autoregressive conditionally heteroskedastic (GARCH) model to the fitted innovations. He then generates artificial innovation sequences according to the estimated GARCH process, where the innovations have a conditional normal distribution. Using this approach, Hodrick also finds evidence of predictability in stock returns, both for short and long horizons.

Nelson and Kim (1993) employ a similar method, simulating from a VAR model under the null hypothesis. However, they randomize fitted innovations for the artificial innovation sequences in order to better match the dispersion of the true marginal distribution of the innovations. The disadvantage of this method is that it destroys any potential dependence in the innovation sequence. The study reports that the simulated distributions of the regular t-statistics are upward biased and that these biases should be taken into account when making inferences. However, even after a bias correction, the authors find some evidence for predictability, especially when looking at postwar data.

Unlike the GMM method, the VAR approach tries to capture the finite sampling distribution of $\hat{\beta}_k$ by generating artificial data having the sample size as the observed data. It succeeds in correcting for both upward biases and skewness to some extent, as demonstrated in Nelson and Kim (1993) and Goetzmann and Jorion (1993). However, for many financial data, using GARCH innovations with a conditional normal distribution tends to underestimate the tails of the true sampling distribution (see also Remark 13.4.1). Underestimating the tails will result in overstating the significance of observed $\hat{\beta}_k$ values again. This might explain why the finding of Nelson and Kim (1993), who randomize fitted innovations, are not quite as significant as those of Hodrick (1992). On the other hand, the small sample effect of destroying the correlation in the second moments of the innovations is not clear.

The obvious shortcoming of the VAR approach is the use of a structural model. Asymptotic consistency will only be assured if VAR(1) is the true model. This is doubtful. Of course, how big the asymptotic mistake is depends on how far the true mechanism is away from VAR(1). The problem is magnified if a parametric model for the innovations, such as GARCH(1,1), is used. In addition, it is noteworthy that the VAR model is estimated from monthly, nonoverlapping data. Small mistakes for $k = 1$ will therefore be magnified for long horizons, such as $k = 48$, via adding up k 1-month returns to construct a k-month return. Another shortcoming of the VAR approach, as pointed out by Goetzmann and Jorion (1993), for example, is that it only indirectly models the serial dependence from the lagged price effect in dividend yield regressions: the variable P_t appears both on the right hand side and on the left hand side of equation (13.3). This motivated Goetzmann and Jorion (1993) to develop a bootstrap approach, designed to fix this shortcoming (see below).

Note that it is very awkward to judge the small-sample properties of the VAR method via simulation studies. Hodrick (1992) presents a simulation study that paints a very favorable picture. The problem is that he uses a VAR model with GARCH innovations as the data-generating mechanism in the study, that is, he pretends to know the true mechanism. Such a study is bound to be overly optimistic.

13.2.3 A Bootstrap Approach

As an alternative to the VAR method, Goetzmann and Jorion (1993) use a bootstrap approach to generate artificial data sequences under a model that satisfies the constraints of the null hypothesis (see Section 1.8 for a general reference on this approach). The motivation is that a model-free method such as the bootstrap should avoid any mistakes due to a potentially misspecified structural model. Their particular bootstrap works as follows.

1. Form the empirical distribution of monthly total stock returns R_t and their associated income returns R_t^I, as defined in (13.2), from the observed data.
2. Generate a pseudo return sequence R_t^* i.i.d. according to the empirical distribution of the observed total returns $R_1 \ldots R_n$.
3. Subtract the contemporaneous income returns $R_t^{I,*}$ to create a pseudo capital-return series $R_t^{C,*}$: $R_t^{C,*} = R_t^* - R_t^{I,*}$. Compound these to create a pseudo price series P_t^*.
4. Create a pseudo dividend yield sequence D_t/P_t^*, in which the D_t are the actual annual dividend flows.

It is obvious that some custom-tailoring is employed here in the attempt to capture the relationship between price levels and dividends. The key problem with this approach is seen in the fact that total returns are resampled at random, implying that returns are i.i.d. according to some unknown distribution, while dividend flows remain fixed, implying that dividend payments are completely nonstochastic. By implicitly imposing these two assumptions, this bootstrap is not really model-free anymore. While the first assumption is slightly troublesome—the null hypothesis of a random walk is stronger than the null hypothesis of no predictability—the second one seems unrealistic. For example, in the bootstrap world, dividend payments are completely independent of prices. For this reason, the asymptotic consistency of this bootstrap approach is doubtful. Goetzmann and Jorion (1993) do not discuss the asymptotic properties of their method.

Note that Goetzmann and Jorion come to basically the opposite conclusion of all previous studies. They do not find strong statistical evidence in favor of predictability of stock returns. P-values of the observed $\hat{\beta}_k$ values are typically slightly above 10%, even for long-term horizons.

13.3 The Subsampling Approach

As an alternative inference method, we propose the subsampling method. The test H_0: $\beta_k = 0$ can be carried out implicitly by computing a subsampling confidence interval for β_k and checking whether or not zero is contained in the interval. Subsampling intervals for regression parameters have been shown to have asymptotically correct coverage probability under very weak conditions (see Example 4.4.3). Since the OLS estimator for β_k is an asymptotically linear statistic with a limiting normal distribution, we can use the calibration method of Subsection 9.3.1 to deal with the choice of the block size b.

Remark 13.3.1. Another feature of the subsampling method is that it can be used as a simple and model-free tool to describe the sampling distribution of an estimator. In our application, the distribution of $\hat{\beta}_k - \beta_k$ is approximated by the empirical distribution of the scaled subsample values $(b/n)^{1/2}(\hat{\beta}_{k;n,b,t} - \hat{\beta}_k)$, $t = 1, \ldots, n - b + 1$. Note that in this notation the dependence on n is suppressed for the OLS estimator $\hat{\beta}_k$ based on the entire sample, while $\hat{\beta}_{k;n,b,t}$ denotes the OLS estimator based on the block of size b of the data starting at index t and ending at index $t + b - 1$. By constructing a histogram of these values, one gets a visual picture of the approximating distribution. In particular, the histogram allows us to judge properties such as bias, skewness, departure from normality, and so on. Previous studies have reported that the sampling distribution of $\hat{\beta}_k$ is upward biased and skewed to the right. However, these studies rely on some parametric model as the VAR(1) model (e.g., Hodrick, 1992; Nelson and Kim, 1993) or on some custom-tailored bootstrap (Goetzmann and Jorion, 1993). Since all subsample values come from the true probability mechanism, the corresponding histogram provides a truly model-free alternative. Some results are shown in Section 13.5.

Note that subsample histograms can be smoothed by any of the conventional techniques to get a smoother estimate of the sampling distribution.

The use of subsampling for the description of a sampling distribution, although not its use for the construction of confidence intervals, was also recognized by Sherman and Carlstein (1996).

Remark 13.3.2. A frequent concern in stock return regressions is the high persistence of dividend yield, at least when sampled at monthly intervals. From a statistical viewpoint and the given sample sizes, a unit root process often cannot be discarded. It has been noted in the literature that standard asymptotic methods based on mean-reverting regressors, such as GMM, can give poor small sample approximations when a regressor is nearly-integrated (for example, see Elliot and Stock, 1994, and Cavanagh, Elliot, and Stock, 1995).

An alternative approach, designed to yield improved small sample performance, makes use of *local-to-unity asymptotics* (see, for example, again the two references above). Roughly speaking, this assumes that the largest root of the regressor process is in a $1/n$ neighborhood of one, allowing for a nearly integrated process. Viceira (1997) derives the asymptotic distribution for $\hat{\beta}_1$ under local-to-unity assumptions. Applying his method to four data sets, he does not find evidence for predictability at the 1-month horizon. Note that his particular method could not be applied to long-horizon regression, since the theory requires the regression residuals to be a martingale difference sequence. Even if this is true for $k = 1$, it cannot be true for bigger k due to moving-average-like behavior of the residuals. However, it seems possible to extend the theory to serially correlated error terms.

The subsampling theory presented in this paper also allows for nearly integrated regressors—in the sense of having a *fixed* root near one—and is intended to be an improvement over standard methods in terms of small-sample performance. It would not work, on the other hand, for exactly integrated regressors. We do not feel that this is a very serious restriction, since economic theory speaks against this case. Also, if dividend yield was an integrated process, the same would have to be true for log returns in case β_k differs from zero, by cointegration theory. Since there is strong evidence against a unit root in log returns, this implies that either β_k equals zero or that dividend yield is not exactly integrated. As this paper will not make a case for predictability, we feel justified in focusing on the latter scenario.

13.4 Two Simulation Studies

As noted before, we use two criteria to judge inference methods for β_k, asymptotic consistency and small-sample properties. It was mentioned that both GMM and subsampling give asymptotically correct results under reasonable assumptions, while the VAR method and the Goetzmann and Jorion (1993) bootstrap only work under restrictive conditions. In this section, we compare the small sample performance of GMM and subsampling via simulation studies.

For the simulations, a data-generating mechanism that jointly models log returns and dividend yields is needed. While the true data-generating mechanism will always be unknown, we aim for a reasonable approximation that includes at least two important features. First, the bias of $\hat{\beta}_k$ due to the predetermined predictor should be modeled. Second, the increasing autocorrelation of the residuals with the return horizon k should be addressed. Both features are captured by the VAR model and the Goetzmann and Jorion (1993) bootstrap. Note that it is not a contradiction to employ models for a *simulation study* that were criticized earlier when used for making an *inference*. By definition, some model is needed to generate data in a simulation study. It does not have to be the true model, it only has to capture its important features, to be useful for comparing model-free inference methods, such as GMM or subsampling. On the other hand, if used for making an inference, it must approximate the true model with a high degree of accuracy. For example, getting the marginal distributions right is crucial for making an inference, while it is of relatively minor importance for a simulation study.

13.4.1 Simulating VAR Data

We use a first-order VAR model as the data-generating mechanism, jointly modeling log return and dividend yield as the vector $X_t \equiv (\ln(R_t), D_t/P_t)$. Let

$$Z_t \equiv [\ln(R_t) - E(\ln(R_t)), D_t/P_t - E(D_t/P_t))]^T.$$

Then, the VAR(1) is given by

$$Z_{t+1} = AZ_t + u_{t+1}, \tag{13.5}$$

where A is a 2×2 matrix and u_t is a white noise innovation process. This model is fitted to the observed data by least squares. The null hypothesis can be enforced by setting the first row of A equal to zero.

For the purpose of the simulation study, the overall mean does not matter and can be set equal to zero. Three different data sets are considered, the NYSE equal-weighted and value-weighted indices and the S&P 500 index, all starting in December 1947. Both of the NYSE data sets consist of 480 basic observations (12/1947 to 12/1986), the S&P 500 data set consists of 577 observations (12/1947 to 01/1995). The fitted VAR parameters for the data sets under consideration are presented in Table 13.1.

To generate artificial X_t^* sequences, GARCH(1,1) vector innovation sequences u_t^* are used. Let $H_t = E_t(u_{t+1}u_{t+1}^T)$ be the conditional covariance matrix of the first order VAR in (13.5) with typical element $h_{ij,t}$. The conditional variances and covariance follow ARMA(1,1) processes:

$$h_{ij,t} = \omega_{ij} + \beta_{ij}h_{ij,t-1} + \alpha_{ij}u_{i,t}u_{j,t}, \quad i = 1, 2. \tag{13.6}$$

Equation (13.6) is known as the *Diagonal Vech* model proposed by Bollerslev, Engle, and Woolridge (1988); of course, $\omega_{12} = \omega_{21}$, and so on. It is not the most general multivariate GARCH(1,1) model, since it assumes that the conditional covariance of variables u_1 and u_2 depends on past realizations of the product u_1u_2 only. Nonetheless, it is popular because it is a reasonable way to reduce the number of free parameters to a manageable size. In particular, note that it is more general than the constant correlation model by Bollerslev (1990), which assumes that the conditional correlation is constant over time; this model is used by Hodrick (1992), for example.

The nine parameters of the model (13.6) are estimated by maximum likelihood, assuming conditional normality. The parameter estimates for our three different data sets are reported in Table 13.2. To judge the size of the ω parameters, note that the models were fitted on the percentage scale; that is, a typical monthly return was on the order of 0.5 to 0.8 rather than 0.005 to 0.008. Of course, the α and β parameters do not depend on the choice of scale. Artificial innovation sequences u_t^* as input to the VAR model (13.5) are generated by computing the Cholesky decomposition of the conditional covariance matrix, $C_t^T C_t = H_t$, and setting $u_t^* = C_t^T \epsilon_t$, where ϵ_t is a sequence of independent bivariate standard normal random variables. When generating those sequences, the first 100 observations are discarded to avoid start-up effects. Long-horizon return data X_t^* can be created by feeding the artificial innovation sequences u_t^* into the fitted VAR models, after imposing the null hypothesis by setting the first row of the VAR matrix equal to zero. The first 100 observations are discarded in this step as well. Finally, the artificial long-horizons returns are compounded according to the formula $\ln(R_{t+k,t}^*) = \ln(R_{t+1}^*) + \cdots + \ln(R_{t+k}^*)$. When generating the artificial data, one obviously needs to match the original sample sizes, which are bigger for the S&P 500 data.

For every scenario, we generate 1000 artificial sequences and compute a 95% calibrated subsampling interval for β_k for each sequence. For comparison, we also compute confidence intervals using the GMM method, employing the quadratic spectral (QS) kernel. This kernel was found to have some optimality properties by Andrews (1991). The bandwidth for the kernel was chosen according to the automatic selection procedure of Andrews (1991). The percentages of intervals which contain the true parameter zero are reported in Table 13.3. Note that we carried out the same simulations using conditional innovations having a (scaled) t-distribution with 4 degrees of freedom. The results were essentially the same and are therefore not reported.

One can see that the GMM intervals undercover consistently. In other words, the GMM method is biased towards falsely rejecting the null hypothesis. The undercoverage is already around 5% at the 1-month horizon and it increases in roughly linear fashion to about 30 to 35% at the 48-month horizon! The subsampling intervals perform much better, although they too undercover significantly at long horizons. The improvment of subsampling over GMM intervals is around 5% at the 1-month horizon and it increases to about 20% at the 48-month horizon.

Remark 13.4.1. In Subsection 13.2.2, we commented on the danger of simulating from a VAR model using GARCH innovation sequences in order to compute a P-value for an *observed* statistic such as $\hat{\beta}_k$. Even in the case the fitted VAR model is a good approximation, if the tails of the artificial GARCH sequences are too light, then one overestimates the significance of observed statistics. It is therefore of interest to test whether the estimated innovations are consistent with the corresponding fitted GARCH distribution. Our test statistic is the empirical 0.95 quantile of the estimated innovations. To compute the P-value, we generate 1000 innovation sequences of the corresponding GARCH(1,1) model and compute the empirical 0.95 quantile for each of them. We then calculate the percentage of GARCH quantiles greater than or equal to the test statistic and the percentage of GARCH quantiles less than or equal to the test statistic. The P-value is two times the smaller of these two percentages. To provide some additional information, we also characterize the sampling distribution of the GARCH quantiles by computing the mean, the median, the 0.01 quantile, and the 0.99 quantile of the 1000 numbers. The results are presented in Table 13.4. Except for the log return innovations of the model for the equal-weighted NYSE data, all empirical 0.95 quantiles of the fitted innovations are too big to be compatible with the corresponding GARCH distribution. Four of the two-sided P-values are equal to zero; the other one is around 0.05.

13.4.2 Simulating Bootstrap Data

While the VAR model captures some of the bias of $\hat{\beta}_k$ due to the predetermined predictor, it fails to directly model the lagged priced effect in dividend yield regressions: the variable P_t appears both on the right-hand side and on the left-hand side of equation (13.3). The Goetzmann and Jorion (1993) bootstrap, on the other hand, explicitly models this effect. Therefore, we use it as a second data-generating mechanism for our simulation studies. By design, it imposes the null hypothesis. This approach requires more knowledge than just the two-dimensional series of log returns and dividend yields; it also requires the split of total returns in capital returns and income returns (see Subsection 13.2.3). Of the three data sets available, this information is included only for the S&P 500 data, so we restrict the bootstrap simulations accordingly.

The method is analogous to the one used before. We generate 1000 sequences for each return horizon, compute a subsampling and a GMM 95% confidence interval for each sequence, and check how frequently the true parameter zero is contained in the intervals. The results are reported in Table 13.5, and are very different from the ones of the VAR simulations. GMM intervals exhibit horrendous coverage, as high as 0.77 for $k = 1$ and as low as 0.46 for $k = 48$. Subsampling intervals, on the contrary, have almost constant coverage between 0.97 and 0.98. These numbers are very surprising indeed and do not conform with our intuition that as the return horizon increases, due to the increasing correlation of the residuals, the performance of both GMM and subsampling should get worse. This behavior was noticed in the VAR simulations and also in some related simulations in Section 9.5.2. It may be that the unrealistic relationship of prices and dividend payments in the bootstrap world is the underlying cause.

13.5 A New Look at Return Regressions

In this section, the subsampling methodology is applied to real-life stock return regression data. There is strong consensus in the literature that the time series properties of stock data differ significantly in the pre- and postwar periods. In particular, predictability seems to be mostly a postwar phenomenon (e.g., Hodrick, 1992; Nelson and Kim, 1993). Therefore, only post-war results are provided in this chapter. Another reason to focus on subperiods is that recent work in finance suggests that conditioning on samples from surviving markets renders the *ex post* observed process non-ergodic. For example, Goetzmann and Jorion (1995) show how this problem can bias stock return regressions toward false rejection of the null hypothesis.

We use three different data sets that have been previously analyzed in the literature. Fama and French (1988) and Nelson and Kim (1993) report regressions of log returns for value-weighted and equally weighted stock portfolios based on the CRSP files for NYSE stocks. Goetzmann and Jorion (1993) use monthly data on the S&P 500 index. In accordance with the majority of the literature, return horizons of 1, 12, 24, 36, and 48 months are considered. Both of the NYSE postwar data sets consist of 480 basic observations (12/1947 to 12/1986); the S&P 500 data set consists of 577 observations (12/1947 to 01/1995).

The strategy is to construct 95% confidence intervals for the regression parameter β_k and to check whether or not zero is contained in the intervals. We use two-sided symmetric confidence intervals (see Section 4.2) in conjunction with the calibration technique described in Section 9.3.1. For the reader interested in the details of the implementation, some remarks are in order. They refer to Algorithm 9.3.1.

i. We used the moving blocks bootstrap with block size $b_{MB} = 100$ to generate the pseudo sequences in Step 1.

ii. To find reasonable block sizes for the subsampling method we used Algorithm 9.3.2. The so chosen block sizes were between $b = 40$ and $b = 120$, with the great majority of them between $b = 60$ and $b = 100$.

iii. A practical issue is the number K of bootstrap samples that we generate in order to estimate the calibration function $h(\cdot)$. We chose $K = 1000$.

The resulting confidence intervals are listed in Table 13.6. The table shows that there is no evidence for predictability for horizons of 1, 12, and 24 months, as zero is contained in all corresponding confidence intervals. For the horizon of 36 and 48 months, the findings are inconclusive. The equal-weighted NYSE index intervals contain zero, while the other two do not. For comparison, GMM confidence intervals, using the QS kernel, were also computed (again see Table 13.6). Note that they are much shorter and that only three of them contain zero. Employing GMM would suggest a strong case for predictability.

At this point, it is natural to ask two questions. First, the simulation study in Subsection 13.4.1 suggests some undercoverage of subsampling intervals at long horizons, due to the very strong correlation of the residuals. How much does this evidence take away from the (weak) case for predictability that could be made at the three- and four-year horizons? Second, the five return horizons $k = 1, 12, 24, 36$, and 48 are always considered simultaneously. If focus is on the overall null hypothesis of no predictability rather than in individual hypotheses concerning particular horizons, it seems preferable to derive a test for the joint null hypothesis of $\beta_1 = \beta_{12} = \beta_{24} = \beta_{36} = \beta_{48} = 0$. This avoids the usual pitfalls of multiple testing. Both issues will be addressed in the following section.

In Remark 13.3.1, we discussed how to use subsampling to describe the sampling distribution of $\hat{\beta}_k - \beta_k$. Figure 13.1 displays the corresponding histograms for the return horizons $k = 1$ and $k = 24$ and all three data sets. Note that the histograms are somewhat ragged, since the approximations are based on $n - b + 1$ values, which is on the order of 500 here. Two well-known features of the sampling distribution of $\hat{\beta}_k$, namely, an upward bias and skewness to the right, are clearly visible. The corresponding point estimates $\hat{\beta}_k$ are indicated by vertical lines. Since the lines are well within the histograms for all six scenarios, $\beta_k = 0$ is a reasonable value. This is in accordance with the confidence intervals of Table 13.6.

13.6 Additional Looks at Return Regressions

13.6.1 A Reorganization of Long-Horizon Regressions

Since the compounded k-period return is simply the sum of k one-period returns, the numerator of the regression coefficient β_k in equation (13.3) is given by

$$Cov[\ln(R_{t+1}) + \cdots + \ln(R_{t+k}), (D_t/P_t)]. \tag{13.7}$$

Under the assumption of stationarity, the covariance (13.7) is identical to

$$Cov[\ln(R_{t+1}), (D_t/P_t) + \cdots + (D_{t-k+1}/P_{t-k+1})], \tag{13.8}$$

which is the numerator of β_k^* in the following, reorganized regression

$$\ln(R_{t+1}) = \alpha_k^* + \beta_k^*[(D_t/P_t) + \cdots + (D_{t-k+1}/P_{t-k+1})] + u_{t+1,k}. \tag{13.9}$$

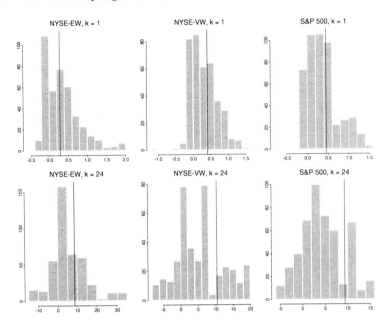

FIGURE 13.1. Histograms of the scaled subsampling values $(b/n)^{1/2}(\hat{\beta}_{k;n,b,t}-\hat{\beta}_k)$, approximating the sampling distribution of $\hat{\beta}_k - \beta_k$. The corresponding point estimates $\hat{\beta}_k$ are indicated by vertical lines.

The test H_0: $\beta_k = 0$ is therefore equivalent to the test H_0: $\beta_k^* = 0$. This fact has been recognized by Hodrick (1992), among others. The advantage of the latter test is that, under the null hypothesis, the stochastic behavior of the error terms $u_{t+1,k}$ in (13.9) is determined by the behavior of the one-period returns $\ln(R_{t+1})$ only, regardless of the horizon k. Hence, the problem of increasing correlation in the error terms due to an increasing return horizon is eliminated.

Hodrick (1992) carries out this alternative test, using critical values obtained by simulating from a VAR model that imposes the null hypothesis. He still finds evidence for predictability at horizons of one year and beyond. The problem is that the critical values might be too small, since the conditionally normal GARCH innovations of the VAR model tend to underestimate the tails of the true sampling distribution.

To provide an alternative answer, we apply the subsampling method. The method of inference for β_k^* is analogous to the method of inference for β_k. The results are reported in Table 13.7. Notice that all subsampling confidence intervals contain zero, and therefore not even at the four-year horizon can a case for predictability be made. On the other hand, all but two of the GMM intervals exclude zero.

Remark 13.6.1. An alternative approach to deal with long return horizons has been developed in Richardson and Stock (1989). Instead of thinking of the return horizon k fixed, while the sample size n tends to infinity, they derive asymptotic theory under the assumption of k/n tending to a positive constant less than one. The idea is to get a better small sample approximation when the return horizon is large compared to the sample size. However, their theory only applies to autocorrelations but not regressions of log returns on another variable such as dividend yield.

13.6.2 A Joint Test for Multiple Return Horizons

We now turn to the problem of making a joint inference for the collection of all β_k considered, that is, the vector $\underline{\beta} = (\beta_1, \beta_{12}, \beta_{24}, \beta_{36}, \beta_{48})'$. The null hypothesis of interest is that $\underline{\beta} = \underline{0} \equiv (0,0,0,0,0)'$. The joint estimation is easily done by combining the individual estimates for each horizon into a vector. However, the joint inference is more complicated. Under reasonable conditions, the vector $\underline{\hat{\beta}} = (\hat{\beta}_1, \hat{\beta}_{12}, \hat{\beta}_{24}, \hat{\beta}_{36}, \hat{\beta}_{48})'$ will have a limiting normal distribution, centered at $\underline{\beta}$. Obviously, the limiting covariance matrix is not diagonal and therefore cannot be estimated by simply combining the individual variance estimates. To make matters worse, the explicit estimation of the 5×5 covariance matrix of $\underline{\hat{\beta}}$ requires along the way the estimation of the 10×10 covariance matrix of $(\hat{\alpha}_1, \hat{\beta}_1, \ldots, \hat{\alpha}_{48}, \hat{\beta}_{48})'$. This does not appear a promising endeavor with sample sizes on the order of 500. Hodrick (1992) runs into this problem when he tries to estimate the limiting covariance matrix by GMM but finds that "simultaneous estimation of the five equations ... results in failure of the GMM matrix to be positive definite". Since, in this instance, the limiting covariance matrix cannot be estimated, Hodrick is unable to test the null hypothesis of $\underline{\beta} = \underline{0}$.

Fortunately, the subsampling methodology can handle multivariate parameters without much difficulty and avoids the problem of *explicit* estimation of the limiting distribution (see Sections 3.3 and 4.3). Here, we will briefly outline how to handle the case $\underline{\hat{\beta}}$. The unknown (multivariate) sampling distribution of $n^{1/2}(\underline{\hat{\beta}} - \underline{\beta})$ is estimated by the empirical distribution of the $n - b + 1$ subsample statistics $b^{1/2}(\underline{\hat{\beta}}_{n,b,t} - \underline{\hat{\beta}})$, $t = 1, \ldots, n-b+1$. With the help of a norm $\|\cdot\|$ on $\mathbb{R}^p$, a confidence region for $\underline{\beta}$ can be found quite easily. Suppose a $1 - \alpha$ confidence region for $\underline{\beta}$ is desired. An asymptotically correct region is given by the collection of all vectors $\underline{\beta}^\dagger$ that satisfy

$$n^{1/2} \left\| \underline{\hat{\beta}} - \underline{\beta}^\dagger \right\| \leq c_{n,b,\|\cdot\|}(1-\alpha). \tag{13.10}$$

Here, $c_{n,b,\|\cdot\|}(1 - \alpha)$ is an $1 - \alpha$ quantile of the univariate "normed" subsampling distribution $L_{n,b,\|\cdot\|}$ having distribution function

$$L_{n,b,\|\cdot\|}(x) = \frac{1}{n-b+1} \sum_{t=1}^{n-b+1} 1\{b^{1/2} \left\| \underline{\hat{\beta}}_{n,b,t} - \underline{\hat{\beta}} \right\| \leq x\}. \tag{13.11}$$

While it would be cumbersome to exhibit explicitly such a confidence region, it is trivial to check whether a specific vector $\underline{\beta}^{\dagger}$ is contained in the region via examining condition (13.10).

The problem of choosing the block size b is analogous to the univariate case and again the calibration technique described in Subsection 9.3.1 can be used The modifications of the algorithm outlined there should be obvious. Note that for Step 2 the explicit computation of the confidence region in Step 1a is not really needed. It is only necessary to check whether $\hat{\underline{\beta}}$ is contained in the region.

When applying this method to three data sets under consideration, one has to be concerned with the magnitudes of the individual coefficients. Note that $\hat{\beta}_k$ naturally will increase with the return horizon k, as a k-horizon compounded return is predicted. It therefore seems sensible to standardize by dividing each estimated regression coefficient by its respective return horizon. Thus, the following modified Euclidean norm is employed.

$$\|(\beta_1, \beta_{12}, \ldots, \beta_{48})'\|_{mod} = \sqrt{\beta_1^2 + (\beta_{12}/12)^2 + \cdots + (\beta_{48}/48)^2}. \quad (13.12)$$

The results are reported in Table 13.8. For all three data sets, a block size of $b = 80$ was used. Since the reader might wish some more details rather than simply whether the vector $(0, 0, 0, 0, 0)'$ is contained in the corresponding confidence region, we give the following information. The observed norm gives the numerical value of $\|(\hat{\beta}_1 - 0, \hat{\beta}_{12} - 0, \ldots, \hat{\beta}_{48} - 0)'\|_{mod}$, with $\|\cdot\|_{mod}$ as defined in (13.12). The observed P-value reports the percentage of subsample statistics $b^{1/2}\|\hat{\underline{\beta}}_{n,b,t} - \hat{\underline{\beta}}\|_{mod}$ exceeding the scaled observed norm $n^{1/2}\|\hat{\underline{\beta}} - \underline{0}\|_{mod}$. Finally, the cut-off point says how small the observed P-value has to be to be deemed significant at the 5% level by the calibration technique of Subsection 9.3.1. In other words, if the observed P-value is bigger than the cut-off point, then the vector $(0, 0, 0, 0, 0)'$ is contained in the 95% confidence region.

For all three data sets, the observed P-value is substantially bigger than the cut-off point at the 5% level. Therefore, in none of the three cases, the null hypothesis can be rejected.

Remark 13.6.2. A related usage of a multiple horizon test is discussed in Richardson (1993). Instead of regressing stock returns on dividend yields, he considers autocorrelations of stock returns. Note that autocorrelations can be thought of regressing stock returns on past stock returns. Similarly to our findings, when Richardson compares the evidence for individual return horizons to the joint evidence for all horizons together, the latter turns out to be weaker. To carry out his test, Richardson uses a Wald test statistic, where he is able to compute the limiting covariance matrix theoretically under the null hypothesis. Thereby he avoids the problem of having to explicitly estimate the matrix. Note that this approach would not work for regressions involving other variables such as dividend yield.

13.7 Conclusions

In this chapter, the subsampling method was utilized to discuss the predictability of stock returns from dividend yields. When comparing subsampling with previous approaches for testing the predictability of stock returns, it was found to be more trustworthy than the VAR approach and Goetzmann and Jorion's (1993) bootstrap on grounds of consistency. A simulation study suggested that subsampling has better small sample properties than GMM, which is a valid competitor in terms of asymptotic properties.

The subsampling method was applied to three different postwar data sets, the NYSE equal- and value-weighted indices and the S&P 500 index, and included five return horizons ranging between one month and four years. No evidence was found for predictability for short and medium horizons, but findings at the long horizons appeared significant. However, some potential undercoverage of subsampling confidence intervals for long horizons due to very strong dependencies in the residuals, as well as the issue of multiple testing, cast some doubt on this evidence.

A reorganization of long-horizon returns, avoiding increasing correlation in the residuals by means of summing dividend yields rather than returns, resulted in insignificant outcomes for all horizons. Moreover, a joint test for all five return horizons also failed to find any evidence. The conclusion therefore is that no convincing case for the predictability of stock returns from dividend yields can be made.

13.8 Tables

TABLE 13.1. Parameter estimates for VAR matrix. This table presents least squares estimates for the VAR matrix A of the following first-order vector-autoregressive model: $Z_{t+1} = AZ_t + u_{t+1}$. Here, Z_t is the joint vector of log return and dividend yield having their respective means subtracted, that is, $Z_t = [\ln(R_t) - E(\ln(R_t)), D_t/P_t - E(D_t/P_t))]^T$, and u_t is white noise. To simplify the notation within the table, we denote the mean zero variables by $\ln(R_t^0)$ and D_t^0/P_t^0, respectively.

NYSE equal-weighted, 12/1947 to 12/1986		
	Coefficients on regressors	
Dependent variable	$\ln(R_t^0)$	D_t^0/P_t^0
$\ln(R_{t+1}^0)$	0.154	0.325
D_{t+1}^0/P_{t+1}^0	-0.004	0.985
NYSE value-weighted, 12/1947 to 12/1986		
	Coefficients on regressors	
Dependent variable	$\ln(R_t^0)$	D_t^0/P_t^0
$\ln(R_{t+1}^0)$	0.062	0.423
D_{t+1}^0/P_{t+1}^0	-0.002	0.984
S&P 500, 12/1947 to 01/1995		
	Coefficients on regressors	
Dependent variable	$\ln(R_t^0)$	D_t^0/P_t^0
$\ln(R_{t+1}^0)$	0.023	0.450
D_{t+1}^0/P_{t+1}^0	0.000	0.985

TABLE 13.2. Parameter estimates for GARCH model. This table presents parameter estimates for the GARCH(1,1) model for the white noise innovation sequence u_t of the VAR. Let $H_t = E_t(u_{t+1} u_{t+1}^T)$ be the conditional covariance matrix of the VAR(1) in (13.5) with typical element $h_{ij,t}$. The conditional variances and covariance follow ARMA(1,1) processes: $h_{ij,t} = \omega_{ij} + \beta_{ij} h_{ij,t-1} + \alpha_{ij} u_{i,t} u_{j,t}$, $i = 1, 2$. All parameters are estimated simultaneously via maximum likelihood, assuming conditional normality.

NYSE equal-weighted, 12/1947 to 12/1986			
Element	ω_{ij}	α_{ij}	β_{ij}
$h_{11,t}$	0.398	0.084	0.894
$h_{12,t}$	-0.017	0.057	0.913
$h_{22,t}$	0.0007	0.044	0.903

NYSE value-weighted, 12/1947 to 12/1986			
Element	ω_{ij}	α_{ij}	β_{ij}
$h_{11,t}$	0.283	0.036	0.937
$h_{12,t}$	-0.012	0.036	0.932
$h_{22,t}$	0.0005	0.040	0.927

S&P 500, 12/1947 to 01/1995			
Element	ω_{ij}	α_{ij}	β_{ij}
$h_{11,t}$	0.257	0.047	0.926
$h_{12,t}$	-0.012	0.058	0.908
$h_{22,t}$	0.0005	0.072	0.891

TABLE 13.3. Estimated coverage probabilities under VAR model. This table presents estimated coverage probabilities of nominal 95% confidence intervals. The data-generating process is a VAR(1) with GARCH(1,1) innovations. The null hypothesis of no predictability is enforced by setting the first row of the VAR matrix equal to zero. Two types of confidence intervals are considered, GMM intervals and calibrated symmetric subsampling intervals. The GMM uses the quadratic spectral kernel with the automatic bandwidth selection procedure of Andrews (1991). Estimated coverage probabilities are based on 1000 simulations for each scenario.

NYSE equal-weighted, 12/1947 to 12/1986		
Horizon	GMM	Subsampling
$k = 1$	0.90	0.95
$k = 12$	0.83	0.93
$k = 24$	0.75	0.91
$k = 36$	0.69	0.86
$k = 48$	0.64	0.81

NYSE value-weighted, 12/1947 to 12/1986		
Horizon	GMM	Subsampling
$k = 1$	0.90	0.97
$k = 12$	0.81	0.93
$k = 24$	0.72	0.90
$k = 36$	0.66	0.85
$k = 48$	0.59	0.80

S&P 500, 12/1947 to 01/1995		
Horizon	GMM	Subsampling
$k = 1$	0.91	0.96
$k = 12$	0.83	0.93
$k = 24$	0.77	0.91
$k = 36$	0.72	0.87
$k = 48$	0.67	0.84

TABLE 13.4. VAR innovation 0.95 quantile: GARCH model vs. observed quantile. This table compares the empirical 0.95 quantile of the estimated VAR innovations with the sampling distribution of the empirical 0.95 quantile from the corresponding GARCH(1,1) model (based on 1000 simulated GARCH(1,1) innovation sequences). The GARCH(1,1) model was obtained via maximum likelihood from the estimated innovations. The sampling distribution is characterized by the 0.01 quantile, the mean, the median, and the 0.99 quantile. The 2-sided P-value tests the null hypothesis that the GARCH(1,1) model gave rise to the estimated innovations.

NYSE equal-weighted, 12/1947 to 12/1986						
	0.01 quant.	Mean	Median	0.99 quant.	Observed	P-value
Log return	4.909	6.751	6.628	10.156	5.554	0.179
Dividend yield	0.176	0.211	0.211	0.252	0.339	0
NYSE value-weighted, 12/1947 to 12/1986						
	0.01 quant.	Mean	Median	0.99 quant.	Observed	P-value
Log return	4.279	5.211	5.179	6.356	7.764	0
Dividend yield	0.170	0.209	0.208	0.255	0.355	0
S%P 500, 12/1947 to 01/1995						
	0.01 quant.	Mean	Median	0.99 quant.	Observed	P-value
Log return	4.073	5.033	5.010	6.324	5.982	0.054
Dividend yield	0.155	0.193	0.191	0.247	0.333	0

TABLE 13.5. Estimated coverage probabilities under bootstrap model. This table presents estimated coverage probabilities of nominal 95% confidence intervals. The data generating process is the Goetzmann and Jorion (1993) bootstrap, as described in Subsection 2.3. Estimated coverage probabilities are based on 1000 simulations for each scenario.

S&P 500, 12/1947 to 01/1995		
Horizon	GMM	Subsampling
$k = 1$	0.77	0.97
$k = 12$	0.60	0.98
$k = 24$	0.54	0.98
$k = 36$	0.50	0.97
$k = 48$	0.46	0.97

TABLE 13.6. 95% confidence intervals for β_k. This table presents 95% confidence intervals for the return regression coefficient β_k, together with the estimated coefficient $\hat{\beta}_k$. We use monthly data, and various return horizons k are considered. The confidence intervals are GMM intervals using the QS kernel and calibrated symmetric subsampling intervals.

NYSE equal-weighted, 12/1947 to 12/1986			
Horizon	$\hat{\beta}_k$	GMM	Subsampling
$k=1$	0.28	[-0.05, 0.61]	[-1.11, 1.67]
$k=12$	4.54	[-0.32, 9.40]	[-11.99, 21.07]
$k=24$	8.70	[-0.41, 17.81]	[-24.55, 41.94]
$k=36$	11.32	[0.72, 21.92]	[-17.90, 40.55]
$k=48$	13.24	[2.59, 23.88]	[-10.55, 37.02]

NYSE value-weighted, 12/1947 to 12/1986			
Horizon	$\hat{\beta}_k$	GMM	Subsampling
$k=1$	0.41	[0.11, 0.72]	[-0.45, 1.28]
$k=12$	5.67	[2.39, 8.95]	[-5.22, 16.57]
$k=24$	10.40	[3.72, 17.07]	[-5.54, 26.34]
$k=36$	13.81	[6.62, 20.99]	[0.08, 27.53]
$k=48$	17.37	[10.07, 24.67]	[7.50, 27.24]

S&P 500, 12/1947 to 01/1995			
Horizon	$\hat{\beta}_k$	GMM	Subsampling
$k=1$	0.45	[0.18, 0.72]	[-0.49, 1.39]
$k=12$	5.47	[3.10, 7.84]	[-5.11, 16.05]
$k=24$	9.41	[5.21, 13.60]	[-2.27, 21.09]
$k=36$	12.38	[7.85, 16.90]	[2.66, 22.10]
$k=48$	15.38	[10.82, 19.95]	[6.88, 23.89]

TABLE 13.7. 95% confidence intervals for β_k^*. This table presents 95% confidence intervals for the return regression coefficient β_k^* of the reorganized regression (13.9), which avoids additional correlation in the residuals for long return horizons. Also, the estimated coefficient $\hat{\beta}_k^*$ is presented. We use monthly data, and various return horizons k are considered. The confidence intervals are GMM intervals using the QS kernel and calibrated symmetric subsampling intervals.

NYSE equal-weighted, 12/1947 to 12/1986			
Horizon	$\hat{\beta}_k^*$	GMM	Subsampling
$k = 1$	0.281	[-0.052, 0.614]	[-1.109, 1.671]
$k = 12$	0.035	[0.006, 0.064]	[-0.195, 0.266]
$k = 24$	0.019	[0.004, 0.035]	[-0.237, 0.276]
$k = 36$	0.011	[0.001, 0.022]	[-0.125, 0.147]
$k = 48$	0.007	[-0.002, 0.016]	[-0.065, 0.078]

NYSE value-weighted, 12/1947 to 12/1986			
Horizon	$\hat{\beta}_k^*$	GMM	Subsampling
$k = 1$	0.412	[0.107, 0.718]	[-0.453, 1.278]
$k = 12$	0.043	[0.017, 0.069]	[-0.150, 0.235]
$k = 24$	0.022	[0.008, 0.036]	[-0.104, 0.147]
$k = 36$	0.013	[0.003, 0.023]	[-0.134, 0.160]
$k = 48$	0.009	[0.001, 0.017]	[-0.036, 0.054]

S&P 500, 12/1947 to 01/1995			
Horizon	$\hat{\beta}_k^*$	GMM	Subsampling
$k = 1$	0.450	[0.187, 0.713]	[-0.485, 1.385]
$k = 12$	0.040	[0.017, 0.062]	[-0.069, 0.148]
$k = 24$	0.019	[0.007, 0.030]	[-0.057, 0.094]
$k = 36$	0.011	[0.003, 0.019]	[-0.053, 0.075]
$k = 48$	0.008	[0.001, 0.014]	[-0.031, 0.046]

TABLE 13.8. Joint test for $\beta_1 = \beta_{12} = \ldots = \beta_{48} = 0$. This table presents results for the joint test of all individual regression coefficients being equal to zero. The observed norm gives the numerical value of $||(\hat{\beta}_1 - 0, \hat{\beta}_{12} - 0, \ldots, \hat{\beta}_{48} - 0)'||_{mod}$. The observed P-value reports the percentage of subsample statistics $b^{1/2}||\hat{\underline{\beta}}_{b,a} - \hat{\beta}||_{mod}$ exceeding the scaled observed norm $n^{1/2}||\hat{\beta} - \underline{0}||_{mod}$. Finally, the cut-off point says how small the observed P-value has to be to be deemed significant at the 5% level by the calibration technique. In other words, if the observed P-value is bigger than the cut-off point, then the vector $(0, 0, 0, 0, 0)'$ is contained in the 95% confidence region.

NYSE equal-weighted, 12/1947 to 12/1986		
Observed Norm	Observed P-value	Cut-off point for 0.05 test
0.727	0.518	0.027
NYSE value-weighted, 12/1947 to 12/1986		
Observed Norm	Observed P-value	Cut-off point for 0.05 test
0.927	0.385	0.060
S&P 500, 12/1947 to 01/1995		
Observed Norm	Observed P-value	Cut-off point for 0.05 test
0.850	0.268	0.081

Appendix A
Some Results on Mixing

In what follows, $\{\ldots, X_{-1}, X_0, X_1, \ldots\}$ is a sequence of random variables taking values in an arbitrary sample space S, and defined on a common probability space. Denote the joint probability law governing the infinite sequence by P. The sequence $\{X_t\}$ will be assumed to satisfy a certain weak dependence condition. To make this condition precise, we introduce the concept of strong mixing coefficients. The original definition, by Rosenblatt (1956), applies to stationary sequences and a modification for arbitrary sequences is needed for our purposes.

Definition A.0.1. Given a random sequence $\{X_t\}$, let $\mathcal{F}_n^m$ be the σ-algebra generated by $\{X_t, n \leq t \leq m\}$, and define the corresponding α-mixing sequence by

$$\alpha_X(k) = \sup_n \sup_{A,B} |P(A \cap B) - P(A)P(B)|, \tag{A.1}$$

where A and B vary over the σ-fields $\mathcal{F}_{-\infty}^n$ and $\mathcal{F}_{n+k}^\infty$, respectively. Note that in case the sequence $\{X_t\}$ is strictly stationary, the $\sup_n$ in this definition becomes redundant. The sequence $\{X_t\}$ is called α-*mixing* or *strong mixing* if $\alpha_X(k) \to 0$ as $k \to \infty$.

The α-mixing coefficient is basically a measure of how much dependence can exist between two events that are at least k (time) units apart. With a strong mixing sequence there is associated a certain notion of asymptotic independence, because $\alpha(k) = 0$ implies that two events that are at least k (time) units apart have to be independent. There exist some other, but less commonly used, mixing coefficients, namely, the β-mixing, ϕ-mixing,

ψ-mixing, and ρ-mixing coefficients. Since they are not used in this book, the reader is referred to Doukhan (1994) for the definitions. However, a special case of α-mixing is known by its particular name:

Definition A.0.2. The sequence $\{X_t\}$ is called *m-dependent* if $\alpha_X(k) = 0$, for $k > m$.

The following two mixing inequalities are of great practical importance. Here, and throughout the Appendices, we use the well-known concept of the *p*-norm of a random variable X:

$$\|X\|_p = (E|X|^p)^{\frac{1}{p}}, \ p \geq 1.$$

Lemma A.0.1. *Let $\{X_t\}$ be a random sequence with corresponding mixing sequence $\alpha_X(\cdot)$. Let the random variables ξ and ζ be $\mathcal{F}_{-\infty}^n$ and $\mathcal{F}_{n+k}^\infty$ measurable, respectively, with $E|\xi|^p$, $E|\zeta|^q < \infty$ for some $p, q > 1$ such that $\frac{1}{p} + \frac{1}{q} < 1$.*

Then, $|Cov(\xi, \zeta)| \leq 8 \|\xi\|_p \|\zeta\|_q \alpha_X(k)^{1 - \frac{1}{p} - \frac{1}{q}}$.

In case the random variables ξ, ζ are bounded with probability one, a sharper result is true.

Lemma A.0.2. *Let $\{X_t\}$ be a random sequence with corresponding mixing sequence $\alpha_X(\cdot)$. Let the random variables ξ and ζ be $\mathcal{F}_{-\infty}^n$ and $\mathcal{F}_{n+k}^\infty$ measurable, respectively, with $|\xi| \leq C_1, |\zeta| \leq C_2$ almost surely.*

Then, $|Cov(\xi, \zeta)| \leq 4 C_1 C_2 \alpha_X(k)$.

Lemma A.0.1 is due to Wolkonoski and Rozanov (1959) in the case $p = q = \infty$, and to Davydov (1970) in its present form. Lemma A.0.2 usually is given as a first step in order to prove Lemma A.0.1. Proofs of both results can be found, for example, in the appendix of Hall and Heyde (1980).

If the sequence $\{X_t\}$ is m-dependent with $E|X_t|^r < \Delta < \infty$ for all t, then it is well known that $E|\sum_{t=1}^n X_t|^r = O(n^{r/2})$. It turns out that an extension of this inequality to strong mixing nonstationary random variables is possible. The result is implicitly contained in a theorem of Doukhan (1994). However, the form in which the theorem is presented is not convenient for our purposes; thus we restate it below. Also, we will give more specific bounds for some special cases. A related bound assuming stationarity was given in Yokoyama (1980).

Lemma A.0.3. *Let $\{X_t\}$ be a sequence of mean zero random variables. Denote the corresponding mixing sequence by $\alpha_X(\cdot)$. Define, for $\tau \geq 2$ and $\delta > 0$*

$$C(\tau, \delta) \equiv \sum_{k=0}^\infty (k+1)^{\tau - 2} \alpha_X^{\frac{\delta}{\tau + \delta}}(k), \qquad (A.2)$$

A. Some Results on Mixing 317

$$L(\tau, \delta, d) \equiv \sum_{t=1}^{d} \|X_t\|_{\tau+\delta}^{\tau}, \tag{A.3}$$

$$D(\tau, \delta, d) \equiv Max\{L(\tau, \delta, d), [L(2, \delta, d)]^{\frac{\tau}{2}}\}. \tag{A.4}$$

Then, the following bound holds:

$$E\left|\sum_{t=1}^{d} X_t\right|^{\tau} \leq BD(\tau, \delta, d), \tag{A.5}$$

where B is a constant only depending on τ, δ and the mixing coefficients $\alpha_X(\cdot)$. We will be specific about the constant B for the two special cases τ is an even integer and $\tau = 2 + \delta$.

1. If c is an even integer

$$E\left|\sum_{t=1}^{d} X_t\right|^{c} \leq B(c, \delta) D(c, \delta, d), \tag{A.6}$$

where bounds for the constants $B(c, \delta)$ can be computed recursively. For example, for c up to 4:

$$B(1, \delta) \leq 1,$$
$$B(2, \delta) \leq 18 Max\{1, C(2, \delta)\},$$
$$B(3, \delta) \leq 102 Max\{1, C(3, \delta)\},$$
$$B(4, \delta) \leq 3024 Max\{1, C^2(4, \delta)\}.$$

2. For $\tau = 2 + \delta$,

$$B \leq \left[3024 Max\{1, C^2(4, \delta)\}\right] 2^{4(4(2-\delta)/\delta+1)}. \tag{A.7}$$

Proof. This result is implicitly contained in Theorem 2 of Section 1.4 in Doukhan (1994). For the proof, we will need a little computational lemma.

Lemma A.0.4. *For integers a, b, $D(a, \delta, d) D(b, \delta, d) \leq D(a+b, \delta, d)$.*

This is Lemma 2 in Section 1.4 of Doukhan (1994).

Step 1: We will prove Lemma A.0.3 for the case $\tau = c$ an even integer first. The case for general τ will then follow by the use of an interpolation lemma. Since c is an even integer:

$$E\left|\sum_{t=1}^{d} X_t\right|^{c} \leq c! \sum_{t_1 \leq \ldots \leq t_c} |E(X_{t_1} \cdots X_{t_c})| \equiv c! A(c, d). \tag{A.8}$$

Let $r = r(i_1, \ldots, i_c)$ be the largest interval among successive points in the sequence $\{t_1, \ldots, t_c\}$: $r = Max_{1 \leq j \leq c-1}(i_{j+1} - i_j) \equiv i_{m+1} - i_m$. Set

$$\nu = \frac{c+\delta}{m}, \quad \mu = \frac{c+\delta}{c-m} \quad \text{and} \quad \zeta = \frac{\delta}{c+\delta}.$$

Using Lemma A.0.1 and the Hölder inequality yields

$$\left| E(X_{t_1} \cdots X_{t_c}) - E(X_{t_1} \cdots X_{t_m}) E(X_{t_{m+1}} \cdots X_{t_c}) \right|$$
$$\leq 8\alpha_X^\zeta(r) \|X_{t_1} \cdots X_{t_m}\|_\mu^{m/c} \|X_{t_{m+1}} \cdots X_{t_c}\|_\nu^{(c-m)/c}$$
$$\leq 8\alpha_X^\zeta(r) \prod_{j=1}^{c} \|X_{t_j}\|_{c+\delta}.$$

Applying this inequality repeatedly, we obtain

$$A(c,d) \leq 8C(c,\delta) \sum_{1 \leq t \leq d} \|X_t\|_{c+\delta}^c + \sum_{m=1}^{c-1} A(m,d) A(c-m,d).$$

Let $\tilde{B}(c,\delta,d) = A(c,m)/D(c,\delta,d)$. Dividing the last inequality by $D(c,\delta,d)$ and applying Lemma A.0.4, we get

$$\tilde{B}(c,\delta,d) \leq 8C(c,\delta) + \sum_{m=1}^{c-1} \tilde{B}(c,\delta,d) \tilde{B}(c,\delta,c-m).$$

We can compute bounds for the $\tilde{B}(c,\delta,d)$'s recursively,

$$\tilde{B}(1,\delta,d) \leq 1,$$
$$\tilde{B}(2,\delta,d) \leq 8C(2,\delta) + \tilde{B}(1,\delta,d)^2,$$
$$\tilde{B}(3,\delta,d) \leq 8C(3,\delta) + \tilde{B}(1,\delta,d)\tilde{B}(2,\delta,d),$$
$$\tilde{B}(4,\delta,d) \leq 8C(4,\delta) + \tilde{B}(1,\delta,d)\tilde{B}(3,\delta,d) + \tilde{B}(2,\delta,d)^2.$$

Assuming for the moment that $1 \leq C(1) \leq C(2) \leq C(3)\ldots$, we are able to obtain cruder, but more explicit, bounds:

$$\tilde{B}(1,\delta,d) \leq 1,$$
$$\tilde{B}(2,\delta,d) \leq 9C(2,\delta),$$
$$\tilde{B}(3,\delta,d) \leq 8C(3,\delta) + 9C(2,\delta)$$
$$\leq 17C(3,\delta),$$
$$\tilde{B}(4,\delta,d) \leq 8C(4,\delta) + 17C(3,\delta) + 81C^2(2,\delta)$$
$$\leq 126C^2(4,\delta). \tag{A.9}$$

Fortunately, these bounds do not depend on d. With inequality (A.8) in mind we can therefore choose $B(c,\delta) = c! \tilde{B}(c,\delta,d), c = 1,2,\ldots$, giving rise to the the constants in (A.7).

Step 2: An interpolation lemma due to Uteev (1984) and given as Lemma 1 in Section 1.4 of Doukhan (1994) allows us to use bounds for even integer moments to get the desired result for arbitrary, real τ.

Lemma A.0.5 (Interpolation Lemma). *Let $F = (\mathcal{F}_t)_{1 \leq t \leq d}$ be a family of sub σ-algebras of $\mathcal{A}$, and suppose B is a separable Banach space. A family of random variables $\eta = (\eta_t)_{1 \leq t \leq d}$, defined on a space $(\Omega, \mathcal{A})$ is said to*

A. Some Results on Mixing

be (F, B)-adapted if the random variable η_t is B-valued, $\mathcal{F}_t$-measurable and has expectation zero for $t = 1, \ldots, n$. Set, for $\nu \geq 2$ and $\delta > 0$,

$$M(\nu, \delta, \eta) \equiv \sum_{t=1}^{d} \left(E \|\eta_t\|^{\nu+\delta} \right)^{\frac{\nu}{\nu+\delta}},$$

$$Q(\nu, \delta, \eta) \equiv Max\{M(\nu, \delta, \eta), [M(2, \delta, \eta)]^{\frac{\nu}{2}}\}.$$

Assume that, for some $\nu \geq 2$ and a fixed constant c, any family $\eta = (\eta_t)_{1 \leq t \leq d}$, (F, B)-adapted satisfies

$$E \left\| \sum_{t=1}^{d} \eta_t \right\|^{\nu} \leq c\, Q(\nu, \delta, \eta).$$

Then, for any $2 \leq \tau \leq \nu$, there exists a constant

$$C = C(c, \nu, \delta, \tau) = c\, 2^{4(\nu(\nu-\tau)/\delta + 1)}$$

such that any family $\phi = (\phi)_{1 \leq t \leq d}, (F, B)$-adapted satisfies:

$$E \left\| \sum_{t=1}^{d} \phi_t \right\|^{\tau} \leq CQ(\tau, \delta, \phi).$$

In the interpolation lemma, the inequality for even powers (A.6) is applied to truncated random variables, which are bounded and consequently admit moments of arbitrary order. In applying the lemma, let

$$\phi_t = X_t, \quad \nu = Min_c\{c \geq \tau,\ c \text{ an even integer}\}, \text{ and}$$

$$\eta_t = \text{various truncated random variables based on } X_t.$$

Details can be found in the proof of Doukhan's Lemma 1. For the special case $\tau = 2 + \delta$, we have $\nu = 4$ (assuming without loss of generality $\delta \leq 2$), and inequality (A.7) follows easily with the bounds (A.6) and (A.9). ∎

In case we have a uniform bound on the $2 + 2\delta$ moments of the sequence $\{X_t\}$, a less sharp but more concisely stated bound can be obtained. It is in a form most useful for our purposes in previous chapters.

Corollary A.0.1. *Assume, in addition to the conditions of Lemma A.0.3, that*

$$\|X_t\|_{2+2\delta} \leq \Delta \quad \text{for all } t.$$

Then,

$$E \left| \sum_{t=1}^{d} X_t \right|^{2+\delta} \leq \Gamma d^{1+\frac{\delta}{2}},$$

where Γ is a constant that only depends on Δ, δ and the mixing coefficients $\alpha_X(\cdot)$. More explicitly,

$$\Gamma = \left[3024\,Max\{1, C^2(4,\delta)\}\right] 2^{4(4(2-\delta)/\delta+1)} \Delta^{(2+\delta)(1+\frac{\delta}{2})},$$

where $C(4,\delta)$ is defined as in (A.2).

Appendix B
A General Central Limit Theorem

Extensions of the classical Central Limit theorems for independent variables to the dependent case were introduced in the literature a long time ago. Central Limit theorems for strong mixing random variables have been proved by Rosenblatt (1956), Ibragimov (1962), Oodaira and Yoshihara (1972), White and Domowitz (1984) and many others. A survey of the literature can be found in Doukhan (1994). Note that in many cases strict stationarity was assumed in addition to moment and mixing conditions. We will present a theorem here that further relaxes the conditions of previous results, since it applies to both nonstationary sequences and triangular array settings. The theorem is interesting in its own right. However, we mainly need it in this book to prove the applicability of the subsampling method in various specific situations.

Theorem B.0.1. *Let $\{X_{n,t}, 1 \leq t \leq d_n\}$ be a triangular array of mean zero random variables. Denote the mixing sequence corresponding to the n-th row by $\alpha_n(\cdot)$. Define*

$$S_{n,k,a} \equiv \sum_{t=a}^{a+k-1} X_{n,t},$$

$$T_{n,k,a} \equiv k^{-\frac{1}{2}} \sum_{t=a}^{a+k-1} X_{n,t},$$

$$\sigma^2_{n,k,a} \equiv Var(T_{n,k,a}).$$

B. A General Central Limit Theorem

Assume the following conditions hold: For some $\delta > 0$:

- $\|X_{n,t}\|_{2+2\delta} \leq \Delta$ *for all n, t,* (B.1)
- $\sigma^2_{n,k,a} \to \sigma^2 > 0$ *uniformly in a (*),* (B.2)
- $C_n(4) \equiv \sum_{k=0}^{\infty} (k+1)^2 \alpha_n^{\frac{\delta}{4+\delta}}(k) \leq K$ *for all n,* (B.3)

where Δ and K are finite constants independent of n, k or a.

(*) *This means: For any sequence $\{k_n\}$ that tends to infinity with n, $\sup_a |\sigma^2_{n,k_n,a} - \sigma^2| \to 0$ as $n \to \infty$.*

Then, $T_{n,d_n,1} \stackrel{\mathcal{L}}{\Longrightarrow} N(0, \sigma^2)$, that is, $d_n^{-\frac{1}{2}} \sum_{t=1}^{d_n} X_{n,t} \stackrel{\mathcal{L}}{\Longrightarrow} N(0, \sigma^2)$.

Remark B.0.1. This result exhibits the familiar trade-off between moment and dependence restrictions, as expressed by conditions (B.1) and (B.3). The larger δ is, which corresponds to the largest finite moment, the lower the minimum mixing rate can be. If all moments exist ($\delta = \infty$), the mixing condition becomes essentially $\alpha_n(k) = o(k^{-3})$. For δ close to zero, on the other hand, the processes must be nearly independent.

Remark B.0.2. Condition (B.2) states that the variance of a standardized block sum has to be close to some (positive) limiting value as long as the block size is large enough, independent of row or block index. This condition is not as restrictive as it might appear at first glance. In fact, it is difficult to imagine a reasonable situation where $\sigma^2_{n,1} \to \sigma^2$, but where condition (B.2) is violated. Two examples where such a condition holds, even though the underlying sequence may not be covariance stationary, are a model with seasonal effects and a Markov chain possessing an equilibrium distribution.

Proof of Theorem B.0.1. In the proof of the theorem we will approximate a sum of weakly dependent random variables by a corresponding sum of independent random variables. The following lemma will help us to establish an upper bound on the sup difference of the corresponding characteristic functions. The idea of bounding this sup difference by mixing coefficients goes back to Ibragimov (1962).

Lemma B.0.6. *Let $\{X_t\}$ denote a sequence of random vectors defined on a probability space, and let $\mathcal{F}_a^b \equiv \sigma(X_t; a \leq t \leq b)$. Also denote the mixing sequence corresponding to the Z_t by $\alpha_X(\cdot)$.*

Let Y_1 and Y_2 be random variables measurable with respect to $\mathcal{F}_{-\infty}^n$ and $\mathcal{F}_{n+m}^{\infty}$, respectively. In addition let Y_1' and Y_2' be independent random variables having the same distribution as Y_1 and Y_2, respectively. Denote the characteristic functions of $Y_1 + Y_2$ and $Y_1' + Y_2'$ by φ and φ', respectively.

Then, $\sup_t |\varphi(t) - \varphi'(t)| \leq 16 \alpha_X(m)$.

Proof.
$$|\varphi(t) - \varphi'(t)| = |E(e^{itY_1}e^{itY_2}) - E(e^{itY_1})E(e^{itY_2})|$$
$$\leq |E(\cos tY_1 \cos tY_2) - E(\cos tY_1)E(\cos tY_2)| +$$
$$|E(\cos tY_1 \sin tY_2) - E(\cos tY_1)E(\sin tY_2)| +$$
$$|E(\sin tY_1 \cos tY_2) - E(\sin tY_1)E(\cos tY_2)| +$$
$$|E(\sin tY_1 \sin tY_2) - E(\sin tY_1)E(\sin tY_2)|$$
$$\leq 4 * 4\alpha_X(m) \quad \text{(by Lemma A.0.2)}. \blacksquare$$

We will now prove the theorem. The main idea of the proof is to split the sum $X_{n,1} + \cdots + X_{n,d_n}$ into alternate blocks of length b_n (the big blocks) and l_n (the small blocks). This is the traditional approach to proving central limit theorems for dependent random variables, and is commonly attributed to Markov or Bernstein (1927) ("Bernstein sums"). Define
$$U_{n,t} = X_{n,(t-1)(b_n+l_n)+1} + \cdots + X_{n,(t-1)(b_n+l_n)+b_n}, \quad 1 \leq t \leq r_n,$$
where r_n is the largest integer i for which $(t-1)(b_n+l_n)+b_n < d_n$. Further define
$$V_{n,t} = X_{n,(t-1)(b_n+l_n)+b_n+1} + \cdots + X_{n,i(b_n+l_n)+b_n}, \quad 1 \leq t \leq r_n,$$
$$V_{n,r_n} = X_{n,(r_n-1)(b_n+l_n)+b_n+1} + \cdots + X_{n,d_n}.$$
Then $S_{n,d_n,1} = \sum_{t=1}^{r_n} U_{n,t} + \sum_{t=1}^{r_n} V_{n,t}$, and the technique will be to choose the l_n small enough that $\sum_{t=1}^{r_n} V_{n,t}$ is small in comparison with $\sum_{t=1}^{r_n} U_{n,t}$ but large enough to ensure that the $U_{n,t}$ are nearly independent.

Let $b_n = \lfloor d_n^{\frac{3}{4}} \rfloor$ and $l_n = \lfloor d_n^{\frac{1}{4}} \rfloor$, where $\lfloor \cdot \rfloor$ denotes the integer part of a real number. Since r_n is the largest integer i such that $(t-1)(b_n+l_n)+b_n < d_n$,
$$b_n \sim d_n^{\frac{3}{4}}, \quad l_n \sim d_n^{\frac{1}{4}}, \quad r_n \sim d_n^{\frac{1}{4}}. \tag{B.4}$$

We will now proceed to show that $d_n^{-\frac{1}{2}} \sum_{t=1}^{r_n} V_{n,t}$ converges to zero in probability, as d_n tends to infinity. Since its expected value equals zero, it suffices to check that its variance tends to zero. First note that by Lemma A.0.3 and assumption (B.3), for all n, t
$$E|V_{n,t}|^2 \leq 10K\Delta^2 l_n \equiv Bl_n. \tag{B.5}$$

Therefore,
$$Var^{\frac{1}{2}}(d_n^{-\frac{1}{2}} \sum_{t=1}^{r_n} V_{n,t}) = E^{\frac{1}{2}}(d_n^{-\frac{1}{2}} \sum_{t=1}^{r_n} V_{n,t})^2$$
$$\leq \sum_{t=1}^{r_n} E^{\frac{1}{2}}(d_n^{-\frac{1}{2}} V_{n,t})^2$$
$$\leq [\sum_{t=1}^{r_n - 1}(Bl_n/d_n)^{\frac{1}{2}}] + [B(l_n + b_n)/d_n]^{\frac{1}{2}} \quad \text{(by (B.5))}$$
$$\leq B^{\frac{1}{2}}\{r_n(l_n/d_n)^{\frac{1}{2}} + [(l_n + b_n)/d_n]^{\frac{1}{2}}\}$$

B. A General Central Limit Theorem

$$\leq O(d_n^{-\frac{1}{8}}) \quad \text{(by (B.4))}.$$

By Slutzky's Theorem, it remains to prove that $d_n^{-\frac{1}{2}} \sum_{t=1}^{r_n} U_{n,t} \Rightarrow N(0, \sigma^2)$.

Let $U'_{n,t}$, $1 \leq t \leq r_n$, be independent random variables having the same distributions as $U_{n,t}, 1 \leq t \leq r_n$. By Lemma B.0.6 applied inductively, the characteristic functions of $d_n^{-\frac{1}{2}} \sum_{t=1}^{r_n} U_{n,t}$ and of $d_n^{-\frac{1}{2}} \sum_{t=1}^{r_n} U'_{n,t}$ differ by at most $16 r_n \alpha_n(l_n)$. Note that by assumption (B.3) we may assume—without loss of generality—that $\alpha_n(k) \leq K/k^2$. Therefore,

$$16 r_n \alpha_n(l_n) \leq 16 r_n K/l_n^2$$
$$\leq O(d_n^{-\frac{1}{4}}) \quad \text{(by (B.4))}.$$

Thus, the proof will be completed by showing that

$$d_n^{-\frac{1}{2}} \sum_{t=1}^{r_n} U'_{ni,} \Rightarrow N(0, \sigma^2).$$

This will be accomplished in two steps:

Step 1: $[Var(\sum_{t=1}^{r_n} U'_{n,t})]^{\frac{2}{2+\delta}} \sum_{t=1}^{r_n} E|U'_{n,t}|^{2+\delta} \to 0$.

Step 2: $\frac{1}{d_n} Var(\sum_{t=1}^{r_n} U'_{n,t}) \to \sigma^2$.

The result then follows by Lyapounov's Central Limit theorem and Slutsky's theorem.

Proof of Step 1:

$$\frac{1}{r_n b_n} Var(\sum_{t=1}^{r_n} U'_{n,t}) = \frac{1}{r_n b_n} \sum_{t=1}^{r_n} E(U'_{n,t})^2$$
$$= \frac{1}{r_n b_n} \sum_{t=1}^{r_n} E(U_{n,t})^2$$
$$= \frac{1}{r_n} \sum_{t=1}^{r_n} E(b_n^{-\frac{1}{2}} U_{n,t})^2$$
$$\to \sigma^2 \quad \text{(by assumption (B.2))}. \tag{B.6}$$

Assume without loss of generality that $1 \leq K$. Then, by Corollary A.0.1,

$$E|b_n^{-\frac{1}{2}} U'_{n,t}|^{2+\delta} \leq [3024 \text{Max}\{1, C_n^2(4)\}] \, 2^{4(4(2-\delta)/\delta+1)} \Delta^{(2+\delta)(1+\frac{\delta}{2})}$$
$$\leq [3024 K^2] \, 2^{4(4(2-\delta)/\delta+1)} \Delta^{(2+\delta)(1+\frac{\delta}{2})} \quad \text{(by (B.3))}$$
$$\equiv \Delta. \tag{B.7}$$

Finally,

$$[Var(\sum_{t=1}^{r_n} U'_{n,t})]^{\frac{2}{2+\delta}} \sum_{t=1}^{r_n} E|U'_{n,t}|^{2+\delta}$$

B. A General Central Limit Theorem

$$= (1/r_n)^{\frac{2+\delta}{2}} [\frac{1}{r_n b_n} Var(\sum_{t=1}^{r_n} U'_{n,t})]^{\frac{2}{2+\delta}} \sum_{t=1}^{r_n} E|b_n^{-\frac{1}{2}} U'_{n,t}|^{2+\delta}$$

$$\leq (1/r_n)^{\frac{2+\delta}{2}} [\frac{1}{r_n b_n} Var(\sum_{t=1}^{r_n} U'_{n,t})]^{\frac{2}{2+\delta}} r_n \Delta \quad \text{(by (B.7))}$$

$$\leq (1/r_n)^{\frac{2+\delta}{2}} O(1) O(r_n) \quad \text{(by (B.6))}$$

$$\leq O\left((r_n)^{-\frac{\delta}{2}}\right).$$

Proof of Step 2: We just showed that $\frac{1}{r_n b_n} Var(\sum_{t=1}^{r_n} U'_{n,t}) \to \sigma^2$. But, by (B.4), $\frac{r_n b_n}{d_n} \to 1$. ∎

References

Anderson, T. (1993). Goodness of fit tests for spectral distributions. *Annals of Statistics* **21**, 830–847.

Andrews, D.W.K. (1984). Non-strong mixing autoregressive processes. *Journal of Applied Probability* **21**, 930–934.

Andrews, D.W.K. (1991). Heteroskedasticity and autocorrelation consistent covariance matrix estimation. *Econometrica* **59**, 817–858.

Andrews, D.W.K. and Monahan, J.C. (1992). An improved heteroskedasticity and autocorrelation consistent covariance matrix estimator. *Econometrica* **60**, 953–966.

Arcones, M. (1991). On the asymptotic theory of the bootstrap. Ph.D. thesis, The City University of New York.

Arcones, M. and Giné, E. (1989). The bootstrap of the mean with arbitrary bootstrap sample size. *Annals of the Institute Henri Poincaré* **25**, 457–481.

Arcones, M. and Giné, E. (1991). Additions and correction to "the bootstrap of the mean with arbitrary bootstrap sample size." *Annals of the Institute Henri Poincaré* **27**, 583–595.

Arcones, M. and Yu, B. (1994). Central limit theorems for empirical and U-processes of stationary mixing sequences. *Journal of Theoretical Probability* **7**, 47–71.

Arunkumar, S. (1972). Nonparametric age replacement policy. *Sankhya*, Series A **34**, 251–256.

Athreya, K. (1985). Bootstrap of the mean in the infinite variance case, II. Technical Report 86-21, Department of Statistics, Iowa State University.

Athreya, K. (1987). Bootstrap of the mean in the infinite variance case. *Annals of Statistics* **15**, 724–731.

Babu, G. (1984). Bootstrapping statistics with linear combinations of chi-squares as weak limit. *Sankhya Series A* **56**, 85–93.

Babu, G. (1992). Subsample and half-sample methods. *Annals of the Institute of Statistical Mathematics* **44**, 703–720.

Babu, G. and Singh, K. (1985). Edgeworth expansions for sampling without replacement for finite populations. *Journal of Multivariate Analysis* **17**, 261–278.

Barbe, P. and Bertail, P. (1995). The weighted bootstrap. *Lecture Notes in Statistics* **98**. Springer, New York.

Bartlett, M.S. (1946). On the theoretical specification of sampling properties of autocorrelated time series. *Journal of the Royal Statistical Society Supplement* **8**, 27–41.

Bartlett, M.S. (1950). Periodogram analysis and continuous spectra. *Biometrika* **37**, 1–16.

Basawa, I.V., Mallik, A.K., McCormick W.P., Reeves, J.H., and Taylor, R.L. (1991). Bootstrapping unstable first-order autoregressive processes. *Annals of Statistics* **19**, 1098–1101.

Becker, R.A., Chambers, J.M., and Wilks, A.R. (1988) *The New S Language: A Programming Environment for Data Analysis and Graphics.* Wadsworth & Brooks/Cole, Monterey.

Beran, J. (1994). *Statistical Methods for Long-Memory Processes.* Chapman & Hall, New York.

Beran, R. (1984). Bootstrap methods in statistics. *Jahresberichte des Deutschen Mathematischen Vereins* **86**, 14–30.

Beran, R. (1986). Simulated power function. *The Annals of Statistics* **14**, 151–173.

Beran, R. and Ducharme, G. (1991). *Asymptotic theory for bootstrap methods in statistics.* Les Publications CRM, Université dé Montréal, Montréal.

Beran, R., LeCam, L., and Millar, P. (1987). Convergence of stochastic empirical measures. *Journal of Multivariate Analysis* **23**, 159–168.

Beran, R. and Millar, W. (1986). Confidence sets for a multivariate distribution. *Annals of Statistics* **14**, 431–443.

Berger, J.O. (1993). The present and future of Bayesian multivariate analysis, in *Multivariate Analysis: Future Directions.* C.R. Rao (ed.), Elsevier Science Publishers, Amsterdam, 25–53.

Bernstein, S. (1927). Sur l'extension du théorème du calcul des probabilités aux sommes de quantités dependantes. *Mathematische Annalen* **97**, 1–59.

Bertail, P. (1997). Second order properties of an extrapolated bootstrap without replacement: the i.i.d. and the strong mixing cases. *Bernoulli* **3**, 149–179.

Bertail, P., Gamst, A., and Politis, D.N. (1998). Moderate deviations in subsampling distribution estimation. Preprint, Department of Mathematics, UCSD.

Bertail, P. and Politis, D.N. (1996). Extrapolation of subsampling distributions in the i.i.d. and strong mixing cases, Technical Report 9604, INRA-CORELA, Ivry, France.

Bertail P., Politis, D.N., and Rhomari, N., (1996). Subsampling continuous parameter random fields, Technical Report TR/18/1996, Department of Mathematics and Statistics, University of Cyprus.

Bertail, P., Politis, D.N., and Romano, J.P. (1999). On subsampling estimators with unknown rate of convergence. *Journal of the American Statistical Association*, to appear June 1999.

Bhattacharya, R. and Ghosh, J. (1978). On the validity of the formal Edgeworth

expansion. *Annals of Statistics* **6**, 434–451.
Bickel, P. and Freedman, D. (1981). Some asymptotic theory for the bootstrap. *Annals of Statistics* **9**, 1196–1217.
Bickel, P., Götze, F., and van Zwet, W.R. (1997). Resampling fewer than n observations: Gains, losses, and remedies for losses. *Statistica Sinica* **7**, 1–31.
Bickel, P., Klaassen, C., Ritov, Y., and Wellner, J. (1993). *Efficient and Adaptive Estimation for Semiparametric Models.* The John Hopkins University Press, Baltimore, MD.
Bickel, P. and Millar, W. (1992). Uniform convergence of probability measures on classes of functions. *Statistica Sinica* **2**, 1–15.
Bickel, P. and Ren, J. (1996). The m out of n bootstrap and goodness of fit tests with doubly censored data. *Robust Statistics, Data Analysis, and Computer Intensive Methods.* Lecture Notes in Statistics, Springer, 35–47.
Bickel, P. and Ren, J. (1997). On choice of m for the m out of n bootstrap in hypothesis testing. Preprint, Department of Statistics, University of California, Berkeley.
Bickel, P.J. and Yahav, J.A. (1988). Richardson extrapolation and the bootstrap. *Journal of the American Statistical Association* **83**, 387–393.
Billingsley, P. (1968). *Convergence of Probability Measures.* John Wiley, New York.
Billingsley, P. (1986). *Probability and Measure*, 2nd edition. John Wiley, New York.
Bingham, N.H., Goldie, C.M., and Teugels, J.L. (1987). *Regular Variation.* Cambridge University Press.
Bollerslev, T. (1990). Modeling the coherence on short run nominal exchange rates: a multivariate GARCH model. *Review of Economics and Statistics* **72**, 498–505.
Bollerslev, T., Engle, R.F, and Woolridge, J.M. (1988). A capital asset pricing model with time varying covariances. *Journal of Political Economy* **96**, 116–131.
Bolthausen, E. (1982). On the central limit theorem for stationary random fields, *Annals of Probability* **10**, 1047–1050.
Booth, J.G. and Hall, P. (1993). An improvement of the jackknife distribution function estimator. *Annals of Statistics* **21**, 1476–1485.
Bose, A. (1988). Edgeworth correction by bootstrap in autoregressions. *Annals of Statistics* **14**, 1709–1722.
Bose, A. and Politis, D.N. (1995), A review of the bootstrap for dependent samples. In *Stochastic Processes and Statistical Inference*, B.R. Bhat and B. L. S. Prakasa Rao (editors). New Age International Publishers, New Delhi, 1995, 39–51.
Bosq, D. (1996). Nonparametric statistics for stochastic processes: estimation and prediction. *Lecture Notes in Statistics* **110**. Springer, New York.
Bradley, R.C. (1991). Equivalent mixing conditions for random fields. Technical Report No. 336, Center for Stochastic Processes, Department of Statistics, University of North Carolina, Chapel Hill.
Bradley, R.C. (1992). On the spectral density and and asymptotic normality of weakly dependent random fields, *Journal of Theoretical Probability* **5**, 355–373.
Bretagnolle, J. (1983). Limites du bootstrap de certaines fonctionnelles. *Annals of the Institute Henri Poincaré* **3**, 281–296.

Brillinger, D. (1975). *Time Series: Data Analysis and Theory*. Holt, Rinehart and Winston, New York.

Brockwell, P.J. and Davis, R.A. (1991). *Time Series: Theory and Models*, 2nd edition. Springer, New York.

Bühlmann, P. (1994). Blockwise bootstrapped empirical process for stationary sequences. *Annals of Statistics* **22**, 995–1012.

Bühlmann, P. and Künsch, H.R. (1994). Block length selection in the bootstrap for time series. Research Report No. 72, Seminar für Statistik, ETH Zürich.

Bühlmann, P. and Künsch, H.R. (1995). The blockwise bootstrap for general parameters of a stationary time series. *Scandinavian Journal of Statistics* **22**, 35–54.

Campbell, J.Y. and Shiller, R.J. (1988a). Stock prices, earnings, and expected dividends. *Journal of Finance* **43**, 661–676.

Campbell, J.Y. and Shiller, R.J. (1988b). The dividend ratio model and small sample bias: A Monte Carlo study. *Economics Letters* **29**, 325–331.

Campbell, J.Y., Lo, A.W., and MacKinley, A.C. (1997). *The Econometrics of Financial Markets*. Princeton University Press.

Carlstein, E. (1986). The use of subseries values for estimating the variance of a general statistic from a stationary time series. *Annals of Statistics* **14**, 1171–1179.

Cavanagh, C.L., Elliot, G., and Stock, J.H. (1995). Inference in models with nearly integrated regressors. *Econometric Theory* **11**, 1131–1147.

Chambers, J.M, Mallows, C.L., and Stuck, B.W. (1976). A method for simulating stable random variables. *Journal of the American Statistical Association* **71**, 340–344.

Cressie, N. (1993). *Statistics for Spatial Data*, 2nd edition. John Wiley, New York.

Csörgö, S. and Mason, D. (1989). Bootstrap empirical functions. *Annals of Statistics* **17**, 1447–1471.

Dahlhaus, R. (1985). On the asymptotic distribution of Bartlett's U_p-statistic. *Journal of Time Series Analysis* **6**, 213–227.

Datta, S. (1995). On a modified bootstrap for certain asymptotically nonnormal statistics. *Statistics and Probability Letters* **24**, 91–98.

Davison, A.C. and Hinkley, D.V. (1997). *Bootstrap Methods and their Application*. Cambridge University Press, Cambridge, England.

Davydov, Y.A. (1970). The invariance principle for stationary processes. *Theory of Probability and its Applications* **14**, 487–498.

Deo, C. (1973). A note on empirical processes of strong-mixing sequences. *Annals of Probability* **1**, 870–875.

DiCiccio, T. and Romano, J.P. (1988). A review of bootstrap confidence intervals (with discussion). *Journal of the Royal Statistical Society, Ser. B* **50**, 338–370.

Dickey, D.A. and Fuller, W.A. (1979). Distribution of the estimators for autoregressive time series with a unit root. *Journal of the American Statistical Association* **74**, 427–431.

Dobrushin, R.L. (1968). The description of a random field by means of conditional probabilities and conditions of its regularity. *Theory of Probability and its Applications* **13**, 197–224.

Doukhan, P. (1994). Mixing: properties and examples. *Lecture Notes in Statistics* **85**. Springer, New York.

Dudley, R. (1989). *Real Analysis and Probability*. Wadsworth, Belmont.

Dudley, R. (1990). Nonlinear functions of empirical measures and the bootstrap. In *Probability in Banach spaces* **7**, 63–82, E. Eberlein, J. Kuelbs, and M. Marcus (editors). Birkhäuser, Boston.

Efron, B. (1969). Student's *t*-test under symmetry conditions. *Journal of the American Statistical Association* **64**, 1278–1302.

Efron, B. (1979). Bootstrap methods: Another look at the jackknife. *Annals of Statistics* **7**, 1–26.

Efron, B. (1981). Nonparametric standard errors and confidence intervals. *The Canadian Journal of Statistcs* **9**, 139–172.

Efron, B. (1982). *The jackknife, the bootstrap and other resampling plans.* CBMS-NSF Regional Conference Series in Applied Mathematics, Society for Industrial and Applied Mathematics, Philadelphia.

Efron, B. (1994). Comment on the paper "Bootstrap: More than a stab in the dark?" by G.A. Young. *Statistical Science* **9**, 396–398.

Efron, B. and Tibshirani, R.J. (1993). *An Introduction to the Bootstrap.* Chapman & Hall, New York.

Elliot, G. (1998). On the robustness of cointegration methods when regressors almost have unit roots. *Econometrica* **66**, 149–158.

Elliot, G. and Stock, J.H. (1994). Inference in time series regression when the order of integration of a regressor is unknown. *Econometric Theory* **10**, 672–700.

Engle, R.F., Granger, C.W.J., Rice, J., and Weiss, A. (1986). Semiparametric estimates of the relation between weather and electricity sales. *Journal of the American Statistical Association* **81**, 310–320.

Fama, E. and French, K. (1988). Dividend yields and expected stock return. *Journal of Financial Economics* **22**, 3–25.

Feller, W. (1957). *An Introduction to Probability Theory and its Applications*, Vol. 1, 3rd edition. John Wiley, New York.

Feller, W. (1971). *An Introduction to Probability Theory and its Applications*, Vol. 2, 2nd edition John Wiley, New York.

Fernholz, L.T. (1983), Von Mises calculus for statistical functionals. *Lecture Notes in Statistics* **19**, Springer, New York.

Ferson, W.E. and Foerster, S. (1994). Finite sample properties of the generalized method of moments in tests of conditional asset pricing models. *Journal of Financial Economics* **36**, 29–55.

Fitzenberger, B. (1997). The moving blocks bootstrap and robust inference for linear least squares and quantile regressions. *Journal of Econometrics* **82**, 235–287.

Franke, J. and Härdle, W. (1992). On bootstrapping kernel spectral estimates. *Annals of Statististics* **20**, 121–145.

Freedman, D.A. (1981). Bootstrapping regression models. *Annals of Statistics* **9**, 1218–1228.

Freedman, D.A. (1984). On bootstrapping two-stage least-squares estimates in stationary linear models. *Annals of Statistics* **12**, 827–842.

French, K., Schwert, W., and Stambaugh, R. (1987). Expected stock returns and volatility. *Journal of Financial Economics* **25**, 23–50.

Fukuchi, J. (1997). Subsampling and model selection in time series analyis. Technical Report No. 97-2, Hiroshima University, Japan.

Gallant, A.R. and White, H. (1988). *A Unified Theory of Estimation and*

Inference for Nonlinear Dynamic Models. Basil Blackwell, New York.
Gastwirth, J.L. and Rubin, H. (1975). The behavior of robust estimators on dependent data. *Annals of Statistics* **3**, 1070–1100.
Giakoumatos, S.G., Vrontos, I.D., Dellaportas, P., and Politis, D.N. (1999). An MCMC Convergence Diagnostic using Subsampling. *Journal of Computational and Graphical Statistics*, in press.
Gill, R., Vardi, Y. and Wellner, J. (1988). Large sample theory of empirical distributions in biased sampling models. *Annals of Statistics* **16**, 1069–1112.
Giné, E. (1997). *Lectures on Some Aspects of the Bootstrap*. École d'Été de Calcul de Probabilités de Saint-Flour.
Giné, E. and Zinn, J. (1989). Necessary conditions for the bootstrap of the mean. *Annals of Statistics* **17**, 684-691.
Giné, E. and Zinn, J. (1990). Bootstrapping general empirical measures. *Annals of Probability* **18**, 851–869.
Götze, F. and Hipp, C. (1983). Asymptotic expansions for sums of weakly dependent random vectors. *Zeitschrift für Wahrscheinlichkeitstheorie und verwandte Gebiete* **64**, 211–239.
Götze, F. and Künsch, H.R. (1996). Second order correctness of the blockwise bootstrap for stationary observations. *Annals of Statistics* **24**, 1914–1933.
Goetzmann, W.N. and Jorion, P. (1993). Testing the predictive power of dividend yields. *Journal of Finance* **48**, 663–679.
Goetzmann, W.N. and Jorion, P. (1995). A longer look at dividend yields. *Journal of Business* **68**, 483–508.
Gray, H., Schucany, W. and Watkins, T. (1972). *The Generalized Jackknife Statistics*. Marcel Dekker, New York.
Györfi, L., Härdle, W., Sarda, P., and Vieu, P. (1989). Nonparametric curve estimation from time series. *Lecture Notes in Statistics* **60**. Springer, New York.
de Haan, L. and Resnick, S.I. (1980). A simple asymptotic estimate for the index of a stable distribution. *Journal of the Royal Statistical Society, Ser. B* **42**, 83–87.
Härdle, W. and Vieu, P. (1992). Kernel regression smoothing of time series. *Journal of Time Series Analysis* **13**, 209–232.
Hall, P. (1982). On simple estimates of an exponent of regular variation. *Journal of the Royal Statistical Society, Ser. B* **44**, 37–42.
Hall, P. (1985). Resampling a coverage pattern. *Stochastic Processes and their Applications* **20**, 231–246.
Hall, P. (1986). On the bootstrap and confidence intervals. *Annals of Statistics* **14**, 1431–1452.
Hall, P. (1988). On symmetric bootstrap confidence intervals. *Journal of the Royal Statistical Society, Ser. B* **50**, 35–45.
Hall, P. (1990). Asymptotic properties of the bootstrap for heavy-tailed distributions. *Annals of Probability* **18**, 1342–1360.
Hall, P. (1992). *The Bootstrap and Edgeworth Expansion*. Springer, New York.
Hall, P. and Heyde, C.C. (1980). *Martingale Limit Theory and its Application*. Academic Press, New York.
Hall, P., Horowitz J.L., and Jing, B.-Y. (1996). On blocking rules for the bootstrap with dependent data. *Biometrika* **50**, 561–574.
Hall, P. and Jing, B.-Y. (1996). On sample re-use methods. *Journal of the Royal Statistical Society, Ser. B,* **58**, 727–738.

Hall, P., Lahiri, S.N., and Jing, B.-Y. (1996). On the sampling window method for long-range dependent data, in press.

Hamilton, J.D. (1994). *Time Series Analysis.* Princeton University Press, Princeton.

Hansen, L. (1982). Large sample properties of generalized method of moments estimation. *Econometrica* **50**, 1029–1054.

Hansen, L. and Hodrick, R. (1980). Forward exchange rates as optimal predictors of future spot rates. *Journal of Political Economy* **88**, 829–853.

Hartigan, J. (1969). Using subsample values as typical values. *Journal of the American Statistical Association* **64**, 1303–1317.

Hartigan, J. (1975). Necessary and sufficient conditions for asymptotic joint normality of a statistic and its subsample values. *Annals of Statistics* **3**, 573–580.

Hausman, J. (1978). Specification tests in econometrics. *Econometrica* 46, 1251–1271.

Heinrich, L. (1982). A method for the derivation of limit theorems for sums of m-dependent random variables. *Zeitschrift für Wahrscheinlichkeitstheorie und verwandte Gebiete* **60**, 501–515.

Heinrich, L. (1984). Non-uniform estimates and asymptotic expansions of the remainder in the central limit theorem for m-dependent random variables. *Mathematische Nachrichten,* **115**, 7-20.

Hill, B. (1975). A simple approach to inference about the tail of the distribution. *Annals of Statistics* **3**, 1163–1174.

Hodrick, R.J. (1992). Dividend yields and expected stock returns: alternative procedures for inference and measurement. *Review of Financial Studies* **5**, 357–386.

Hotelling, H. (1961). The behavior of some standard statistical tests under non-standard conditions. *Proceedings of the Fourth Berkeley Symposium of Mathematical Statistics* Prob. **1**, 319–360.

Hu, I. (1985). A uniform bound for the tail probability of Kolmogorov–Smirnov statistics. *Annals of Statistics* **13**, 851-856.

Huang, J., Sen, P., and Shao, J. (1996). Bootstrapping a sample quantile when the density has a jump. *Statistica Sinica* **6**, 299–309.

Ibragimov, I. (1962). Some limit theorems for stationary processes. *Theory of Probability and its Applications* **7**, 349–382.

Ibragimov, I.A. and Rozanov, Y.A. (1978). *Gaussian Random Processes.* Springer, New York.

Ivanov, A.V. and Leonenko, N.N. (1989). *Statistical Analysis of Random Fields.* Kluwer Academic Publishers, The Netherlands.

Jing, B.Y. (1997). On the relative performance of the block bootstrap for dependent data. *Communications in Statistics—Theory and Methods* **26**, 1313–1328.

Jolivet, E. (1981), Central limit theorem and convergence of empirical processes for stationary point processes, in *Point Processes and Queueing Problems*, P. Bártfai, and J. Tomkó (editors). North-Holland, Amsterdam.

Karr, A.F. (1986). Inference for stationary random fields given Poisson samples. *Advances in Applied Probability* **18**, 406–422.

Karr, A.F. (1991). *Point Processes and their Statistical Inference*, 2nd edition. Dekker, New York.

Kendall, M.G. (1954). Note on bias in the estimation of autocorrelation. *Biometrika* **41**, 403–404.

Kinateder, J. (1992). An invariance principle applicable to the bootstrap. In *Exploring the Limits of Bootstrap*, 157–181, R. LePage and L. Billard (editors). John Wiley, New York.

Knight, K. (1989). On the bootstrap of the sample mean in the infinite variance case. *Annals of Statistics* **17**, 1168–1175.

Kocherlakota, N.R. and Savin, N.E. (1995). Confidence intervals for the sample mean of overdifferenced data. Preprint, Department of Economics, The University of Iowa.

Krickeberg, K. (1982). Processus ponctuels en statistique, École d' été de probabilités de Saint-Flour X – 1980, P.L. Hennequin (Ed.), *Lecture Notes in Mathematics* **929**. Springer, New York, 205–313.

Künsch, H. (1984). Infinitesimal robustness for autoregressive processes. *Annals of Statistics* **12**, 843–863.

Künsch, H.R. (1989). The jackknife and the bootstrap for general stationary observations. *Annals of Statistics* **17**, 1217–1241.

Kutoyants Yu.A. (1984). *Parameter estimation for stochastic processes*. Heldermann Verlag, Berlin.

Lahiri, S.N. (1992). Edgeworth correction by 'Moving block' bootstrap for stationary and nonstationary data. In *Exploring the Limits of Bootstrap*, 183–214, R. LePage and L. Billard (editors). John Wiley, New York.

Lahiri, S.N. (1998). Effects of block lengths on the validity of block resampling methods. Preprint, Department of Statistics, Iowa State University.

Leblanc, F. (1994). Estimation of marginal density of a continuous time stochastic process by wavelets and application to diffusion processes. *CREST working paper* No. **9456**.

Léger, C. and Cléroux, R. (1990). Nonparametric age replacement: bootstrap confidence interval for the optimal cost. Publication 731, Département d'informatique et de recherche opérationnelle, Université de Montréal.

Léger, C., Politis, D.N., and Romano, J.P. (1992). Bootstrap technology and applications. *Technometrics* **34**, 378–399.

Li, H. and Maddala, G.S., (1996). Bootstrapping time series models, *Econometric Reviews* **15**, 115–158.

Lii, K-S., and Masry, E. (1994). Spectral estimation of continuous time stationary processes from random sampling. *Stochastic Processes and their Applications* **52**, 39–64.

Lii, K-S. and Tsou, T-H. (1995). Bispectral analysis of randomly sampled data. *Journal of Time Series Analysis* **16**, 43–66.

Liu, R.Y. (1988). Bootstrap procedures under some non-iid models. *Annals of Statistics* **16**, 1696–1708.

Liu, R.Y. and Singh, K. (1992). Moving blocks jackknife and bootstrap capture weak dependence. In *Exploring the Limits of Bootstrap*, 225–248, R. LePage and L. Billard (editors). John Wiley, New York.

Lo, A.W. and MacKinley, A.C. (1988). Stock market prices do not follow random walks: Evidence from a simple specification test. *Review of Financial Studies* **1**, 41–66.

Lo, A.W. and MacKinley, A.C. (1989). The size and power of the variance ratio test in finite samples: A Monte Carlo investigation. *Journal of*

Econometrics **40**, 203–238.

Logan, B.F., Mallows, C.L., Rice, S.O., and Shepp, L.A. (1973). Limit distributions of self-normalized sums. *Annals of Probability* **1**, 788–809.

Loh, W.-Y. (1987). Calibrating confidence coefficients. *Journal of the American Statistical Association* **82**, 155–162.

Mahalanobis, P. (1946). Sample surveys of crop yields in India. *Sankya, Series A* **7**, 269–280.

Mammen, E. (1992). When does bootstrap work? *Lecture Notes in Statistics* **77**. Springer, New York.

Martin, R.D. and Yohai, V.J. (1986). Influence functionals for time series. *Annals of Statistics* **14**, 781–818.

Masry, E. (1978). Poisson sampling and spectral estimation of continuous-time processes. *IEEE Transactions on Information Theory* **24**, 173–183.

Masry, E. (1983). Spectral and probability density estimation form irregularly observed data. In *Time Series Analysis of Irregularly Observed Data*, 224–250, E. Parzen (editor). Springer, New York.

Masry, E. (1988). Random sampling of continuous-parameter stationary processes: statistical properties of joint density estimators, *Journal of Multivariate Analysis* **36**, 133–165.

McCarthy, P. (1969). Pseudo-replication: half-samples. *Review of the International Statistical Institute* **37**, 239–263.

McLeish, D.L. (1975a). A maximal inequality and dependent strong laws. *Annals of Probability* **3**, 829–839.

McLeish, D.L. (1975b). Invariance principles for dependent variables. *Zeitschrift für Wahrscheinlichkeitstheorie und verwandte Gebiete* **32**, 165–178.

Meketon, M.S., and Schmeiser, B. (1984). Overlapping batch means: something for nothing? *Proceedings Winter Simulation Conference*, 227–230, S. Sheppard, U. Pooch, and D. Pegden (editors). IEEE, Piscataway, New Jersey.

Mittnik, S., Paolella, M.S., and Rachev S.T. (1996). A tail estimator for the index of the stable paretian distribution. Working paper No. 103, Institute of Statistics and Econometrics, Christian Albrechts University at Kiel.

Mykland, P. and Ren, J. (1996). Algorithms for computing the self-consistent and maximum likelihood estimators with doubly censored data. *Annals of Statistics*, **24**, 1740–1764.

Naik-Nimbalkar, U. and Rajarshi, M. (1994). Validity of blockwise bootstrap for empirical processes with stationary observations. *Annals of Statistics* **22**, 980–994.

Nelson, C.R. and Kim, M.J. (1993). Predictable stock returns: the role of small sample bias. *Journal of Finance* **48**, 641–661.

Oodaira, H. and Yoshihara, K. (1972). Functional central limit theorems for strictly stationary processes satisfying the strong mixing condition. *Kodai Math. Sem. Rep.* **24**, 259–269.

Owen, A. (1988). Empirical likelihood ratio confidence intervals for a single functional. *Biometrika* **72**, 45–58.

Pedrosa, A. and Schmeiser, B. (1993). Asymptotic and finite-sample correlations between OBM estimators. *Proceedings of the Winter Simulation Conference*, 481–488. IEEE, Piscataway, New Jersey.

Phillips, P.C.B. and Perron, P. (1988). Testing for a unit root in time series regression. *Biometrika* **75**, 335–346.

Pickands, J. (1975). Statistical inference using extreme order statistics. *Annals of Statistics* **3**, 119–131.

Politis, D.N., Paparoditis, E., and Romano, J.P. (1998). Large sample inference for irregularly spaced dependent observations based on subsampling. *Sankhya, Series A.* **60**, 274–292.

Politis, D.N. and Romano, J.P. (1992a). A general resampling scheme for triangular arrays of α-mixing random variables with application to the problem of spectral density estimation. *Annals of Statistics* **20**, 1985–2007.

Politis, D.N. and Romano, J.P. (1992b). A circular block-resampling procedure for stationary data, in *Exploring the Limits of Bootstrap*, 263–270, R. LePage and L. Billard (editors). John Wiley, New York.

Politis, D.N. and Romano, J.P. (1992c). A general theory for large sample confidence regions based on subsamples under minimal assumptions. Technical Report 399, Department of Statistics, Stanford University.

Politis, D.N. and Romano, J.P. (1993a). Nonparametric resampling for homogeneous strong mixing random fields. *Journal of Multivariate Analysis* **47**, 301–328.

Politis, D.N. and Romano, J.P. (1993b). On the sample variance of linear statistics serived from mixing sequences, *Stochastic Processes and their Applications* **45**, 155–167.

Politis, D.N. and Romano, J.P. (1993c). Estimating the distribution of a studentized statistic by subsampling. *Bulletin of the International Statistical Institute*, 49th Session, Firenze, August 25–September 2, 1993, Book 2, 315–316.

Politis, D.N. and Romano, J.P. (1994a). The stationary bootstrap. *Journal of the American Statistical Association* **89**, 1303–1313.

Politis, D.N. and Romano, J.P. (1994b). Large sample confidence regions based on subsamples under minimal assumptions. *Annals of Statistics* **22**, 2031–2050.

Politis, D.N., and Romano, J.P. (1994c). Limit theorems for weakly dependent Hilbert space valued random variables with applications to the stationary bootstrap. *Statistica Sinica* **4**, 461–476.

Politis, D.N. and Romano, J.P. (1995). Bias-corrected nonparametric spectral estimation. *Journal of Time Series Analysis* **16**, 67–103.

Politis, D.N. and Romano, J.P. (1996a). On flat-top kernel spectral density estimators for homogeneous random fields. *Journal of Statistical Planning and Inference* **51**, 41–53.

Politis, D.N. and Romano, J.P. (1996b). Subsampling for econometric models— Comments on bootstrapping time series models. *Econometric Reviews* **15**, 169–176.

Politis, D.N. and Romano, J.P. (1999). Multivariate density estimation with general flat-top kernels of infinite order. *Journal of Multivariate Analysis* **68**, 1–25.

Politis, D.N., Romano, J.P., and Wolf, M. (1997). Subsampling for heteroskedastic time series. *Journal of Econometrics* **81**, 281–317.

Politis, D.N., Romano, J.P., and Wolf, M. (1999). Weak convergence of dependent empirical measures with application to subsampling and confidence bands. *Journal of Statistical Planning and Inference*, in press.

Politis, D.N., Romano, J.P., and You, L. (1993). Uniform confidence bands for the spectrum based on subsamples. In *Computing Science and Statistics, Proceedings of the 25th Symposium on the Interface*, 346–351.

Politis, D.N. and Sherman, M. (1998). General moment estimation for statistics from marked point processes. Technical Report, Department of Statistics, Texas A&M University.

Pollard, D. (1984). *Convergence of Stochastic Processes*. Springer, New York.

Possolo, A. (1991). Subsampling a random field. In *Spatial Statistics and Imaging*, 286–294, A. Possolo (editor). IMS Lectures notes, Vol. 20, Hayward, CA 1991.

Priestley, M.B. (1981). *Spectral Analysis and Time Series*. Academic Press, New York.

Quenouille, M. (1949). Approximate tests of correlation in time series. *Journal of the Royal Statististical Society, Ser. B* **11**, 68–84.

Radulović, D. (1996). The bootstrap of the mean for strong mixing sequences under minimal conditions. *Statistics and Probability Letters* **28**, 65–72.

Radulović, D. (1998). On the subsample bootstrap variance estimation. *Test* **7**, 295–306.

Raïs, N. (1992). *Méthodes de Reéchantillonage et de sous Echantillonage dans le Contexte Spatial et pour des Données Dépendantes*. Ph.D. thesis, Department of Mathematics and Statistics, University of Montreal, Montreal, Canada.

Raïs, N. and Moore, M. (1990). Bootstrap for some stationary α-mixing processes, Abstract, *INTERFACE '90*, 22nd Symposium on the Interface of Computing Science and Statistics.

Reeds, J. (1976). *On the definition of von Mises functionals*. Ph.D. thesis, Department of Statistics, Harvard University, Cambridge, Massachusetts.

Resnick, S. (1997). Heavy tail modeling and teletraffic data (with discussion). *Annals of Statistics* **25**, 1805–1869.

Richardson, M. (1993). Temporary components of stock prices: A skeptic's view. *Journal of Business & Economic Statistics* **11**, 199–207.

Richardson, M. and Stock, J.H. (1989). Drawing inferences from statistics based on multiyear asset returns. *Journal of Financial Economics* **25**, 323–348.

Ripley, B.D. (1981). *Spatial Statistics*. John Wiley, New York.

Ripley, B.D. (1988). *Statistical Inference for Spatial Processes*. Cambridge University Press, Cambridge, England.

Robinson, P. (1994). Semiparametric analysis of long-memory time series. *Annals of Statistics* **22**, 515–539.

Romano, J.P. (1988a). Bootstrapping the mode. *Annals of the Institute of Mathematical Statistics* **40**, 565–586.

Romano, J.P. (1988b). On weak convergence and optimality of kernel density estimates of the mode. *Annals of Statistics* **16**, 629–647.

Romano, J.P. (1988c). A bootstrap revival of some nonparametric distance tests. *Journal of the American Statistical Association* **83**, 698–708.

Romano, J.P. (1989). Bootstrap and randomization tests of some nonparametric hypotheses. *Annals of Statistics* **17**, 141–159.

Romano, J.P. and Siegel A.F. (1986). *Counterexamples in Probability and Statistics*. Wadsworth, Belmont.

Romano, J.P. and Thombs, L.A. (1996). Inference for autocorrelations under weak assumptions. *Journal of the American Statistical Association* **91**, 590–600.

Romano, J.P. and Wolf, M. (1998a). Inference for the mean in the heavy-tailed case. Technical Report 1998-1, Department of Statistics, Stanford University.

Romano, J.P. and Wolf, M. (1998b). Subsampling confidence intervals for the autoregressive root. Technical Report 1998-5, Department of Statistics, Stanford University.

Rosenblatt, M. (1956). A central limit theorem and a strong mixing condition. *Proceedings of the National Academy of Sciences* **42**, 43–47.

Rosenblatt, M. (1984). Asymptotic normality, strong mixing and spectral density estimates, *Annals of Probability* **12**, 1167–1180.

Rosenblatt, M. (1985). *Stationary Sequences and Random Fields*. Birkhäuser, Boston.

Roussas, G.G. and Ioannides, D. (1987). Moment inequalities for mixing sequences of random variables. *Stochastic Analysis and its Applications* **5**, 61–120.

Rozeff, M. (1984). Dividend yields are equity risk premium. *Journal of Portfolio Management* **11**, 68–75.

Samorodnitsky, G. and Taqqu, M.S. (1994). *Stable Non-Gaussian Random Processes*. Chapman & Hall, New York.

Schmeiser, B. (1990). Simulation experiments. In *Stochastic Models*, 295–330, D.P. Heyman and M.J. Sobel (editors). North-Holland, Amsterdam.

Schmid, P. (1958). On the Kolmogorov and Smirnov limit theorem for discontinuous distribution functions. *Annals of Mathematical Statistics* **29**, 1011–1027.

Serfling, R. (1980). *Approximation Theorems of Mathematical Statistics*. John Wiley, New York.

Shao, J. and Wu, C.F. (1989). A general theory for jackknife variance estimation. *Annals of Statistics* **17**, 1176–1197.

Shao, J. and Tu, D. (1995). *The Jackknife and the Bootstrap*. Springer, New York.

Shen, X. and Wong, W. (1994). Convergence of sieve estimates. *Annals of Statistics* **22**, 580–615.

Sherman, M. (1992). *Subsampling and asymptotic normality for a general statistic from a random field*. Ph.D. thesis, Department of Statistics, University of North Carolina, Chapel Hill (Report # 2081, Mimeo Series).

Sherman, M. (1994). Kernel estimation of the density of a statistic. *Statistics and Probability Letters* **21**, 29–36.

Sherman, M. (1996). Variance estimation for statistics computed from spatial lattice data. *Journal of the Royal Statistical Society, Ser. B* **58**, 509–523.

Sherman, M. (1997). Subseries methods in regression. *Journal of the American Statistical Association* **92**, 1041–1048.

Sherman, M. and Carlstein, E. (1994). Nonparametric estimation of the moments of a general statistic computed from spatial data. *Journal of the American Statistical Association* **89**, 496–500.

Sherman, M. and Carlstein, E. (1996). Replicate histograms. *Journal of the American Statistical Association* **91**, 566–576.

Sherman, M. and Carlstein, E. (1997). Omnibus confidence intervals. Technical Report No. 278, Department of Statistics, Texas A&M University.

Shi, X. (1991). Some asymptotic results for jackknifing the sample quantile. *Annals of Statistics* **19**, 496–503.

Singh, K. (1981). On the asymptotic accuracy of Efron's bootstrap. *Annals of Statistics* **9**, 1187–1195.

Stambaugh, R. (1986). Biases in regressions with lagged stochastic regressors. Working paper 156, The University of Chicago.

Stock, J.H. (1991). Confidence intervals for the largest autoregressive root in U.S. macroeconomic time series. *Journal of Monetary Economics* **28**, 435–459.

Tsirel'son, V. (1975). The density of the distribution of the maximum of a Gaussian process. *Theory of Probability and its Applications* **20**, 847–856.

Tu, D. (1992). Approximating the distribution of a general standardized functional statistic with that of jackknife pseudo values. In *Exploring the Limits of Bootstrap*, 279–306, R. LePage and L. Billard (editors). John Wiley, New York.

Tukey, J.W. (1958). Bias and confidence in not quite large samples (abstract). *Annals of Mathematical Statistics* **29**, p. 614.

Uteev S. (1984). Inequalities and estimates of the convergence rate for the weakly dependent case. *Advances in Probabability Theory. 1985*, Novosibirsk. §1.4, §1.5.

van der Vaart, A. and Wellner, J. (1996). *Weak Convergence and Empirical Processes*. Springer, New York.

Varadarajan, V.S. (1958). Weak convergence of measures on separable metric spaces. *Sankhya, Ser. A* **19**, 15–22.

Veretennikov, A. (1994). On large deviation for diffusion processes under minimal smoothness conditions. *Comptes Rendus Academie Science Paris,* **319**, Serie I, 727–732.

Viceira, L.M. (1997). Testing for structural change in the predictability of asset returns. Preprint, Harvard University, Cambridge, Massachusetts.

Welch, P.D. (1967). The use of fast fourier transform for estimation of spectra: A method based on averaging over short, modified periodograms. *IEEE Transactions on Electronics* AU-15, 70.

Welch, P.D. (1987). On the relationship between batch means, overlapping batch means and spectral estimation. *Proceedings Winter Simulation Conference*, 320–323, A. Thesen, H. Grant, and W.D. Kelton, (editors).

West, K. (1988). Asymptotic normality, when regressors have a unit root. *Econometrica* **56**, 1397–1417.

White, H. (1980). A heteroskedasticity-consistent covariance matrix estimator and a direct test for heteroskedasticity. *Econometrica* **48**, 817–838.

White, H. and Domowitz, I. (1984). Nonlinear regression with dependent observations. *Econometrica* **52**, 143–161.

Wolf M. (2000). Stock returns and dividend yields revisited: A new way to look at an old problem. *Journal of Business and Economic Statistics*, in press.

Wolkonoski, V.A. and Rozanov, Y.A. (1959). Some limit theorems for random functions, Part I. *Theory of Probability and its Applications* **4**, 178–197.

Woodroofe, M. and Van Ness, J. (1967). The maximum deviation of sample spectral densities. *Annals of Mathematical Statistics* **38**, 1558–1570.

Wu, C.F. (1986). Jackknife, bootstrap and other resampling methods in regression analysis. *Annals of Statistics* **14**, 1261–1343.

Wu, C.F. (1990). On the asymptotic properties of the jackknife histogram. *Annals of Statistics* **18**, 1438–1452.

Wu, J., Carlstein, E., and Cambanis, S. (1993). Bootstrapping the sample mean for data with infinite variance. Technical Report No. 296, Department of Statistics, University of North Carolina.

Yadrenko, M.I. (1983). *Spectral Theory of Random Fields*. Optimization Software, Inc., New York.

Yokoyama, R. (1980). Moment bounds for stationary mixing sequences.

Zeitschrift für Wahrscheinlichkeitstheorie und verwandte Gebiete **52**, 45–57.

Yoshihara, K. (1975). Weak convergence of multidimensional empirical processes for strong mixing sequences of stochastic vectors. *Zeitschrift für Wahrscheinlichkeitstheorie und verwandte Gebiete* **33**, 133–137.

Zhurbenko, I.G. (1986). *The Spectral Analysis of Time Series*. North-Holland, Amsterdam.

Zolotarev, V.M. (1986). *One-dimensional Stable Distributions*. Vol. 65 of "Translation of mathematical monographs," American Mathematical Society.

Index of Names

Anderson, T., 169, 327
Andrews, D.W.K., 204, 205, 284, 300, 310, 327
Arcones, M., 15, 16, 49, 50, 167, 253, 263, 265, 327
Arunkumar, S., 46, 327
Athreya, K., 14, 15, 50, 253, 263, 327

Bühlmann, P., 100, 166, 168, 192, 193, 330
Babu, G., 14, 50, 51, 219, 327, 328
Barbe, P., 181, 223, 328
Bartlett, M.S., 97, 189, 193, 206, 241, 328
Basawa, I.V., 270, 284, 328
Becker, R.A., 328
Beran, J., 328
Beran, R., ix, 3, 4, 11, 22, 31–33, 36, 38, 44, 48, 51, 72, 161, 176, 185, 186, 328
Berger, J.O., 118, 328
Bernstein, S., 323, 328
Bertail, P., ix, 133, 149, 175, 181, 187, 216, 217, 223, 225–227, 233–235, 245, 328
Bhattacharya, R., 219, 328

Bickel, P.J., viii, 3, 13, 48, 49, 51, 55, 161, 175, 217, 218, 221–223, 329
Billingsley, P., 12, 76, 329
Bingham, N.H., 144, 329
Bollerslev, T., 299, 300, 329
Bolthausen, E., 122, 134, 329
Booth, J.G., 149, 216, 219, 222, 223, 225, 251, 329
Bose, A., 65, 329
Bosq, D., 93, 135, 136, 329
Bradley, R.C., 122, 127, 135, 329
Bretagnolle, J., 16, 49, 329
Brillinger, D., 169, 330
Brockwell, P.J., 67, 241, 330

Cambanis, S., 50, 253, 263, 339
Campbell, J.Y., 291, 294, 330
Carlstein, E., viii, 41, 50, 65, 66, 68, 87, 96–98, 117, 126, 131, 138, 174, 175, 187–189, 253, 263, 297, 330, 338, 339
Cavanagh, C.L., 288, 298, 330
Chambers, J.M., 328, 330
Cléroux, R., 46, 334
Cranger, C.W.J., 331
Cressie, N., 139, 330

Csörgö, S., 15, 330

Dahlhaus, R., 168, 169, 330
Datta, S., 175, 330
Davis, R.A., 67, 241, 330
Davison, A.C., 38, 330
Davydov, Y.A., 316, 330
de Haan, L., 261, 332
Dellaportas, P., 332
Deo, C., 89, 167, 330
DiCiccio, T., 38, 330
Dickey, D.A., 271, 281, 287, 330
Dobrushin, R.L., 122, 330
Domowitz, I., 321, 339
Doukhan, P., 121, 122, 284, 316–319, 321, 330
Ducharme, G., 38, 328
Dudley, R., 25, 28, 163, 330

Efron, B., vii, 3, 38, 40, 62, 65, 66, 98, 100, 176, 197, 253, 263, 265, 331
Elliot, G., 95, 288, 298, 330, 331
Engle, R.F., 181, 299, 329, 331

Fama, E., 291, 294, 302, 331
Feller, W., 14, 253–257, 260, 331
Fernholz, L.T., 26, 331
Ferson, W.E., 294, 331
Fitzenberger, B., 110, 115, 331
Foerster, S., 294, 331
Franke, J., 65, 331
Freedman, D.A., 3, 13, 32, 48, 49, 65, 175, 329, 331
French, K., 291, 294, 302, 331
Fukuchi, J., 97, 118, 331
Fuller, W.A., 271, 281, 287, 330

Götze, F., viii, 48, 51, 217, 218, 222, 229, 230, 232, 235, 239, 329, 332
Gallant, A.R., 284, 331
Gamst, A., 245, 328
Gastwirth, J.L., 89, 332
Ghosh, J., 219, 328
Giakoumatos, S.G., 119, 332
Gill, R., 54, 332
Giné, E., 14–16, 22, 38, 50, 55, 253, 263, 265, 327, 332

Goetzmann, W.N., 291, 293–299, 301, 302, 307, 311, 332
Goldie, C.M., 144, 329
Gray, H., 217, 332
Györfi, L., 89, 332

Härdle, W., 65, 331, 332
Hall, P., 15, 38, 51, 72, 138, 149, 176, 193, 194, 216, 219, 220, 222, 223, 225, 235, 251, 262, 316, 329, 332, 333
Hamilton, J.D., 113, 271, 279, 281, 333
Hansen, L., 293, 333
Hartigan, J., vii, 40, 51, 333
Hausman, J., 86, 333
Heinrich, L., 248, 249, 251, 333
Heyde, C.C., 316, 332
Hill, B., 261, 333
Hinkley, D.V., 38, 330
Hipp, C., 232, 332
Hodrick, R.J., 291–297, 300, 302, 304, 305, 333
Hoeffding, 44, 45, 60
Horowitz, J.L., 193, 332
Hotelling, H., 263, 333
Hu, I., 5, 333
Huang, J., 31, 333

Ibragimov, I., 333
Ibragimov, I.A., 85, 122, 321, 322
Ioannides, D., 125, 131, 143, 338
Ivanov, A.V., 135, 333

Jing, B.-Y., 176, 193, 194, 235, 250, 332, 333
Jolivet, E., 139, 145, 333
Jorion, P., 291, 293–299, 301, 302, 307, 311, 332

Künsch, H.R., 66, 89, 96–100, 131, 166, 176, 192, 193, 195, 229, 230, 235, 239, 330, 332, 334
Karr, A.F., 139, 145, 155, 156, 333
Kendall, M.G., 334
Kim, M.J., 291, 294, 295, 297, 302, 335
Kinateder, J., 50, 253, 334
Klaassen, C., 329

Knight, K., 14, 50, 253, 334
Kocherlakota, N.R., 90, 334
Krickeberg, K., 139, 145, 334
Kutoyants, Yu.A., 136, 334

Léger, C., 46, 190, 334
Lahiri, S.N., 70, 110, 176, 193, 194, 198, 229, 235, 333, 334
Leblanc, F., 136, 334
LeCam, L., 161, 328
Leonenko, N.N., 135, 333
Li, H., 334
Lii, K-S., 139, 145, 156, 157, 334
Liu, R.Y., 65, 66, 98, 166, 195, 235, 244, 334
Lo, A.W., 86, 88, 89, 116, 117, 294, 330, 334
Logan, B.F., 254, 263, 264, 335
Loh, W.-Y., 194, 335

MacKinley, A.C., 86, 88, 89, 116, 117, 294, 330, 334
Maddala, G.S., 334
Mahalanobis, P., vii, 40, 335
Mallik, A.K., 328
Mallows, C.L., 13, 330, 335
Mammen, E., 335
Martin, R.D., 89, 335
Mason, D., 15, 330
Masry, E., 139, 141, 145, 156, 334, 335
McCarthy, P., 40, 335
McCormick, W.P., 328
McLeish, D.L., 284, 335
Meketon, M.S., 97, 98, 335
Millar, W., 22, 161, 328, 329
Miller, P., 328
Mittnik, S., 262, 335
Monahan, J.C., 204, 327
Moore, M., 131, 337
Mykland, P., 55, 335

Naik-Nimbalkar, U., 166, 168, 335
Nelson, C.R., 291, 294, 295, 297, 302, 335

Oodaira, H., 321, 335
Owen, A., 37, 335

Paolella, M.S., 262, 335
Paparoditis, E., 140, 146, 336
Pedrosa, A., 239, 335
Perron, P., 271, 281, 335
Phillips, P.C.B., 271, 281, 335
Pickands, J., 261, 336
Politis, D.N., vii, viii, 39, 54, 66, 68, 80, 86, 87, 94, 96, 98, 100, 101, 123, 127, 128, 130, 131, 133, 138, 140, 146, 149, 151, 155, 157, 160, 170, 175, 176, 182–184, 186, 187, 189–192, 217, 220, 226, 235, 237–240, 242–245, 328, 329, 332, 334, 336, 337
Pollard, D., 20–22, 161, 167, 337
Possolo, A., 131, 337
Priestley, M.B., 67, 84, 85, 97, 337

Quenouille, M., vii, 40, 93, 129, 337

Raïs, N., 131, 337
Rachev, S.T., 262, 335
Radulović, D., 77, 97, 99, 110, 337
Rajarshi, M., 166, 168, 335
Reeds, J., 23, 337
Reeves, J.H., 328
Ren, J., 55, 329, 335
Resnick, S.I., 261, 262, 332, 337
Rhomari, N., 133, 328
Rice, J., 331
Rice, S.O., 335
Richardson, M., 294, 304, 306, 337
Ripley, B.D., 139, 337
Ritov, Y., 329
Robinson, P., 186, 337
Romano, J.P., vii, viii, 36–39, 50, 51, 54, 66, 68, 80, 84–87, 94, 96, 98, 100, 101, 123, 127, 128, 130, 131, 138, 140, 146, 149, 151, 157, 160, 170, 175, 176, 182–184, 186, 187, 189–192, 217, 220, 235, 237–244, 253, 255, 272, 280, 328, 330, 334, 336, 337
Rosenblatt, M., 85, 121, 122, 160, 176, 315, 321, 338
Roussas, G.G., 125, 131, 143, 338

Rozanov, Y.A., 85, 122, 316, 333, 339
Rozeff, M., 291, 338
Rubin, H., 89, 332

Samorodnitsky, G., 254, 256, 338
Sarda, P., 332
Savin, N.E., 90, 334
Schmeiser, B., 97, 98, 239, 335, 338
Schmid, P., 22, 338
Schucany, W., 217, 332
Schwert, W., 331
Sen, P., 31, 333
Serfling, R., 5, 19, 28, 44, 51, 338
Shao, J., viii, 31, 38, 41, 42, 62, 63, 138, 219, 333, 338
Shen, X., 197, 338
Shepp, L.A., 335
Sherman, M., viii, 41, 80, 126, 131, 138, 155, 174, 175, 187, 297, 337, 338
Shi, X., 338
Shiller, R.J., 291, 294, 330
Siegel, A.F., 255, 337
Singh, K., 3, 66, 98, 166, 195, 219, 235, 244, 328, 334, 338
Stambaugh, R., 293, 331, 338
Stock, J.H., 271, 286–290, 294, 298, 304, 330, 331, 337, 339
Stuck, B.W., 330

Taqqu, M.S., 254, 256, 338
Taylor, R.L., 328
Teugels, J.L., 144, 329
Thombs, L.A., 84, 85, 241, 280, 337
Tibshirani, R.J., 38, 65, 197, 331
Tsirel'son, V., 22, 170, 339
Tsou, T-H., 139, 145, 157, 334
Tu, D., 38, 63, 138, 217, 219, 338, 339
Tukey, J.W., vii, 40, 62, 339

Uteev, S., 318, 339

van der Vaart, A., 22, 29, 161, 167, 339
Van Ness, J., 170, 339
van Zwet, W.R., viii, 48, 51, 217, 218, 222, 329
Varadarajan, V.S., 160, 161, 170, 339
Vardi, Y., 54, 332
Veretennikov, A., 136, 339
Viceira, L.M., 298, 339
Vieu, P., 332
Vrontos, I.D., 332

Watkins, T., 217, 332
Weiss, A., 331
Welch, P.D., 97, 98, 339
Wellner, J., 22, 23, 29, 54, 161, 167, 329, 332, 339
West, K., 281, 339
White, H., 80, 284, 321, 331, 339
Wilks, A.R., 328
Wolf, M., 80, 101, 160, 253, 272, 292, 336, 337, 339
Wolkonoski, V.A., 316, 339
Wong, W., 197, 338
Woodroofe, M., 170, 339
Woolridge, J.M., 299, 329
Wu, C.F., viii, 41–43, 45, 50, 62, 65, 338, 339
Wu, J., 253, 263, 339

Yadrenko, M.I., 135, 339
Yahav, J.A., 217, 221, 223, 329
Yohai, V.J., 89, 335
Yokoyama, R., 316, 339
Yoshihara, K., 167, 321, 335, 340
You, L., 128, 149, 170, 182–184, 186, 336
Yu, B., 167, 327

Zhurbenko, I.G., 98, 122, 340
Zinn, J., 14, 15, 22, 55, 332
Zolotarev, V.M., 254, 340

Index of Subjects

α-mixing, *see* Strong mixing
m-dependence, 239, 240, 286

AR model, 270
ARMA model, 243

Bias reduction
  for variance estimation, 236
  general theory, 93, 129
Block size
  choice of, 188
    data-dependent, 59, 92, 117
    for estimation of a c.d.f., 193
    for hypothesis testing, 200
    for variance estimation, 189, 240
    via calibration method, 194, 200, 297
    via minimum volatility method, 197, 201, 265, 284
Bootstrap
  and differentiability, 23
  asymptotic approximation, 6
  asymptotic pivotal method, 6
  averaged, 263
  comparison with subsampling, 47
  consistency of, 9

counterexamples, 49, 50, 270
examples
  p-th quantile, 28
  correlation model, 32
  goodness of fit, 36
  Kolmogorov–Smirnov bands, 6
  linear functional, 16
  linear regression, 31
  location and scale families, 5
  median, 30
  multivariate mean, 17
  smooth functions of means, 18
  square of mean, 27
  testing independence, 37
  testing the mean, 35
  trimmed mean, 7, 28
  univariate mean, 6
for calibration method, 195
for empirical process, 19
for hypothesis testing, 33
for mean, 11
for mean-like statistics, 16
general theory, 3
moving blocks, *see* Moving blocks bootstrap
pivotal method, 4

Bootstrap (*continued*)
  with smaller resample size, 48, 49, 55, 181, 263, 265

Calibration method
  for choice of block size, 194, 200, 297
Central limit theorem
  for triangular array of mixing sequences, 321
Comparison with the bootstrap, 47
Comparison with the moving blocks bootstrap, 98
Confidence interval
  equal-tailed vs. symmetric, 72, 105
  general theory, 41, 68, 102
    based on studentized roots, 52, 257, 274
  for studentized roots, 74
Confidence region
  general theory, 53, 74, 106
Convergence rate
  estimation of, 177
  examples when unknown, 175, 176
  subsampling when unknown, 173, 226, 260

Differentiability, 23

Edgeworth expansion
  for general statistic, 216, 225
  for sample mean, 219, 228, 232
  of subsampling distribution
    for general statistic, 216, 225
    for sample mean, 219, 229
Efron, B., 166
Empirical measure, 160
Empirical process, 19, 166
Examples
  autocorrelations, 83
  autocovariance, 156
  bispectrum, 157
  density estimation, 50
  diffusion process, 136
  extreme observation, 175
  extreme order statistic, 49
  first marginal distribution, 89

  integral, 156
  kernel regression, 135
  linear regression, 80, 113
  long memory, 175
  MCMC simulation, 118
  mean with infinite variance, 50
  multivariate mean, 78, 111
  nonnormal limit distribution, 175
  nonparametric regression, 176
  optimal replacement time, 45
  overdifferenced data, 90
  probability, 156
  robust statistics, 89, 116
  smooth function of the mean, 78, 111
  spectral density, 156
  spectral density function, 85
  superefficient estimator, 50
  test for normality, 157
  U-statistics, 49
  univariate mean, 77, 108, 134, 155
  variance ratio, 86, 116
Extrapolation, 213, 221

Finite population correction, 218, 235

Higher-order accuracy, 213
Hypothesis testing
  choice of block size, 200
  general theory, 54, 90, 117

Interpolation, 213, 216, 224

Jackknife
  delete-*d*, 62, 219
  generalized, 217, 239

MA model, 191, 241
Marked point processes, 138
Minimum volatility method
  for choice of block size, 197, 201, 265, 284
Mixingale, 284
Moderate deviations, 244
Moment bound
  for mixing sequence, 316

Monte Carlo approximation, *see* Stochastic approximation
Monte Carlo study, *see* Simulation study
Moving blocks bootstrap
  choice of block size, 194
  comparison with subsampling, 98
  for calibration method, 195
  for empirical process, 166
  for linear regression, 115
  for mean, 109
  studentization of, 239

Parameter
  general, 53, 159
  multivariate, 53, 74, 106, 127
  univariate, 67, 102, 123

Random fields, 120
Rate of convergence, *see* Convergence rate

Self-normalizing sum, 263, 283
Simulation study
  for autoregressive parameter, 286
  for linear regression, 204, 298
  for univariate mean, 203, 266
Spectral density
  relation to variance estimation, 237, 239
  uniform confidence bands for, 170
Spectral measure, 168
Stable distribution
  as limit law, 260
  general definition, 254
Stochastic approximation, 51, 151
Strong mixing
  general CLT for triangular arrays, 321
  general definition, 315
  general results, 315
Studentized statistics
  for sample mean, 219, 228, 263
  general theory, 52, 74, 257, 274

Time series
  m-dependent, 239, 240, 286
  AR model, 270
  ARMA model, 243
  MA model, 191, 241
  nonstationary, 101
  stationary, 65, 159, 213

Variance estimation
  bias-corrected, 236
  choice of block size, 189, 240
  for general statistic, 95
  for sample mean, 97
  general theory, 95, 117, 129
  negativity issues, 241

Springer Series in Statistics

(continued from p. ii)

Kotz/Johnson (Eds.): Breakthroughs in Statistics Volume II.
Kotz/Johnson (Eds.): Breakthroughs in Statistics Volume III.
Kres: Statistical Tables for Multivariate Analysis.
Küchler/Sørensen: Exponential Families of Stochastic Processes.
Le Cam: Asymptotic Methods in Statistical Decision Theory.
Le Cam/Yang: Asymptotics in Statistics: Some Basic Concepts.
Longford: Models for Uncertainty in Educational Testing.
Manoukian: Modern Concepts and Theorems of Mathematical Statistics.
Miller, Jr.: Simultaneous Statistical Inference, 2nd edition.
Mosteller/Wallace: Applied Bayesian and Classical Inference: The Case of the Federalist Papers.
Parzen/Tanabe/Kitagawa: Selected Papers of Hirotugu Akaike.
Politis/Romano/Wolf: Subsampling.
Pollard: Convergence of Stochastic Processes.
Pratt/Gibbons: Concepts of Nonparametric Theory.
Ramsay/Silverman: Functional Data Analysis.
Rao/Toutenburg: Linear Models: Least Squares and Alternatives.
Read/Cressie: Goodness-of-Fit Statistics for Discrete Multivariate Data.
Reinsel: Elements of Multivariate Time Series Analysis, 2nd edition.
Reiss: A Course on Point Processes.
Reiss: Approximate Distributions of Order Statistics: With Applications to Non-parametric Statistics.
Rieder: Robust Asymptotic Statistics.
Rosenbaum: Observational Studies.
Ross: Nonlinear Estimation.
Sachs: Applied Statistics: A Handbook of Techniques, 2nd edition.
Särndal/Swensson/Wretman: Model Assisted Survey Sampling.
Schervish: Theory of Statistics.
Seneta: Non-Negative Matrices and Markov Chains, 2nd edition.
Shao/Tu: The Jackknife and Bootstrap.
Siegmund: Sequential Analysis: Tests and Confidence Intervals.
Simonoff: Smoothing Methods in Statistics.
Singpurwalla and Wilson: Statistical Methods in Software Engineering: Reliability and Risk.
Small: The Statistical Theory of Shape.
Stein: Interpolation of Spatial Data: Some Theory for Kriging
Tanner: Tools for Statistical Inference: Methods for the Exploration of Posterior Distributions and Likelihood Functions, 3rd edition.
Tong: The Multivariate Normal Distribution.
van der Vaart/Wellner: Weak Convergence and Empirical Processes: With Applications to Statistics.
Vapnik: Estimation of Dependences Based on Empirical Data.
Weerahandi: Exact Statistical Methods for Data Analysis.
West/Harrison: Bayesian Forecasting and Dynamic Models, 2nd edition.
Wolter: Introduction to Variance Estimation.
Yaglom: Correlation Theory of Stationary and Related Random Functions I: Basic Results.
Yaglom: Correlation Theory of Stationary and Related Random Functions II: Supplementary Notes and References.